ABOUT THE AUTHORS

Paula Wagner, MBA, PMP, has more than 20 years of business experience in technology, strategy, and planning. As a long time Turner Broadcasting employee, her experience encompasses leading multimillion-dollar cutting-edge technology projects, research and development, long-range technology planning, business planning and development, financial management and forecasting, and process engineering. Ms. Wagner teaches project, program, and portfolio management at DeVry/Keller Graduate School of Management.

Bruce T. Barkley is a former federal senior executive, vice president of The Learning Group Corporation, and program analyst in the federal Office of Management and Budget. He is a former faculty member at the University of Maryland University College and now serves on the senior faculty at DeVry/Keller Graduate School of Management. Mr. Barkley is the author of *Project Risk Management, Integrated Project Management, Customer-Driven Project Management* with James H. Saylor (translated into Chinese), and *Project Management in New Product Development,* all published by McGraw-Hill.

Global Program Management

Paula Wagner, MBA, PMP

Bruce T. Barkley

New York   Chicago   San Francisco   Lisbon   London
Madrid   Mexico City   Milan   New Delhi   San Juan
Seoul   Singapore   Sydney   Toronto

The **McGraw·Hill** Companies

Cataloging-in-Publication Data is on file with the Library of Congress

Copyright © 2010 by The McGraw-Hill Companies, Inc. All rights reserved. Printed in the United States of America. Except as permitted under the United States Copyright Act of 1976, no part of this publication may be reproduced or distributed in any form or by any means, or stored in a data base or retrieval system, without the prior written permission of the publisher.

1 2 3 4 5 6 7 8 9 0   DOC/DOC   0 1 5 4 3 2 1 0 9

ISBN 978-0-07-162183-0
MHID 0-07-162183-0

Sponsoring Editor	**Copy Editor**
Judy Bass	Lisa McCoy
Editorial Supervisor	**Proofreader**
Stephen M. Smith	Shivani Arora, Glyph International
Production Supervisor	**Indexer**
Richard C. Ruzycka	Robert Swanson
Acquisitions Coordinator	**Art Director, Cover**
Michael Mulcahy	Jeff Weeks
Project Manager	**Composition**
Preeti Longia Sinha, Glyph International	Glyph International

Printed and bound by RR Donnelley.

McGraw-Hill books are available at special quantity discounts to use as premiums and sales promotions, or for use in corporate training programs. To contact a representative, please e-mail us at bulksales@mcgraw-hill.com.

This book is printed on acid-free paper.

Contents

Introduction

Global program management involves directing the design, development, delivery, and monitoring of major, long-term business or agency initiatives in countries and communities that are foreign to the home country. The setting for global program management is worldwide. To be successful, the process requires an extraordinary degree of leadership, organization, communication, and technical and managerial competence and agility, along with a keen sensitivity to foreign markets, cultures, and values—and a little luck.

Two major developments spurred us on in the development of this book: the Project Management Institute's (PMI) recent publication of the first *Standard for Program Management*—expanding the narrow concept of "projects" to a more realistic and longer-term context for achieving major benefits—and the expansion and dynamics of the global economy generated by new communication technologies and international trade policies. PMI's new standard is broad and comprehensive, and provides a new framework for generating and nesting projects in a more strategic and global program framework. And the forces of the global economy now exhibit themselves in a wide variety of products and services as the world becomes flat, as described by Thomas Friedman in *The World Is Flat*, and barriers to global marketing and business development disappear.

Flow of This Book

Our book reflects the movement of the field to a broader and more enterprise-wide perspective.

Chapter 1, "The Global Setting for Program Management," provides an initial look at the organizational and global setting for program management, including the process of strategic business and agency planning for global success.

Chapter 2, "The Standard for Program Management According to PMI," is a specific and targeted discussion of the new PMI *Standard for Program Management,* a useful reference for those who seek and/or maintain the new PMI program certification or who simply want to know about the program standard for their own career and/or company development. This chapter also includes a detailed discussion of the role of the program management office as an integral part of the program standard.

Chapter 3, "Global Portfolio Management Strategies," discusses the development of strategies, programs, and candidate projects into a strategic portfolio of projects. The chapter addresses the portfolio development process to ensure that the *right* projects are generated, chosen, funded, and delivered in the context of broad program benefits and goals worldwide.

Chapter 4, "Global Program Management Strategies," further addresses the process through which companies align their strategies and programs to ensure success, profitability, and program benefits. The chapter also discusses two generic cases: Western Inc. and Eastern Inc.

Chapter 5, "Global Program Risk Management," provides a discussion of program-level risk and how risk is assessed, monitored, and mitigated on a program scale.

Chapter 6, "The Global Program Manager," addresses the characteristics and leadership competencies of program managers who do business globally.

Chapter 7, "Partnerships, Contracts, and Procurement," emphasizes the process through which global companies find and secure partners to provide program support and supply chain benefits.

Chapter 8, "Federal Program Management," a specialized chapter, discusses applications or project management in the federal sector, including a proposed new Office of Federal Innovative Management in the Executive Office of the President.

Chapter 9, "New Global Program Development," a special-case application, is a focused discussion of how new products can be encouraged and developed in a program framework.

Appendix A, "Program Risk Management Checklist," is a practice checklist for program managers to ensure that risks are anticipated and managed. Appendix B, "Microsoft Applications for Program Management," provides a short discussion of how Microsoft Project can support global program management. Finally, Appendix C, "Global Project Management Strategies in a Program Management World," supplies refresher information on project management concepts and also how they integrate with program management.

The Global Setting

Program managers face challenges globally in five distinct categories: technical, regulatory/legal, political, economic, and social/cultural.

Technical. Technology is linking economies and customers together so that communication is no longer a major barrier to doing business in foreign countries.

Regulatory/legal. Local laws and regulations govern the design and development of markets and programs, thus complicating the process of program management across national borders.

Political. Varying levels of political stability and unrest generate risks and threats to successful program development and delivery, thus adding uncertainty to programs already characterized by a creative and changing development process.

Economic. Economic conditions, currency values, and pricing issues are faced by program managers as they attempt to estimate program costs and deliver on time and within budget.

Social/cultural. Language and cultural differences generate major challenges as programs are delivered in countries whose values and lifestyles vary widely.

The Program Management Setting

The setting of this book is *the program,* not individual projects. This means that one must conceptualize at a higher level or abstraction and a higher level of business/agency planning and management to understand broader, program-level strategies, goals, objectives, and projects. It is the difference between NASA's shuttle program and a particular part of the program to develop alternative fuel safety systems. It is the difference between Alcatel's global cable development program and a component project to develop a seagoing vehicle for laying oceanic cable. It is the difference between targeting a company's resources in a particular part of the world to increase market share and a specific marketing project to elicit local values and needs.

The setting of this book is *benefits,* not project outputs. Programs create long-term benefits—e.g., NASA creates a Space Shuttle competency and technology transfer program that makes major contributions to space science and domestic science and technology; a platform for Internet operations around the world is created and the groundwork for flexible expansion of cable-based or satellite-based communications and data flows is established; jobs for poor people in Africa are created by establishing a series of individual training and technology projects throughout the continent.

The Business Setting

Program management is a part of a total business strategy and support system; thus, successful program management starts with a solid business structure and competency. Good global business planning creates programs and projects that can deliver on longer-term business goals and objectives, and good operational planning provides the production facilities and program and project support systems so that businesses can be delivered anywhere in the world. Program managers are business officers who see the big picture in their companies or agencies, understand where the organization is going, and can link planning to solid operations and program development.

Seven Principles

We identify seven key principles that seem to be associated with successful global program management at the operational level:

1. *Tight central control of quality and cost.* Because so many forces are working against consistency and working toward fragmentation of an agency's global outreach, the global enterprise administering programs must ensure discipline and tight control of quality, and it must exercise tight cost control to ensure long-term profitability.

2. *Loose guidance on local operations, team relationships, and schedules.* Here, the global enterprise ensures that headquarters decisions and biases do not dominate local delivery of programs. Because those closest to program delivery in foreign environments are better able to translate programs to local conditions and needs, central program management must loosen the reins on team operations, schedules, and contracting. The key is leadership and vision, but not micromanagement.

3. *Liberal approach to changes and shifts in program and project scopes of work.* Central program managers must create agile programs; they have to let go in a way that is not characteristic of the traditional project management process. Because scopes of work and other program components are subject to more complex and unanticipated forces in a global context, local changes and views must not only be tolerated, but also encouraged. This will come hard to some program managers, who are accustomed to locking into up-front schedules and costs as if they were the gospel.

4. *Complete understanding of local regulatory, safety, and public policy dynamics.* Global program managers must know the local lay of the land and provide useful guidance for local program team members to interpret and translate program interests into local initiatives. A thorough understanding of local governmental and regional (e.g., European Union) policies is essential to good global program management.

5. *Establishment of locally empowered program and project teams.* Teams must be set up to perform locally and have enough delegation of power and responsibility to make necessary changes to program and project elements to succeed locally. Program managers in New York cannot run their program to create a local food processing plant and delivery system in Munich; they must have people in Munich who understand the local context for program development.

6. *In-depth risk assessment of key program and project activities in the context of local conditions.* Risk potentials are higher and more consequential in global program operations; implications of program and project failure are more severe. Things can get bad very fast, driven by forces often out of control of the program manager and even the local team. Of course, with more risk comes more opportunity, since successful mitigation of risk often opens up new markets and business opportunity.

7. *Direct liaison with home-based governmental and diplomatic circles to ensure consistency with national intents.* Central program managers have to keep on top of changes and dynamics in diplomatic circles in their own home countries so that they can reflect those forces—especially the goals of their home country—in program delivery. Sometimes, the biggest obstacles to doing business globally are in the home country.

Change and Program Management

Project changes are placed in the context of overall program benefits. In a field characterized by command and control tools—e.g., tight scopes of work, linked and controlled scheduling structure, tight cost controls, etc.—the demands of global competition require new, more agile approaches. The shift in thinking from project to program is from tight, centralized notions of management to loose, decentralized, open-loop program delivery, from internally driven corporate strategies to open, partner-driven, web-based collaboration and peer decision-making. Program managers must sometimes give away traditional controls to generate local teams that respond to local talent and local conditions to satisfy customers and client needs as they change.

So the reader versed in traditional project management tools will see in this book some of the paradoxes inherent in the move from project to program. While program managers must be able to call up traditional, centrally administered project management tools to address major issues, the reality of today's global markets requires program managers to focus on building local coalitions and partnerships, and seeking out and trusting local talent on-site, rather than on running things from a central control point with classic project tools. The new approach recognizes that programs are actually creative and dynamic processes to produce long-term benefits that require teams throughout the world to continuously build coalitions locally and to redefine and redesign program products, benefits, and outcomes as their local constituents see them.

The new emphasis on long-term benefits and broad program outcomes and on building lasting international partnerships changes the job description of tomorrow's program manager, whether in the private or public sector. Now the role becomes one of leadership and personal diplomacy, of packaging outsourced relationships to allow partners and contractors to own the program and to contribute product ideas and systems that make the program better and more responsive. This sea change does not mean that traditional project management tools and techniques are rendered obsolete, but rather that these tools are seen more as guideposts and criteria for making locally staged decisions as the process evolves.

Striking changes beginning in the global environment in the 2000s suggest that program management will get more challenging. Unstable political settings, restructuring of economic and financial systems, and major shifts in how governments interact with the business community all suggest the importance of environmental monitoring and risk assessment. The design, planning, and management of a program in tomorrow's global economy must take into account the changes going on in the financial and market world. National economies are going through major contraction and restructuring, governments are intervening in business investment and finances to spur growth, and business is reeling from the deepest recession in many decades. How does a program manager approach this kind of global challenge, and how does classic program and project management change as a result?

1. Design programs that can generate early impacts on generating jobs.
2. Limit programs to confined goals and objectives with realistic benefits.
3. Target programs on economic and technology goals that align with appropriate foreign government objectives and recovery programs.
4. Target technologies that stimulate local development, manufacturing, and marketing competencies.

This book also captures and defines a major new trend in business management—*core program management,* a concept that defines program management as the mainline business function. In other words, programs target resources in a given global market. They are not simply oriented to new systems or products as much as they are focusing resources locally.

At the same time, many recurring and administrative business services are being outsourced. This trend toward putting the best resources of the company in programs and projects has focused them on strategic change and innovation, new product development, system upgrades, and research and development in order to stay competitive. The trend has been accelerated by increasing global pressure to reduce cycle times for new products and services, and by the need to focus the business workforce on innovation and change. One iteration of a product is now typical, requiring companies to redefine their products and services constantly.

The Global Setting for Program Management

Introduction

The purpose of this chapter is to explore and set the context for global program management. What makes global management different from domestic management, and how do program managers—as differentiated from project managers—handle such conditions?

Global conditions increase both risk and opportunity at the same time. While global marketing and program development open up major new markets for business development, global forces increase risk and uncertainty. These risks and uncertainties require that businesses structure themselves to be global in their thinking and actions. They embed the global theme in their business planning, and they continuously conduct environmental scanning—the process of monitoring key program success indicators globally—in order to understand and anticipate forces that will affect program success. And they characterize themselves in global terms in their mission, vision, values, and strategic plans.

Program management capacity, that is a corporation's core competency in designing and implementing long-term programs globally, is a strategic differentiator in itself. Foreign sponsors and clients look to a company's infrastructure and process to support multiple projects involving many countries and many cultures. Foreign owners and customers depend on the management and administrative capacity of global program managers to deliver on complex programs such as: (1) the development of a major global information network; (2) the installation of a major, global communication system; (3) the development of major power and/or utility systems in foreign countries; and (4) rehabilitating buildings all over the world to make them more energy efficient. They look for proven executive and management talent with the experience to build relationships across countries and to build international partnerships.

Global Forces

The global forces that translate to success factors can be categorized as follows: technical, regulatory/legal, political, economic, and social/cultural.

1. *Technical*. Technology is linking economies and customers together so that communication is no longer a major barrier to doing business in foreign countries. But the cycle of change in technology is so rapid that major investments in technology-oriented programs can be endangered midstream.

2. *Regulatory/legal*. Local laws and regulations govern the design and development of markets and programs, thus complicating the process of program management across national borders. For instance, one country might employ major safety restrictions on oil discovery and exploitation, while another might have no safety restrictions. Legal issues include intellectual property protection, licensing, and trademark protection, complicated enough in the home country but more complex in a global context.

3. *Political*. Varying levels of political stability and unrest generate risks and threats to successful program development and delivery, thus adding uncertainty to programs already characterized by a creative and changing development process. Terrorism has become a major risk in various parts of the world, and danger is no longer restricted to one part of the world. With governments going through major upheavals, programs can be endangered by the threat of terrorist actions. One country might suddenly move to expropriate or nationalize an industry related to an ongoing program, while an adjoining country might be encouraging more private enterprise and deregulating its markets.

4. *Economic*. Economic conditions, currency values, and pricing issues face program managers as they attempt to estimate program costs and deliver on time and within budget. Major unemployment in parts of the world creates both risk and opportunity. While global programs can generate jobs, they can also threaten local economies with change and job loss as new technologies overtake old ones.

5. *Social/cultural*. Language and cultural differences generate major challenges as programs are delivered in countries whose values and lifestyles vary widely. Language and culture differences create differences in customer demand and how markets behave. A program designed to create a quick market for a given line of products or system enhancements can be stalled because local values interpret those new products or systems as being contrary to local standards. A good example would be a project to introduce a new mobile communication system in a country where women are not encouraged to participate actively in business decisions, or even to use new technologies.

Thus, the global business landscape is changing and evolving. No longer are we faced with the dimension of local concerns, but rather multidimensional forces of a global backdrop. In this rapidly changing world, where the economy in one country affects the economies in other countries, for a company to remain buoyant and even prosper, it is essential for an organization to remain agile. The organization needs the flexibility of a clipper ship even if they are an aircraft carrier. A business must sustain a clear understanding of underlining components that propels them forward, while keeping a hand on the pulse of the global atmosphere. For a company to thrive in the frenzied waters of the global market, it has to make tough decisions. Where is the business going to optimize and invest its limited resources? How is the company going to capitalize on strategic partnerships? Where is the next frontier for growth? How is the organization going to maintain its core business while developing new markets?

Setting the groundwork for good global program management requires the business or agency to establish its mission and values and to *understand itself* first. Then, the core to any business project venture is the strategic investment decisions the organization makes. This includes decisions on where a number of core business functional areas can be outsourced, such as human resources, accounting, and manufacturing. It is possible that the left-brain process-oriented activities can be outsourced, whereas the right-brain creative process for strategic decision-making becomes central to the operation of the organization. In the project portfolio process, an organization determines where to invest to maintain its business and power it forward. The project portfolio process allows for identification, analysis, optimization, and execution of strategic programs and plans. Programs leverage both the strategic elements of portfolios and the tactical capability of projects to align the program's efforts to produce business benefits. Projects provide for the tactical ability to deliver on business objectives.

Strengthening the Business for Global Project Delivery

Businesses targeting global markets and programs increasingly restructure their core mission, vision, and values in the context of global change and opportunity. In many cases, businesses turn their attention to global markets through a redefinition of their mission and values, fundamentally shifting the focus of the business. While there is value in the products of such an exercise, the real benefit is in the changes in mindsets that occur during the process, widening horizons about what is possible around the world.

Global Mission

The process of developing a new, globally oriented mission for a business involves redefining what the business is all about in a global context. Global opportunity may even change the core nature of the business. The mission statement for an organization states the purpose of the business. This statement allows the business to clearly define what business they are operating. A mission statement can be used for a number of purposes. It could be a tool used to signal the intent to go global. It can inform the public of what the company provides, or it can be used internally to serve as a gauge for decision-making by executive leaders. The mission could be a powerful tool or motto used to drive performance by having the staff focus on a common purpose; if the staff know that programs are going global, they are likely to respond with more energy and enthusiasm because of the opportunities worldwide. Some companies publish their mission statements in their annual reports for their investors and potential clients.

The mission statements can be concrete and tangible, or conceptual and belief-driven. Concrete areas would center on:

- Products and services
- Main customers or markets
- Areas the business operates within

Belief areas—or values—would include:

- Philosophical and social concerns of the business
- Competitive advantage
- Image or quality standards
- Reasons for existence

The concrete areas might be considered the "hard" areas, while belief areas could resonate with people and be considered the "soft" areas.

Here are some examples of mission statements from various companies using both hard and soft components:

- The Dow Chemical Company
  - Slogan/motto:
    Living. Improved daily.
  - Description:
    The Dow Chemical Company is a chemical company whose specialty is performance plastics, such as engineering plastics, polyurethanes, etc. Aside from this, Dow is also into performance chemicals, agricultural sciences, hydrocarbons and energy, etc.
  - Mission statement:
    To constantly improve what is essential to human progress by mastering science and technology.
- Global Gillette
  - Slogan/motto:
    Welcome to everyday solutions.
  - Description:
    Global Gillette is a manufacturer of shaving equipment, specializing particularly in razors and blades, aside from being a manufacturer of batteries

as well. Its brands include Sensor, Trac II, Mach3, M3Power, Fusion, and Duracell for its batteries product.

- Mission statement:
  We will provide branded products and services of superior quality and value that improve the lives of the world's consumers. As a result, consumers will reward us with leadership sales, profit, and value creation, allowing our people, our shareholders, and the communities in which we live and work to prosper.

- Harley-Davidson, Inc
  - Slogan/motto:
    Define your world in a whole new way.
  - Description:
    Harley-Davidson, Inc., is the manufacturer of a line of motorcycles, with over 32 models of touring and custom Harleys. Aside from their line of motorcycles, Harley-Davidson also offers motorcycle accessories, motorcycle clothing apparel, and engines.
  - Mission statement:
    We fulfill dreams through the experience of motorcycling by providing to motorcyclists and to the general public an expanding line of motorcycles and branded products and services in selected market segments.

- The Walt Disney Company
  - Description:
    The Walt Disney Company operates a global entertainment portfolio of media networks, parks and resorts, studio entertainment, and consumer products. This wide array reaches out to the world through its television broadcasts, Internet businesses, theme parks, and the many ventures of The Walt Disney Company's subsidiaries.
  - Mission statement:
    The mission of The Walt Disney Company is to be one of the world's leading producers and providers of entertainment and information. Using our portfolio of brands to differentiate our content, services, and consumer products, we seek to develop the most creative, innovative, and profitable entertainment experiences and related products in the world. (www.missionstatements.com)

The Gillette mission statement is indicative of the intent to go global. Reference to the *world's consumers* is a good example of how companies signal their intent in their mission statements.

To construct a global mission statement, you have to understand what is unique about your organization and how it will compete globally. What makes this organization different from another one? Why would a world customer or client select you above your competitor? What would be the few tangible success measures? The mission statement goes through a number of iterations before it is finally complete. It should provide information on how this organization is unique, as well as an identity and purpose for the organization.

When constructing a global mission statement, it is important to consider who the customers are. What products, services, and/or technologies are needed to satisfy that customer group? Is the business customer-driven or market-driven? Are the products, services, and technologies specialized, integrated with other systems, or commodities?

Mission statements should be well crafted and specific. The statement serves as a boundary to what the organization will and will not pursue. It defines a business position. If the mission statement is ill-defined, not specific, or unclear, it will not aid as a tool in decision-making. The mission should provide a strong identity for the organization.

Mission statements can be created for all levels in the company. The approach to the creation would be top-down. The mission from the corporation would be created first, then departments would describe how they contribute to the firm's mission. The department mission would also include the department's role and responsibilities within the business and where the department is headed in relation to the big picture of the business.

Global Vision

There is often confusion between mission and vision statements. Both mission and vision statements are important and complement one another; however, there are distinctive differences. Where mission statements concentrate on the current state of affairs, the vision statement has its sight set on the future state of the organization. The mission is a reminder of why the business exists. The vision inspires propagation of a new harvest. And that new harvest for a global company is worldwide in dimension. As it expresses its intent to deliver programs worldwide, the vision gives a message not only to its competitors, but also to its workforce. A vision statement is the emotional component of the business plan: "This is where we get excited about what we want to become."

The vision statement is more than a picture of the future of the company; it also provides a framework for strategic planning on a global scale. The vision can be constructed to concentrate at the corporate, division, department, or business unit level. It can answer the question of where do we want to go and how I am going to contribute to getting there. The vision statement keeps to the forefront what we are trying to build. Though the vision statement doesn't tell you how to get there, it does provide a destination to aim toward for your planning purposes. A global vision can create a sense of differentiation and confidence in the workforce as they begin to think of themselves as part of their company's response—and perhaps even their nation's response—to the global economy. Programs with worldwide reach are exciting in themselves.

Vision statements allow you to dream and capture your passion on paper. Generally, they are produced for "in-house consumption" and are not necessarily publicized outside the organization. A vision statement is a tool used to help with decision-making in determining where the organization should invest the organization's resources, including people's talents, technical resources, facilities, and money to name a few.

To create a vision statement, it is important to understand the organization's core competencies in a global context. What business are you in? The vision goes beyond the mission and describes the desired future outcome, and inspires and energizes the staff to reach that target. The outlook of the vision statement is often five to ten years into the future. Some organization's vision statements look out further than ten years.

At the beginning of a vision statement, you should have a powerful first paragraph capturing the essence of the vision in a memorable phrase, picture, or metaphor. The beginning of the statement will trigger the rest of the vision explanation. Visions are unique to the organization and are not a one-size-fits-all statement. Vision statements tend to be longer than mission statements. The vision should be as long as it needs to be to create a complete picture. The picture that is described in the statement should feature the best possible outcome—a vision that may be beyond what is thought currently achievable. A vision is not used as a measuring stick for success, but rather it is used to spur creativity and ingenuity. It stretches expectations and generates leaping out of old shells.

Some questions can be used to help create a vision statement.

- What will our global customers want in the future?
- What are the expectations of our global stakeholders and shareholders?
- Who will be our global competitors, partners, or suppliers?
- How will emerging technology affect the industry?
- What are the global or governmental factors that might affect our business?

Some might not feel comfortable creating "unrealistic" visions. However, if you set a path for the clouds, you will never reach the stars. If needed, two versions of the vision statement could be produced: one more realistic and the other more ambitious. In the early 80s, it might have seemed highly unrealistic to have the vision of a PC in every household, but Microsoft stayed true to this ambitious vision. It is true that even today, not every home in the world has a PC, or even runs Microsoft products; however, that just means Microsoft is still in the process of achieving their vision. The vision provides a cohesive document describing the business the organization plans to pursue. It also provides long-term direction and the big-picture perspective of where the company is headed.

Vision statements keep the mind open to what is possible. When employees are committed to the vision, they will make decisions that are aligned with it. Employees will not be confused on the business's direction or try to move in too many directions. A vision helps provide a foundation for improved execution of the organization's strategy.

Global Strategic Management

The global strategic management process answers the questions of what we are going to do, where in the world are we going to do it, and how we are going to get it done. Strategy takes a holistic as well as a component view of the situation. Strategic management is the process of planning, analyzing both the internal and global environments, clearly defining the objectives and direction of the company, implementing that strategy, evaluating the success of the strategy, and then adjusting the plan to achieve the greatest program benefits and rewards for the organization.

Strategy examines the current business and its environment, and then analyzes potential options based on how they will provide a competitive advantage, and then executes the decided-on plan. A strategic plan drives ideas to action. The process then repeats itself with the examination of the business and the environment, and determines what has changed and what is emerging. Strategy is the path to competitive advantage. The strategic plan allows for decisions to be made that affect the investment of the organization's resources. It focuses on what is inside and outside of the organization's business perimeters. When developing a strategy, thought is given to markets, products, services, and internal competencies. To formulate the strategic plan, executives take a step out of normal business operation mode to focus on the broader perspective of the organization.

First, the company needs to determine its business objectives on a global scale and then decide upon the best way to achieve them. The organization then determines what resources are required to execute the plan. The business objectives are housed in the organization's mission statement. What is the purpose of the organization? What values does the organization hold? The core ideals of an organization are its foundation and usually remain constant over time. The ideals provide decision-making guidance. The mission statement focuses on what the organization is, and is composed of three main components: business purpose, values, and goals.

The business purpose describes why the entity exists. The purpose of a company generally remains unchanged over time. The purpose answers the question, "How will the business be profitable?" The organization should never define the purpose too narrowly by focusing on a single project or service. As an example, consider the railroad industry. A railroad organization's mission could be something like:

To operate railroads in this territory and support railroads worldwide and their customers' needs.

Railroads in the United States reigned supreme until the highway system and airline industry gained much of railroad's market. If the rail industry defined itself as in the transportation industry rather than the rail industry, during their strategic management process they would have made different investment decisions in the beginning, which could have led to different opportunities. The execution on the strategy could have led to the players in the transportation business being different than what they are today. Having a too narrow a mission can stunt long-term growth for a business.

Some companies create impressive think tanks and develop wondrous products, but in order not to disrupt its core business, they fail to capitalize on the investment. As reported in the article "How Xerox Forfeited the PC War" by Rob Landley, September 18, 2000, Xerox narrowly defined their business to the super computer market. The super computer market had large margins compared to the personal computer market. As the article states:

> . . . Xerox PARC invented modern desktop computing. Windows, icons, mice, pull-down menus, "What You See Is What You Get" (WYSIWYG) printing, networked workstations, object-oriented programming—the works. Xerox the copier company feared the paperless office and formed a think tank to invent it before anybody else could, but once its commandos had succeeded, it simply couldn't bring itself to disrupt its core business of making copiers.
>
> Xerox could have owned the PC revolution, but instead, it sat on the technology for years. Then, in exchange for the opportunity to invest in a hot, new pre-IPO start-up called Apple, the Xerox PARC commandos were forced—under protest—to give Apple's engineers a tour and a demonstration of their work. The result was the Apple Macintosh, which Microsoft later copied to create Windows.

A well-crafted strategic plan can launch a company into greater market domains. A poorly crafted strategic plan can be like putting your fist into a bucket of water, taking it out, and expecting it to leave an impression.

After the organization answers the how, it moves on the question of why. The answer to the "why" question takes the purpose statement to the next level of clarity. The purpose of a business should not only describe the area it operates within, but also should capture the minds and hearts of the employees.

Global Values

Most companies and agencies generate key values that go to the heart of the organization. These values usually remain unchanged even if the business environment changes. If the market tastes change and conflict with core values, the business can migrate to new markets to stay true to its core values. Examples of core values are innovation, superior customer service, collaboration, and sustained growth on a global scale. Value statements can be structured in a number of ways. To illustrate, here are the value statements from the IBM and Microsoft websites:

> IBM's value . . .
> - Dedication to every client's success
> - Innovation that matters—for our company and for the world
> - Trust and personal responsibility in all relationships
>   www.ibm.com/ibm/us/en
>
> Microsoft's value . . .
> As a company, and as individuals, we value integrity, honesty, openness, personal excellence, constructive self-criticism, continual self-improvement, and

mutual respect. We are committed to our customers and partners and have a passion for technology. We take on big challenges, and pride ourselves on seeing them through. We hold ourselves accountable to our customers, shareholders, partners, and employees by honoring our commitments, providing results, and striving for the highest quality.

www.microsoft.com/About/default.mspx#values

Global Goals

The goals of an organization describe the future state of the business. The goal is the vision of the potential opportunities for the company, e.g., "It is our goal to dominate the world market for mobile phone systems, especially in developing countries." The visionary goals are long-term objectives rather than tactical. These goals could be focused on a target market or on becoming number one or two in the markets that you are currently operating in. Visionary goals are often stretch goals. However, most companies seem to achieve their goals because of the inspiration they cause within the staff. The vision provides direction, purpose, alignment, motivation, and inspiration.

Ted Turner, a true visionary, had the inspiration to create a *"New York Times* of the airways." Critics of the 24-hour news service called the venture crazy and instead of referring to the organization by its proper name, Cable News Network (CNN), they dubbed it Chicken Noodle News. Turner was committed to serious journalism and soon gained loyalty from viewers and respect from the journalism industry. The ratings for the network were known to surpass those of the three major networks at the time, beginning as early as the breaking-news reporting of the Persian Gulf War in the early 1990s. Turner motivated his staff to do more with less, to achieve what was thought unachievable, and to redefine the creation of network television, all through sheer force of will and the desire to get it done. Turner's vision and inspiration created a new segmentation: the 24-hour niche broadcast market. And the target was clearly global in nature.

Developing a global strategic plan

To begin a strategic plan, a critical look at the current operation needs to take place. The strategic plan is generally undertaken by the business owners or executive members of the company. The focus is a two- to four-year outlook. Some companies stretch their outlook beyond four years; regardless of the duration of the outlook window, the plan is updated periodically to ensure alignment with corporate structure and market conditions. Strategic planning is usually done outside normal business operations. Time has to be allotted to suspend day-to-day business operations and concentrate on the future direction of the organization. The strategic process needs to be a realistic approach to the future of the business, where the cause and effect of decisions are evaluated. The strategic planning process is ongoing and is reviewed periodically. At the conclusion of strategic planning sessions, decisions are documented, action items are reviewed, and plans are structured for the dissemination of decisions.

To begin the strategic planning process, a number of components need to be in place, which provide the foundation for strategic planning. First, there needs to be a clear understanding of the status of the business and its potential to go global. Often, this type of study is done through a SWOT analysis on a global level. SWOT stands for strengths, weaknesses, opportunities, and threats. An organization examines its core internal strengths and weakness. It then looks outside the organization for opportunities and threats. The SWOT analysis provides valuable information for the company to build a strategic plan and direction for global program development.

Global strategy

After the mission and vision, the next step is to create the roadmap that takes you from the current state and drives you to the vision. This roadmap is your corporate global strategy. The corporate strategy maps out specific actions required to successfully implement the plan. It integrates the goals, policies, and tactics into a cohesive whole. The strategy defines business approaches to lead to a successful outcome. This roadmap is the plan on how to satisfy customers, achieve performance targets, strengthen the organization's market position, and outlines the procedures on how the business should be run. To think strategically, you need to understand where you are and where you want to go. The new destination can include financial and strategic objectives.

There are five tasks of strategic management.

1. Create a mission and vision.
2. Set measurable objectives.
3. The objects should be both short- and long-term objectives.
4. Craft a strategy to achieve the objectives.
5. Implement and execute the strategy.
6. Evaluate performance, review the situation and new developments, and initiate corrective action.

Deliverables related to these five tasks are revised and improved upon over time.

Setting global objectives

Global objectives set measurable indicators of success, e.g., we intend to increase our global market share by 12 percent through new product development globally and ensure a profit margin of over 10 percent in the process." It also motivates teams by giving them a way to track their progress and measure their success. Objectives convert mission statements into performance targets. It is critical that objectives be created so that they are both challenging and achievable. Objectives should push organizations to be inventive and focused on their actions to reach desired results. Clear objectives prevent confusion, complacency, and stagnation. Objectives should set specific performance targets and spell out how much of what is going to be achieved by when. Objectives for an organization are the same as in project management; they should be measurable, quantifiable, and timely.

Both financial and strategic objectives should be created. Financial objectives provide targets to improve financial performance. An example of a financial objective would be to increase earnings by a certain percentage over a certain period, say, 15 percent by the end of the year. Strategic objectives focus on increasing market share and improving competitiveness. They can include becoming the number-one business or being the low-cost solution in a particular market. There could also be a focus on diversifying into new businesses, creating strategic alliances, collaborative partnerships, or introducing new products to the market.

Financial performance reflects past decisions and actions, which could be called "lagging indicators." On the other hand, strategic performance signals how well the company is growing in competitiveness and strength in the marketplace, which can be considered a "leading indicator."

Objectives need to reflect both short- and long-term objectives. Short-term objectives focus on quick gains, or "low-hanging fruit." These types of objectives could also be milestones on the way to a long-term objective.

Long-term objectives target three to five years in the future. These types of objectives help the organization make decisions that will lead them on the path to achieving the long-term objective success in the future.

Objectives are needed at every layer of the organization. They should be created first at the corporate level, then move down the chain to the business, then functional, and then finally the individual level. Objectives that are created in a top-down approach allow for objectives that are connected, cohesive, and unified.

Crafting a global strategy

In developing a strategy, the focus is on opportunities—either doing new things or doing the old things even better. When developing a strategy to meet its global objectives, an organization needs to be concerned with the "how" questions—how to achieve the objectives, respond to global factors, defend against threats, and grow the business. To create a strategy, you need to combine planned goals with reactions to market, environmental, and global conditions.

To build a strategy, an organization needs to understand market trends, global conditions, and their competitors' positions. The business must be familiar with their customers' needs and anticipate their taste changing. Emerging technologies need to be analyzed. A determination should be made on the appropriateness and effectiveness of a technological solution and how the company could benefit from it. An organization also looks at ways to increase market share with acquisitions, new products, or increasing competitive capabilities.

The corporate strategy is created by executive directors of the organization. The team that makes up this group generally includes the C-level executives: chief executive officer (CEO), chief financial officer (CFO), chief operating officer (COO), and so on; the corporate board of directors; division heads; and other key executives throughout the company. In the book *The Medici Effect,* Frans Johansson describes the benefits of diversity and innovation. At the intersections of diverse domains, ideas ignite and lead to an explosion of new discoveries. Diversity in the team that builds the corporate strategy will take the business to new frontiers. Diversity includes age, gender, socioeconomic, cultural, and departmental/functional areas, among others. The team of executives could benefit from working with their staffs to generate ideas to be analyzed later by the strategy team. Often, the team charged with creating a strategy is composed of the same individuals charged with implementing it.

Often, when the corporate strategy is complete, each business and functional area takes the plan and crafts how their division strategically contributes to the overall plan. Each individual in the organization is then tasked with how to contribute to the business and steer business decisions and activities that support the plan.

An organization's strategy is a collection of objectives to achieve the business benefits the organization is seeking. The objectives should fit together as a cohesive action plan.

Implement and execute the global strategy through portfolios, programs, and projects

In sports, the team spends countless hours practicing and memorizing plays. On game day, the plays turn to action on the game field. The strategic plan moves from the planning phase to the implementation phase. The plan becomes actionable. Progress is measured. Actions are aligned to the plan.

An organizational culture is created and managed to pursue the strategy. The three Ps for strategy success are portfolio, program, and projects. The strategy is administered through the portfolio process. Resources are allocated to candidate programs and projects that will produce the benefits planned and aligned with organizational objectives. As in a program management office, best practices are created and disseminated to allow for a continuous improvement model. Information is collected on the status of progress. Communications among teams are standardized and expectations set so that comparison of "apple-to-apple" information is presented. Managers inspire, motivate, and coach their staff teams to reach their targeted objectives. The more mature the portfolio, program, and project process, the better positioned the organization will be in achieving the benefits it seeks.

Evaluate and align performance

The final task of strategy, just as in the project process, is to evaluate performance and align or realign tasks accordingly. The strategy process is not an isolated event. The strategic plan should not be created and then dropped on a deserted island. The strategic process is a continuous business function that is exercised regularly. Global situations migrate states of conditions. Customer needs change. Advancements in technology or new opportunities may appear to be valuable. The executive board of the organization evaluates earlier decisions for continued relevance. As new information is gathered and evaluated, updates are made to the strategic plan and its execution. Expectations of performance may also be adjusted.

Industry and global competitive analysis

Part of the work in building a strategy is to understand internal and global environments through a situation analysis. Internal analysis focuses on the core strengths and weaknesses of the organization. Competencies, capabilities, and competitive strengths are examined. External analysis focuses on global factors, industry advancements, and competitive conditions. The results of this analysis are fed into the mission, vision, and strategic plan for the business.

The external factors to an organization within the industry include buyers, suppliers, competitors, alternative products or services, and new entrants to the market. When you step beyond the industry, you have other conditions to examine, including technology, government regulations, economic conditions, society, and culture. There are a number of key areas to focus on when analyzing an industry and competitive environment.

- Dominant economic traits for the industry
- Global competitor analysis
- Drivers of industry change
- Competitive forces
- Success factors
- Industry attractiveness

Dominant economic traits for the industry.   When you evaluate the economic traits for an industry, you look at a number of factors. How many competitors are there in the industry globally? What is the industry's size and growth rate? What are the customers' needs and requirements? What is the pace of technological advancements and product innovation? How different are the products within the industry? How can economies of scale be used to improve a market position?

Global competitor analysis.   In the analysis process, the competitors in the target industry and global market are analyzed for differences. A strategic group mapping can be done to graphically depict the organizations with a similar competitive market position. Strategic groups can be identified as having a similar price or quality range, using identical technology, offering similar services, and/or covering the same geographic territories. When an organization analyzes a strategic group map, they look for driving forces and pressure from other organizations. They also evaluate the potential for differences among the groups.

Competitor analysis examines the strategies of the rivals. How do the rivals position themselves? How do the rivals rank now, and what are the estimates for their ranking in the future? What is the likelihood that the rival would enter a new market or produce a new product?

Drivers of industry change. Industries change due to factors that motivate and alter the actions of those involved in that industry. Drivers can affect the demand for products to increase or decrease, make competition more or less intense, or lead to higher or lower profit margins. Types of driving forces that have changed industries include the Internet, globalization, government regulation, innovation, and advancements in technology. Customer tastes, attitudes, concerns, and lifestyle changes can also be driving factors for industry change.

Competitive forces. Research shows that different industries can sustain different levels of profitability due to their structure. An analysis of competitive forces examines the forces within the industry and their strength. A key tool used to analyze forces is Porter's Five Forces Model of Competition. This model depicts how an industry is influenced by five forces. It allows for a better understanding of an industry in which a business operates.

- *Power of the customers.* The force of the power of the customers examines the buyer. What type of impact does the customer have on the industry? Are the buyers well informed about the products? Is there a lot of differentiation between products? How strong is the buyer bargaining power? What is the relationship between buyer and seller? Are there incentives? Is there price sensitivity and/or brand identification?

- *Power of the suppliers.* All organizations are in need of suppliers for the supplies to build their product. This need leads to buyer/supplier relationships. Powerful suppliers can exert influence on an industry. How strong is the bargaining power from the suppliers of raw materials, components, parts, and other resources?

- *Competition from substitutes.* What type of pressure do other industry competitors place on the industry being analyzed? What types of alternative products or services could compete with the organization's products or services?

- *Barriers/threats to potential entry.* Barriers prevent new entrants into the market. One barrier for an entrant could be the large start-up costs. Another barrier could be extreme uncertainty of the industry. Barriers can be created by governments, patents and proprietary knowledge, or economies of scale. There is always the possibility for new entrants to the industry. What advantages would a new entrant have over a longtime contender in the marketplace?

- *Rivalry among competing sellers.* What competition is presented by industry competitors jockeying for more market share or position? Some ways to combat rivals would be to change price, improve or differentiate the product, improve distribution channels, and enhance relationships with suppliers.

To use the Five Forces model, first you need to identify the issues with each of the forces. Then evaluate the strength of each item under all the forces. Then determine a mitigation plan to capitalize on or alleviate each issue within each force. This analysis could decipher an unattractive industry. An industry could be too competitive if rivalry is vigorous, substitutes are strong, barriers are low to entrants, and suppliers and customers have enormous bargaining power. On the other hand, if barriers to entrants are high, there are no good substitutes, suppliers and customers have a weak bargaining position, and rivalry is moderate, the competitive environment could be considered ideal from a profit-making standpoint.

Global success factors.   To understand the success factors for an industry, an organization examines the access to skilled labor, technological research in the industry, ability to manufacture with access to materials and economies of scale, and the distribution chain to get products to market.

Industry attractiveness.   Industry attractiveness refers to the conditions of the organization in relation to the industry to determine suitability. Some factors help determine the attractiveness of an industry: market size and growth potential, degree of risk or uncertainty, the organization's capitalization of strengths within the industry, and the strength of the organization in relation to the competition. The opportunities available within an industry depend on an organization's ability to capitalize on them.

Organization's internal factors

External factors that affect an organization help determine market position and areas that a business might want to capitalize on worldwide. Internal factors help an organization understand how well they can take advantage of opportunities and eliminate threats in the marketplace. The organization examines how the present strategy is faring. Have the organization's strengths and weaknesses changed over time?

The company could have built strengths in research and development by creating new products or more efficient processes. Improvements could have been in the supply chain strategy or manufacturing processes. Gains could have been made in marketing, promotion, or sales. The organization's talent pool and financial strategy may have improved. All these strengths, over time, could flip and become weaknesses.

You can see this trend most prominently in the retail industry. Take McDonald's, for instance. McDonald's initially introduced their "happy meal," originally called Circus Wagon Train, in 1977. This product went national in 1979, and was a huge success. In the 90s, childhood obesity statistics were alarming, and one of the culprits cited for the rise in childhood obesity was fast food. This report caused much anxiety for parents, which led to the happy meal not being a popular meal choice for concerned parents. This criticism slowed sales, but over time, McDonald's repaired their image by introducing healthier items for children.

Some industries don't recover as well when a strength turns into a weakness. The American auto industry is known for their large gas guzzling automobiles, which fall in and out of fashion as the price of gasoline and oil rises and falls.

When analyzing the organization's competitive strategies, you have to examine the competitive approach scope. Has the organization strived to be a low-cost leader, focused on a niche market or differentiation of the product or service? Has the organization focused on a specific geographic area?

To determine how well the current strategy is doing, you have to examine both qualitative and quantitative measures. How well has the company met the current strategy? Is the strategy still relevant? Is the company achieving its financial objectives? How does the organization compare to industry benchmarks?

Some of the indicators that can be used to determine if a strategy is successful are:

- Trend in profit margins
- Return on investment (ROI)
- Reputation with customers
- Ability to acquire and maintain customers
- Position in the market (leader in technology, innovation, or quality)

SWOT

For a company to successfully deliver a strategic plan, it needs to be well aligned with the company's strengths and market opportunities. At the same time, it needs to defend against threats in the market and internal weaknesses. A SWOT analysis allows an organization to position itself to its best advantage.

Strengths and weaknesses.   Strengths are what an organization does well and are a benefit for its competitiveness. Strengths can be assets that the company possesses. The organization could have key competencies in areas that surpass their rivals'. The workforce could have certain skills, expertise, and experience that accumulate to a proficiency in an activity. The strengths in an organization could be the ability to get new products to market quickly, accuracy in filling customers' orders, or efficiency in supply chain management.

The value of a strength that a business possesses can be determined by its ability to be copied, repeated, or easily mitigated. Could other organizations easily copy the strength and adapt it into their business? Could it be repeated effectively and efficiently? Could rivals easily mitigate the strength with alternatives?

Weaknesses are the opposite of strengths. These are areas where the organization struggles, does poorly, or lacks expertise. Weaknesses are competitive liabilities.

Opportunities and threats.   Opportunities and threats exist outside the walls of the organization; they exist in the industry or beyond. Opportunities provide potential advantages for an organization. Threats could harm the long-term competitiveness of an organization if not handled appropriately. Numerous conditions could be considered threats to an organization.

- Rise in interest rates
- Demographic shifts
- Political upheaval
- Government regulations
- Entrants to the industry with better price, products, or position
- Emerging technologies

A threat could be turned into an opportunity; however, the opposite is true, too. A prominent scenario for threats and opportunities falls under the emerging technologies category. As innovation reaches an industry, how well a company embraces and capitalizes on the change can lead to the innovation being categorized as an opportunity or a threat. Some threats that most companies face today are souring energy prices worldwide and global climate change. Green technologies like solar and wind energy are improving and becoming a promising solution. How a company embraces the trend to "go green" could be an opportunity or a threat to the future growth of the company.

The SWOT analysis can be a powerful tool in determining the best match for a company's strengths and market opportunities. It could also identify weaknesses that can be corrected and threats that need to be defended against.

Strategic cost analysis

To continue the analysis of a company's strategy, one must examine the organization's costs and how they compare to their competitors. The two key analytical tools used are value chain analysis and benchmarking.

The organization's value chain consists of all the activities required to take a product from concept to execution to the customer's hands. The value chain includes product design, production, marketing, distribution, and ongoing support. Both primary and support activities affect the value chain. The primary activities directly affect the value change and can be compared to the critical path in a project. These activities include operations, distribution, sales and marketing, and service. Support activities indirectly affect the value chain and are activities around research and development, human resources, and general administration.

Value chain analysis looks at the cost of the activities, or what could be called the internal cost structure. It compares the internal costs with those of its competitors. Who is better able to procure raw materials or get a better price from their customers? The analysis can determine if the costs are an advantage or a disadvantage for an organization. There could be many reasons for the differences in costs between an organization and its competitors. Each organization is structured differently, and those differences can lead to varying internal operational costs. The way an organization approaches strategy and how it chooses to execute it are the key.

The value chain can be examined for the entire industry. Key industry activities need to be determined, along with their associated costs. As in project management, activity-based costing is used to analyze costs, and it determines whether the company's costs are aligned with their rivals.

Benchmarking

Benchmarking compares the cost of activities and provides information on the standard or best practice in a given area. It provides guidance for an organization on how it measures against industry standards. This information allows an organization to generate a plan to either meet or exceed industry standards by lowering costs on specific activities. Activities that could be cost-benchmarked for an organization include training of employees, purchasing materials, paying suppliers, and processing payroll. Lowering costs in one or more activity can make a company more competitive.

There are a number of ways to improve the costs of activities. Some cost activities can be eliminated by streamlining the value chain. High-cost activities can be relocated to lower-cost geographic areas or performed by cheaper vendors/suppliers. Best practices can be implemented throughout the company to streamline costs.

A well-managed value chain can provide a competitive advantage for an organization. The organization can better coordinate activities to enhance their capabilities, and economies of scale can be leveraged.

Competitive advantage and strategy

Competitive advantage is when one organization has an advantage over another organization within their industry. The advantage could be how they attract customers with a lower-cost solution with the same type of product. The advantage could be that an organization offers the same cost, but a different or more advanced product or service. Finally, the advantage could be on a narrow focus of the market in a niche area.

A competitive strategy for an organization focuses on enhancements to customer relationships, responses to market and global conditions, and strengthening market position. An organization can use generic strategies. These strategies focus on cost, differentiation, and segmentation. An organization can focus on being the low-cost or best-cost provider of a product or service. The low-cost option is achieved by not only having lower production costs, but all the costs are lower as well. The value chain for the organization outperforms that of its rivals, or it streamlines its operation to offer savings. The organization looks at cost controls or at bypassing costs to produce a low-cost option. Where a low-cost provider provides the same product at a lower cost, a best-cost provider provides more value for the money.

Differentiation offers customers more value that cannot be easily matched by competitors. Differentiated products can build brand loyalty and allow for a price premium. Differentiation can be captured by uniqueness, one-stop shopping, prestige, technology features, wide selection, or superior service.

Niche strategies concentrate the corporate strategy on a narrow piece of the market. The narrow focus allows unique capabilities to be added to the product or service to better target the needs of the customer group. To define a niche market, an organization can analyze the geographic uniqueness, special requirements in a product or service, or an emerging need that is not being met.

There are risks to a focused strategy. Competitors will eventually meet or exceed your competitive advantage, and niche segments may shift preferences. Focused strategies can offer a number of advantages, including better positioning. When selecting the strategy for an organization, it is important to look back at the industry and the internal analysis to determine the best fit. Does the strategy allow the organization to enter new markets, capitalize on its strengths, or defend against threats?

The strategy that gets you to where you want to go will not necessarily be the same strategy that keeps you there.

Alliances and collaborative partnerships

An organization may create a strategy, but how to execute on that strategy can be another challenge. Through analysis, a business may find that they have weaknesses or outside threats that they can't manage. An available option is to develop alliances or collaborative partnerships with other organizations to strengthen their competitive advantage. Alliances can take the form of shared research efforts, technology sharing, or shared manufacturing or facilities costs. Alliances could market each other's products or services. Costs could be lowered and/or access to needed capabilities or expertise can be gained. There can be many advantages and disadvantages to this strategy. Alliances provide participants access to a global platform while preserving each company's structure. They also provide competitive advantages over rivals and enhance market position.

However, alliances can fail due to issues with collaboration and the willingness of parties to adjust to changing conditions, among other issues. Alliances can be time consuming. There could be barriers to effectively working together in the form of corporate cultural differences and mistrust by parties on both sides. It can be difficult to discuss sensitive issues. Another challenge is that one organization may get too dependent upon the other.

When forming a strategic alliance, it is important to select an ally who has complementary strengths. An organization should select a business where there can be mutual benefits, without concern for divulging sensitive information. Alliances tend to be temporary in nature, because the two companies come together for a purpose and after the objective is met, they tend to go their own directions.

Merger and acquisition strategies

Moving to the next level beyond alliances and collaborative partnerships, you have mergers and acquisition strategies. Mergers happen when two equal companies decide to join. An acquisition is when one organization purchases or absorbs another. Here, many of the benefits that you have with alliances and partnerships are realized. The newly formed organization, whether in a merger or acquisition, would benefit by gaining access to new territories, technologies, products, services, skills, and knowledge. Other benefits with the newly formed company can be in areas of cost savings and enhanced competitive advantage.

It can be difficult to combine the operations of two different companies. Conflicts can be found in styles and cultural differences, which can limit or even prevent the synergistic capabilities.

Outsourcing

In a global environment, outsourcing becomes a fundamental strategy for success. In outsourcing, an organization hires a second business to provide needed services, products, activities, or support. An organization should consider an outsourcing strategy when a second party could perform the service more inexpensively, or if local understanding of the culture, government, and policies is needed. It can also be a good strategy if the second party provides higher quality and/or faster delivery of products at more economical prices.

Though outsourcing has many advantages, there are factors to consider. When too many areas are outsourced, it provides the potential of losing touch of the operation and lowering internal expertise.

Global competitive strategies

Going global adds another dimension to the strategic plan. Will the organization deploy the same strategy in all countries or modify it for each area? How will the business be structured, dispersed, or centralized? Are the products and services going to be customized or standardized? How are communication, information, and knowledge going to be transferred to all the locations?

There are a number of reasons to venture beyond one's borders. Global businesses have the opportunity to gain new customers, spread business risk across a larger market, obtain access to resources, capitalize on core competencies, and help lower costs. There is a difference between a global and an international company. Global companies are expansive and operate in 50 or more countries. International companies tend to operate in a few countries and have conservative plans for growth in that area. A global strategy is beneficial if the buyers require little customization and tastes are more homogeneous for the products that are offered. With a global strategy, organizations operate worldwide.

Global businesses face a number of challenges and need to manage them with insight and flexibility. There are cultural and market variances. Materials, manufacturing, and distribution costs vary. Local conditions place different governmental, political, and economic demands on organizations. At the customer level, their tastes in products and their buying habits and capacity vary. There is a condition where one size doesn't always fit all. Organizations are challenged with the decision to offer a standard product abroad or customize it for local tastes.

There are benefits to globalization. The availability of natural resources and cost-effective sources of energy provide a fertile ground for growth for some industries. Wage rates and labor skills, too, can be assets or liabilities in a different geographic location. The value chain for an organization can vary greatly, depending on a particular location's access to the sources that produce the activities in the chain. Governments have different policies on exports and imports. Tariffs, a government tax placed on imports or exports, can add to the overall cost of production. An organization can obtain a competitive advantage with regard to its global strategy by locating activities in the most beneficial locales.

Often within an industry, competitors compete globally or multinationally in the same geographic areas. If one multinational organization "wins" in one country, it doesn't guarantee that they will win in all countries. Global competitors, like the Coca-Cola Company, look for worldwide branding and positioning.

There are many different types of opportunities to compete in internationally. Products and services can be franchised, exported, or licensed.

In a franchising strategy, the franchisor licenses the organization's products or services to a franchisee in an exclusive geographic area to operate the franchisor's business model and carry the brand name. Per the contractual agreement, the franchisee pays the franchisor a fee to operate the franchisor's business model. In return, the franchisor trains and supports the franchisee to maintain the quality and standards that are incorporated into the business model. It may be difficult to maintain the quality standards with franchisees that are geographically dispersed. However, a franchisor is able to expand into new territories with these agreements without capital risk or the need to develop expertise in understanding the conditions in a specific territory.

Exporting provides a way to sell abroad by maintaining the domestic production chain. This strategy allows the ability to test markets without the risk of capital investment in local production, as well as the ability to reach new customers. Exporting can be a challenge when production costs abroad can be more cost effective. Other challenges can be found in export, import, and shipping costs. Variations in exchange rates at times could be an advantage or a disadvantage with exporting products.

Licensing could be a viable strategy if the organization has technical expertise or a patented product that is not available abroad. Licensing grants the licensee permitted access to proprietary information. There could be many limitations, exemptions, and conditions in a licensing contract. The licensee receives access to valuable information, processes, and/or technology. Licensing is a strategy that helps mitigate the risk of committing resources to a geographic area that the company does not have internal expertise in.

Another strategy could be to use alliances and collaborative partnerships with complementary competencies, which could provide additional leverage for penetration into an uncharted market. This strategy allows the competencies of each organization to amalgamate in a synergistic way to provide access to new markets that neither of the business could have undertaken individually.

Multicountry strategies provide a good way to begin a global enterprise. The strategy is matched to the host country's conditions. A multicountry strategy is a good option when countries are diverse, customer tastes require customized products, government regulations prevent a standardized approach, or if demand for the product is not consistent across the globe. It can be difficult to coordinate efforts across countries, and it may not be a competitive advantage to operate in this fashion. Conflicts can occur in transferring competencies across countries, which can turn into a liability for the organization.

Location, location, location

The real estate industry will tell you there are three Ls to their business: location, location, location. Decisions need to be made on how the organization will structure its location strategy. Should activities be concentrated in a central location or dispersed geographically?

A concentrated location strategy is best to achieve economies of scale and enhanced coordination of activities. At times, it is more cost-effective to perform activities centrally. It could be easier to scale an operation if it is centrally located. If there is a concern with access to materials or an extensive learning curve in order to perform the activity, a central location for all activities is the best option.

A dispersed strategy allows activities to be located where there is a cost advantage or a need to be close to the customer. This strategy helps alleviate trade barriers, distribution costs, and buffers exchange rates. It contributes to transferring knowledge and expertise to other locations. An organization can take advantage of favorable costs or conditions. Brand recognition can be enhanced globally if there is a local presence. An organization can choose how, when, and where they challenge their competitors.

Profit sanctuaries

The intent of organizations may vary industrywide. However, when an organization discovers a profit sanctuary, they take full advantage of it. Profit sanctuaries are markets where a business has a protected and strong market share, with sustainable profits afforded to them. The position of the organization often allows for a substantial percentage of the profits in that market. Often, a profit sanctuary will be in the organization's home market, which is strategically crucial to them. An organization is considered a fierce competitor if it possesses a number of profit sanctuaries, with the ability to aggressively compete with a domestic competitor that only has the home market as a profit sanctuary.

Critical markets

Critical markets are in countries that offer profit sanctuaries for competitors. These locations possess sought-after and essential buyers whose business is important to the organization. Sales are generally high and offer good profit margins. Competing in all markets would not be advantageous to the organization. To be a successful competitor in a global industry, it is essential to compete in the critical markets.

Other strategies

Countless business strategies are available for an organization to select from to incorporate in its strategic plan. Industries evolve over time, from emerging and turbulent, to mature, stagnant, and fragmented. The position an organization is in changes over time, from new entrant and rapidly growing to an industry leader or follower. The organization needs to be able to understand the market and internal conditions to manage their strategy over time.

Strategies for emerging industries. Emerging industries are a fertile ground for an organization to be bold and creative. Quickly entering an emerging industry allows you to gain the first-mover advantage positioning. The focus is on pursuing new customers and gaining entry into new locations. There is risk with emerging industries. As the dust settles in an emerging industry, a number of competitors will enter the market.

Strategies for turbulent, high-velocity markets. Turbulent markets are characterized by exponential advancements in technology, short product life cycles, and evolving buyer expectations. The turbulent markets force the need to invest aggressively in research and development. It is critical to be able to shift resources, adapt new competencies, create new capabilities, and match or exceed competitors' abilities to get new products to market. Strategic partnerships could be developed to better maneuver high-velocity markets. Agility, innovativeness, and flexibility are keys to success in a turbulent market.

Strategies for mature industries. Mature industries present challenges in finding ways to be more efficient and effective. The "doing more with less" philosophy is crucial in a mature industry. The value chain should be streamlined. Competitors may be available for purchase. It is also a good time to explore untapped international markets and develop flexible competitive capabilities. Strategies to differentiate the organization become paramount. The focus should be on long-term competitiveness. Capacity should not exceed buyers' needs.

Strategies for stagnant or declining industries.   In a stagnant or declining industry, it is important to not be overly optimistic about the future. Generally, the customer group is shrinking and tastes are changing. Substitute products may have been carved into the market. However, you don't want to divert resources to other opportunities too quickly, though you do want to focus on growing market segments. The focus for a stagnant market is to provide differentiation based on quality or feature sets. There is also the need to drive down costs, which can be managed by streamlining, consolidating, or outsourcing activities.

Strategies for fragmented industries.   A fragmented industry is one where there is no market leader; customer demand is dispersed; this is a young industry with low entry barriers. Fragmented industries need a strategy to become the low-cost provider. In addition, focus can be placed on specializing in customer or product type. Another strategy would be to operate in a limited geographic area.

Strategies for industry leaders.   An industry leader is an organization with a proven strategy, strong reputation, and a powerful market position. When you are the leader in an industry, the focus is on maintaining that position. To keep your present hold on an industry, it is important to protect the competitive advantage and constantly strengthen the current market position. A strategy could be to make it difficult for new entrants to gain access to the market by challenging the positions that they bring to bear. To challenge new entrants, an organization needs to provide high levels of customer service, build strong brands, increase advertising, invest in research and development, and keep prices low while maintaining quality.

Strategies for runner-up firms.   Runner-up organizations fall short of the industry leader position. Runner-ups can focus on niches in the industry, work to gain market share, or maintain status. When prices are dropped, new features are added, or new services are offered, the runner-up organizations need to keep pace with the industry leader. The focus for runner-ups is to build market share or change markets. To build market share, the organization can focus on being the low-cost provider or provide additional features that differentiate itself from the rest of the pack. They can also focus on a niche or specialist strategy. The runner-ups have challenges in brand awareness, funding ability, and weaker economies of scale.

Strategies for weak and crisis-ridden businesses.   There are a number of options for a weak business. If resources permit, they can focus on a turnaround strategy. In this case, the strategy is updated and new objectives are created. To generate funding, assets may be sold.

If funding is not available, the organization may have to pursue a defend strategy. It is also possible to engage in a fast exit or a slow exit in a harvest strategy. The harvest strategy focuses on generating cash flow in the short term and exiting the market in the long term.

Organizations should avoid spreading themselves too thin by pursuing all opportunities. Success comes from well-positioned strategies that leverage core competencies harmonized with industry conditions.

Diversification strategies

Diversification allows an organization to pursue the strategy of growth by increasing profitability through greater sales volumes. A diversified firm is a grouping of a number of businesses under a single organizational umbrella. Organizations generally start as a single entity and then grow and expand to other regions. Growth for an organization can be acquired through market or product development, market penetration, or diversification. Market penetration focuses on the same products and the same market and on increasing the organization's market share. Market development takes the same products and sells them in new markets. Organizations could also develop new products for the same market, which would then be a product development strategy.

Diversification is needed when other growth opportunities have been depleted. Diversification could be a solution when there is strong competition and market growth is slow, or when the organization is in a weak competitive position and there is market growth. Diversification provides an avenue to increase profits and shareholders' value.

Summary

Much has been written on business strategies. Only a sampling of the information was presented in this chapter to provide an overview of the considerations that must be identified, analyzed, and executed by an organization.

There are many strategies for expanding beyond the local borders of the organization's home field. Often, to survive global competition, a company may be forced to abandon its focus on a local business operation. For an organization to be able to enter a new market frontier, it must have assets that are suitable for maneuvering in the diverse market conditions. The business would then need to construct a multinational strategy to thrive. Global companies can undercut domestic-only firms with severe price cuts and offset the losses in other parts of the business. Industry leaders hold an advantage in that they have a strong proven brand. However, leadership in an industry can be eroded quickly if an organization is not agile and is unable to realign its resources and activities to compete more effectively.

An organization must make crafting and executing a strategic plan a top priority to maintain its competitive advantage. The strategic plan must be clear and actionable, and built upon the company's strengths, assets, and market opportunities. The plan must also mitigate a business's weaknesses and defend against market threats. The strategy needs to allow the organization to differentiate itself from the rest of the pack and highlight growth potential. When an organization is competing to be the leader in the industry, they must be aggressive in obtaining that objective. If they have achieved the market leader position, they need to be just as aggressive in defending that position. To manage your success in a global backdrop, strategies should be avoided that allow success in only the best of circumstances. Strategies need to be flexible and buoyant. Understanding the industry and completing a competitive analysis will help an organization prepare for the competitors' reactions.

To compete aggressively in a global market, a firm needs to be well funded, or have a strategy to offset funding shortfalls through opportunities with alliances, corporate partnerships, franchises, and licenses, among other strategies. Building a strong value chain process could help the bottom line, as well as allow for a competitive advantage.

For an organization to be successful, it needs to have a strong portfolio, program, and project management process. With a strong portfolio process, the company's mission, vision, and strategy are kept alive and active. Investments and resources are aligned with the objectives of the organization. Projects tactically produce the deliverables that they were formed to manage. Programs align the activities in projects to that of the strategic plan to produce business benefits.

The Standard for Program Management According to PMI

Introduction

In Chapter 1, we explored the process of "globalizing" a company, equipping the business to think and plan globally and to enable it to design, develop, and deliver programs globally. In effect, the business becomes *global* when its people think, plan, and act globally and its programs extend to worldwide markets. The process involves internal planning and business communication, and sometimes it requires major restructuring. This transition can be difficult for companies that do not have a global culture and a history reflecting a worldwide perspective and global market horizon. However, the process is necessary to successfully deliver global programs.

In that chapter, we addressed the global strategic planning process and the articulation of mission, vision, goals, objectives, programs, and projects in a worldwide context. Our bottom line in that chapter was that if a business or agency intends to deliver programs globally, it must reflect that intent in all of its business planning and strategic thinking. A portfolio of programs and projects that is generated from the strategic plan of a globalized company has a much better chance of succeeding simply because of the analysis and planning involved.

What Is Program Management?

Before we go into the global program management process, we want to describe in this chapter what program management is and how it differs from project management.

The standard discussed in this section has a specific language with valuable processes and has been commonly accepted for creating program success. Awareness of the program management methodology helps those who are interested in the field gain access to the breadth of the practice. An understanding of this practice prepares project managers to collaborate and interact with program managers. This standard also helps program managers understand the nuances of their position. Moreover, it allows portfolio managers, stakeholders, project teams, and anyone involved with a program to understand how they interact with program managers and what they can expect from them. This information is, therefore, essential for any person who has to manage multiple projects that culminate under an umbrella of work to achieve a successful result.

Project Management Institute

The Project Management Institute (PMI) is a nonprofit professional organization dedicated to advancing the project management profession. PMI was founded in 1969 by five working project managers, and has grown into an international organization with members in more than 170 countries. According to the PMI website, "Project management allows an individual to speak with one common language, no matter their industry, geography, or whether they manage projects, programs, or portfolios. This common language steers organizations toward achieving repeatable, predictable results."

The primary goal of PMI is to advance the profession of project management throughout the world by setting standards, conducting research, and providing education to strengthen the profession. PMI offers a number of certifications, which are held by people who have the professional experience and understanding of the standards. These certifications include:

- Certified Associate in Project Management (CAPM)
- PMI Scheduling Professional (PMI-SP)
- PMI Risk Management Professional (PMI-RMP)
- Project Management Professional (PMP)
- Program Management Professional (PgMP)

This last and most recent certification, PgMP, continues the PMI mission of advancing the project management practice. In 2006, PMI published the first edition of *The Standard for Program Management*. This book provides a guideline for managing programs and defines program management and its life cycle. In 2008, the second edition of this book was released with a number of updates. Those who earn a PgMP certification have the experience and proven ability to manage programs. They also have passed the certification test, demonstrating their understanding of the standards.

This chapter explores *The Standard for Program Management* and further explains the concepts in this publication. The standard focuses on all the activities within a program, which may or may not be project-related work.

The PMI *Standard for Program Management* defines a program as "a group of related projects managed in a coordinated way to obtain benefits and control not available from managing them individually." Programs are seen as long-term business initiatives that involve both projects and supporting systems, and that focus on benefits and broad outcomes. Programs involve more than managing many projects at the same time. Program managers oversee the work of multiple projects and project managers, and coordinate their efforts so that, in combination, they can achieve a program goal and business objective in the company's strategic business plan. A program is a synergistic effort. Here, the combined impact of the projects that make up the program is greater than if they stood on their own. Together, as a program, the projects achieve the business result that is desired—and often required—for the short- and long-term success of the company.

The definition of a program involves scale, breadth, and timeline. In terms of scale a program is wider and deeper, involves more of the company's resources, and involves the integration of many projects and supporting systems. Programs are typically global in nature because they tend to involve other countries and markets in both program support and market opportunity. A program is required when the degree of the work involved in a business goal is large enough that a coordinated effort over time is necessary. Once achieved, the program provides an enormous benefit to the business and to its clients and customers by ensuring that not only project deliverables are produced, but also that business goals and program benefits are achieved. When there are dependencies among projects, or if there is a need to consolidate and prioritize issues, it is best to centralize the oversight of the projects within a program structure.

Program management is not about managing the intricacies of individual projects, but about managing the big picture—the global benefits that the business is expecting. With a program, you gain the benefit of integrating all of the project's efforts into a program dashboard of information. This information can be presented to executive management and/or the portfolio governance committee for a streamlined view of the status and progress of the program. Within a program, you combine the triple constraints of cost, schedule, and scope into one system. You optimize the integration of cost, schedule, resources, and efforts over multiple projects in the program.

Usually, programs are managed within one company with shared resources. The program manager ensures optimum use of the business's valuable resources. At the program level, you manage risk that affects multiple projects or affects the program as a whole. The program manager becomes the single point for escalation of issues or scope changes from the projects within his or her domain. The program manager has a strategic focus and understands how a change in one project may affect the other projects in the program or the strategic benefits of the program as a whole. A program allows the ability to produce integrated deliverables across multiple projects.

Overall, program management can be thought of as the centralized, coordinated management of a program to achieve the program's strategic benefits and objectives. It involves aligning multiple projects to achieve the program goals and allows for optimized or integrated cost, schedule, and effort.

Projects can be assembled under a program umbrella if there is a connection or an association with a specific market or customer. For example, a convenience foods business can target a specific market, like an organic foods market, or "go green," meaning to eliminate harmful chemicals, products, or processes and replace them with environmentally friendly alternatives.

According to the Kellogg's website, this company has a strategic focus on reducing their impact on the environment:

> Corporate responsibility has always been the foundation of Kellogg Company and a key part of our heritage and culture. More than a century after our founding, we remain a company consumers rely on to provide consistent, high-quality, great-tasting foods. "Now more than ever, it's important to do this while minimizing environmental impacts and positively addressing global challenges," said David Mackay, president and chief executive officer, Kellogg Company.

Here, the Kellogg Company could undertake various projects to achieve the goal of "minimizing environmental impact" and could group them under a single program to track the cumulative *benefits* of this initiative.

Projects could also be grouped together based on a shared capability. Projects that have collective capability are those that deliver an integrated product. An example of collective capability would be a company that would like to launch a new product. Here, each functional area may have a project within its group, but a program is necessary to tie the work together to achieve the desired collective benefit. The functional areas that would work with a program could include:

- *Research and development.* Generates the product idea or concept and may also test new product feasibility for manufacture.

- *Marketing.* Researches markets to determine if this product has a viable market with no or limited competition. This functional area would later develop a product launch plan, including packaging and other marketing material.

- *Manufacturing.* May need to develop an environment to produce the product.

- *Finance.* Focuses on understanding the profitability of the product as well as the development of a break-even-point analysis for decision-maker review. They would also do a business analysis for selling price based on completion and customer feedback.

Projects could be centralized in a program under many different circumstances. Besides new product development, you could centralize projects based on escalation level, direction change for the organization, resource constraints, or risk mitigation.

- *Escalation level.* Projects could be linked together based on the escalation level for specific components such as quality, risks, or constraint. Escalation is needed when issues cannot be handled by lower levels of staff. Combining the projects into a program ensures that issues are raised to the appropriate person's or group's attention. It allows the appropriate managerial level to resolve the issue in an effective and timely manner. The escalation path could have two parts: one within the program and the other from the program level to the executives in the organization or stakeholders for the program.

- *Organizational change in direction.* The organization may change direction, which will affect projects and how they relate to each other. With a change in direction, the program may no longer support the organizational strategic goals and, therefore, may no longer be needed. On the other hand, the change may cause the deliverables of the projects under the program to be redirected. A program structure allows for better navigation of changes in a business's direction or strategy.

- *Resource constraints.* Grouping projects together according to the resource constraints would provide an optimal use of resources. Constraints are circumstances that place boundaries around a program's ability to maneuver. Program constraints could be related to time, cost, resources, and/or deliverables. The program allows a program manager to identify and define constraints across the projects within the program. Furthermore, it allows a better way to analyze and prioritize resource needs. With the knowledge of resource constraint analysis, a program manager could manage the constraints to provide the best optimization of resource allocation.

- *Risk mitigation.* Projects that need to mitigate or reduce the probability and/or impact of a risk occurring should be linked together. Here, a program focuses on the full risk management process and uses synergistic effort to create contingency plans and to form decisions around risks to resolve them at the program level.

- *Task dependency.* The program focuses on interdependencies of cross-project deliverables and manages them at the appropriate level. Linking projects with task dependencies can provide improvements to schedule and cost management by eliminating redundant efforts and enhancing how dependencies are managed.

- *Global barriers.* Programs allow the ability to manage processes and interfaces across dispersed and global teams to handle language, time, distance, culture, environment, and resource barriers. It is best for projects to roll up into a program with additional expertise in global management to aid the projects with global challenges that they may encounter.

Relationships with Other Project Standards

The project portfolio is the connection between the corporate strategy and the investment in projects and programs. In your portfolio, business executives make the decision as to whether or not they will invest in an initiative and how it will be implemented—whether in a project or a program. The project portfolio is the 50,000-foot view of the business project investments. A project, which is a temporary endeavor to achieve a business deliverable, is at the ground level of work. A program is between these two: the portfolio's 50,000-foot view and the project's ground level view. In a program, a program manager navigates from the ground level at times, and may roll up her sleeves and dive deep into a program issue; other times, she focuses on a higher strategic level. Portfolio management is a continual process and becomes a major functional or operational area within the business. Programs and projects are temporary. At the conclusion of a project or program, after the deliverables or planned strategic results are accomplished, resources are released and the project or program closes.

Projects produce planned deliverables. Programs deliver results, which are capabilities and benefits that a company can use to achieve its goals. Another way to look at it is that *projects deliver outputs and programs produce outcomes*. Portfolios focus on identifying, selecting, prioritizing, and providing resources for projects and programs.

PMI defines a project, a program, and a portfolio as follows:

> Project management is the application of knowledge, skills, tools, and techniques to project activities to meet project requirements. A program is composed of multiple related projects that are initiated during the program's life cycle and are managed in a coordinated fashion. The program manager coordinates the efforts between projects, but does not directly manage the individual projects. A portfolio is a collection of components (i.e., projects, programs, portfolios, and other work, such as maintenance and related ongoing operations) that are grouped together to facilitate the effective management of that work in order to meet strategic business objectives. The projects or programs of the portfolio may not necessarily be interdependent or directly related.

While program managers oversee the efforts of projects, they do not directly manage any of the projects. Program managers administer the interdependencies of the projects and focus on identifying, monitoring, and controlling these interdependencies.

The Role of the Program Manager

The program manager's responsibilities are distinct from that of a project manager.

It is critical for the program manager to understand and abide by the "rules of the road" in the office. Program managers need to adapt their management style to that of the company's culture. When managing the program team, a program manager needs to understand and follow the norms and policies of their company. A program manager will have varying degrees of authority, depending on the structure of the company. If they work in a functional company, they will have little influence, while in a fully projectized organization, they will have more control over the program and how it operates. This is also true for the reporting structure for a program manager. In a functional organization, the program manager would report to a specific functional area, while in a projectized organization, they would report to the director of the project management office (PMO). Most programs will fall between the extremes of functional and projectized organizations and run in a matrix structure.

As a project manager manages a project, a program manager directs the projects within the program and provides guidance and support to the project managers. The program managers manage the project's interdependencies among each other, not the projects themselves. The program manager focuses on the links between projects, which is an iterative process throughout a program. At the beginning of a program, in the planning phase, there may be a top-down approach to planning and organizing the program. The interaction between a program and a project is from the program level, providing information and guidance to the project level. As the program progresses, the top-down approach transforms into a bottom-up approach as information from the project teams is raised to the program level. The project teams provide the program manager with schedules, costs, risks, and issue information and reports. This information gathering is an iterative cycle. As the program manager receives the information, she may redirect the project manager to stay on course for the program to achieve the business benefits that are required.

While a project manager focuses on preventing changes, the program manager focuses on producing benefits and achieving results. Project managers lead a project team by focusing on project deliverables. A program manager works to mitigate conflicts and manage shareholder expectations. Project managers motivate their teams to focus on their task. Program managers set the stage and provide the vision and inspiration for the course of the program. Where a project manager monitors and controls tasks, a program manager oversees projects and non-project-related work. They also have to keep an eye on the broader picture of the governance or portfolio process. Program managers must understand corporate changes of direction, as well as opportunities and threats from their industry.

Project managers have a diverse team to manage. They may manage functional managers, specialists, and technical experts. Program managers, on the other hand, only manage project managers. Project managers create detailed plans. Program managers develop high-level planning and reporting tools, and provide guidance. A project manager's scope of work resides within only one project. A program manager's scope encompasses a broader theme of benefits delivery through the oversight of many projects. Project managers can measure success of their projects by time, cost, and quality. A program manager's success is determined by benefits delivered, enhanced or new capabilities for the company, and the return on investment. When you look at the overall difference between a project manager and a program manager, you can think of a quilt. The project manager focuses on delivering one square on the quilt. The program manager, on the other hand, focuses on the overall technique of execution, all the squares, and how they will be combined to create the overall design or pattern. Program managers also have to focus on other aspects of the quilt, like the batting, the backing, how it is put together (or "tying"), and how the quilt will be used. The program manager focuses on three main areas: benefits management, stakeholder management, and program governance.

Benefits management

The purpose of a program is to deliver benefits. A program manager must focus on the overall benefits that the program delivers. Benefits management is the skill used to define, create, maximize, and sustain benefits from a program. These benefits can be tangible and intangible. The tangible benefits are actual, measurable deliverables. While the intangible benefits can be perceived yet be of an unquantifiable quality, such as employee morale, this could lead to a tangible result. If the business has a portfolio process, the portfolio will determine the expected benefits for the program. If a portfolio process does not exist, the business determines the benefits that should be provided by the program. After the program is formed, the program manager begins the benefits management process.

Benefits management includes:

- Assessing the value of the program with regard to the benefits it will produce and how it will affect the organization.

- Ensuring that the expected benefits are SMART (specific, measurable, achievable, realistic, and within a specific time period) goals.

- Assigning responsibilities for achieving actual benefits expected from the program.

- Analyzing risk, impact, and benefits of any changes to the program.

- Examining interdependencies of the projects within the program.

Planning is core to program management. In the early phases of a program, a benefits realization plan is created and is maintained during the life cycle of the program.

- *Communication plan.* In this plan, you determine the benefit information that needs to be communicated and to which stakeholder based on their level of interest and need. Who needs what information, when do they need it, and how should it be delivered?

- *Benefit register.* This is a document containing the results of the benefits analysis, listing all the benefits, and describing how they will be realized.

- *Benefit mapping.* This is the component of the plan that maps program results to the expected benefits of the program.

- *Benefit metrics.* This component of the plan determines which metrics will be used to measure benefits. There are a number of ways to measure benefits. The decision on how to measure the benefit is based on what type of benefit you are seeking. Is the benefit financial? If so, you could use financial tools and techniques like return on investment (ROI), among others. Earned value can be used to help measure the status of a project and make sure that it is on track to deliver the benefits planned.

- Value realization can be used to measure the benefits of the program. It is a way of understanding the value obtained from the investment.
- Balanced score card is another technique that can be used as a benefit metric. This system, developed by Robert Kaplan and David Norton, uses strategic goals that are segmented into four components: customer, financial, internal operations, and learning and innovation. Metrics can be built around each of the four dimensions.
- Key performance indicators (KPIs) help measure performance at stages of the program process.

- *Roles and responsibilities.* This part of the plan identifies the roles required to facilitate the benefits management and assigns responsibilities to each role.
- *Transition plan.* The transition plan includes how the transition of responsibilities regarding sustainment of benefits to ongoing operations will be performed.

Figure 2-1 maps out the benefits management process. First, benefits that the program will be seeking need to be identified. After the benefits are identified, they need to be analyzed. In the analysis process, benefits are prioritized and metrics are established to best measure the level of benefit realization at later stages. Next, the program manager establishes a plan as to how the benefits will be realized, with ways on how they will monitor the benefits throughout

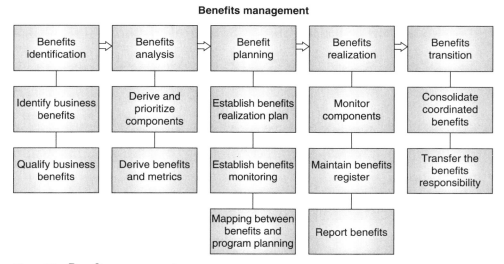

Figure 2-1   Benefits management process.

the program. As benefits are achieved, the benefits register is updated and key information is presented to program stakeholders. At the final stage, benefits are transitioned to the business.

Program stakeholder management

Program managers are responsible for stakeholder management. Anyone who can influence a program positively or negatively is a stakeholder. It is easy to want to avoid negative stakeholders, but it is critical that all stakeholders receive the appropriate attention, information, and involvement in the program. At the end of a program, the program manager is responsible for transferring the benefits to the business.

Program stakeholder management includes understanding and managing the expectations of the program stakeholders. A program stakeholder is a person who is actively involved in the program or whose interest may be affected by the execution or result of the program. Stakeholders can exert influence, both positively and negatively, over program results. They play a critical role in the success of the program. The program manager is responsible for ensuring that stakeholders understand the benefits of the program. It is important to focus on the positive as well as the negative stakeholders; each can greatly affect the success of the program. Program stakeholder management is the second theme of program management. In stakeholder management, you need to:

- Identify stakeholders
- Analyze and resolve their issues
- Manage their needs and expectations

Identifying program stakeholders.   There are many ways to identify program stakeholders. Stakeholders may be internal or external to your company. They can be at different corporate levels and express different levels of influence on a program. Identifying and analyzing stakeholders takes time and is an iterative process. There are some obvious candidates that could be a part of your list.

- *Program manager*. The person who is managing the program.
- *Project managers*. The people who manage the projects that fall under the program.
- *Program team members*. The people responsible for managing program activities.
- *Project team member*. The people responsible for managing project activities that fall under the program.
- *Program sponsor*. The individual or group in the organization providing resources, helping remove barriers, and delivering the program results. This is the person ultimately responsible for delivering program benefits.

- *Program director*. The person with executive ownership of the program. The program directory may also oversee other programs.

- *Program management office (PMO)*. The group responsible for defining and managing program-related governance processes, procedures, standards, templates, histories, and best practices for all programs within the organization.

- *Program office (PO)*. The central office that provides administrative support for all program management team members.

- *Program governance board*. This is the group responsible for ensuring that all program benefits are met and for providing support for managing program risks and issues.

- *Performing organization*. The organization that is performing the work through projects.

- *Customer*. The individual or organization that will receive and make use of the results, benefits, and capabilities that are delivered by the program.

In addition to the key stakeholders, there could be a number of other stakeholders—both internal and external to the organization—that may be less obvious to identify. Examples of external program stakeholders include:

- Vendors and suppliers may be affected by modifications to policies and procedures, or the product of the program.

- Special-interest groups, including those representing the interests of investors, consumers, and the environment.

- Government regulatory agencies may need to ensure that current regulations are being complied with and may create new policies that could affect a program.

- External customers and society at large might be affected by the program outcome.

- Competitors and the marketplace could have an impact on the program.

- Media outlets have an interest in analyzing and reporting on business programs. The information documented could have a positive or negative effect on a program.

Managing stakeholders.   To effectively manage stakeholders, first you need to identify them and how the program will affect them in terms of organizational culture, current issues, and resistance or barriers to change and then develop a communication strategy to manage their expectations and improve their acceptance of the program's objectives. This process includes examining different levels of stakeholders and their broader interdependencies among the projects in the program.

The communication plan centers on taking the initiative to deliver timely, accurate, consistent, and needed information to the appropriate stakeholders. The communication plan delivers a clear vision to help resolve issues and manage expectations. It can be summed up as "delivering the correct message, to the correct the audience, at the correct time."

Program governance

Program governance is a third area that a program manager focuses on within a program environment. Program governance provides the mechanism to ensure alignment between the business strategy and the outcomes of the program. The methods used in program governance provide oversight and control over the program execution. It allows the program manager to assess the state of the program and determine if adjustments are necessary. Program governance is the process of developing, communicating, implementing, and monitoring the program's policies, procedures, structure, and progression. It provides:

- A structure for efficient and effective decision-making
- A consistent method for delivering focus to achieving program goals
- A process for addressing program risks and stakeholder requirements

The team needs to make sure that the program governance complies with the overall organization's governance structure.

- Corporate processes and structures that drive, monitor, and constrain the operation of the organization. This process focuses on the company's mission, vision, and strategy.
  - *Mission.* A mission is a brief statement describing the present purpose of the company. The statement could also include who the key customers are and critical processes.
  - *Vision.* A vision statement defines where the organization is going in the future. It is a source of inspiration and a tool for decision-making for the company.
  - *Strategy.* A strategy is a long-term plan of action to achieve the company's vision and maintain its mission.
- Guidance provided by strategic management on the direction the business is going. The program manager must have knowledge about this process.
- Practices and policies on portfolio and project management.
- Formal methods to capture executive needs and issues can be addressed.

In Fig. 2-2, you see the overall program governance process. The process always starts at the top with the corporate governance team, which usually is your executive management committee. The corporate governance team develops the strategic plan. This is presented to the business operations group, who then implements changes to their functions to meet business goals. Changes that cannot be implemented easily in operations, due to the nature of the strategy, move to the corporate project investment branch, then down to the portfolio

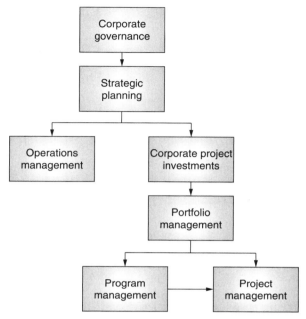

Figure 2-2   Program governance process.

management process. In the portfolio management process, the strategy is broken into project deliverables and program benefits to be realized through processes.

Program management is concerned with managing the business's program investment. As a program progresses, program governance monitors and controls the delivery of benefits. It accomplishes the controlling process with progress reports and program reviews.

The program board can consist of:

- Executive sponsor
- Program director
- Portfolio manager
- Business change manager
- Other necessary stakeholders

The executive sponsor is the primary decision-maker on the board, because they are ultimately responsible for realizing the benefits of the program. Participation on the program board is not a full-time position. A heavy reliance is placed on the program manager and their team.

The skills of a program manager

In order for a program manager to be successful, they must have a blend of skills and talents. Just like functional managers, program managers need to be strong leaders. The program manager's leadership style needs to be aligned with the overall style of the business and focus on vision, strategy, and guidance. It is a benefit to the program manager if they have previous technical knowledge or skills in the field that they are in. People skills are also a basic requirement for a program manager. The program manager works with executives as well as those on the front line. Thus, they need to have a positive attitude and remain flexible, regardless of what comes their way.

As you can see from Fig. 2-3, skills build upon each other to create a strong and productive program manager. At the base of any great program manager is strong communication skills. A program manager needs to communicate effectively with everyone in a timely fashion and while keeping stakeholder needs in mind.

Building upon strong communication skills are the basic skills of technical knowledge and skill in your particular field, people skills, organization skills, and time management skills. These basic skills are valuable for anyone to be successful at the front line of a business.

As you move up the pyramid, project management, political skills, and environmental awareness become more critical. All great program managers have a background in project management and the project management methodology. Having this background provides valuable insight when they manage and provide guidance to the project managers that work for them. Office politics plays a part in any position. As a program manager, having strong political skills provides you with the ability to gain resources and balance requirements in a politically savvy manner. Environmental awareness refers to more than the conditions outside your building. A person who is environmentally aware reads the trade magazines and news sites. Program managers know what is going on inside their company,

Figure 2-3 Program manager's skills.

industry, government, and society. They know if regulations change. When working globally, they understand different cultures and customs.

All great program managers have a strong background in the standards of program management and how best to apply it to their environment. Strategic vision and an ability to communicate that vision to inspire those under you to accomplish the work that is required is a high-level talent.

Finally, at the crown of the skill pyramid are leadership skills. The ability to lead effectively is critical.

External Program Elements

A program's success can be affected by a number of external elements; however, these factors can be internal or external to the company. The company that the program resides in may have other projects, programs, or business functions competing for the same resources. The program manager needs to identify, analyze, plan, and determine how to manage these external elements. Many organizations have established tools, techniques, and policies around best practices, knowledge bases, and risk information, which can be called organizational process assets. These organizational process assets can also include financial management processes, procurement processes, and quality control processes. A program manager needs to understand how the organization's process assets are managed and use them appropriately.

Another element that is external to the program and that could affect it is enterprise environmental factors. When the company changes to a strategy that is no longer in alignment with the benefits of the program, this situation would be considered an enterprise environmental factor. These factors can be a change of company priority or a realignment of resources. A primary stakeholder may leave the company and support is lost. On the other hand, a new stakeholder could join the company and bring a new vision.

Other situations affect a program outside of the boundaries of the company, and these are called enterprise external factors. These elements can be anything that happens in the world that we live in that could affect a program.

Enterprise external factors can be found in any of these areas:

- *Economic conditions.* Changes in interest rates; a downturn in the local, regional, or global economy
- Natural disaster
- *Market needs.* A better product or service enters the market and changes the demand for your program benefits.
- Government regulations
- Availability of raw materials

The program manager must look beyond the borders of their program to the external factors that could affect the successful delivery of the benefits that the program was formed to provide.

Program Life Cycle

As projects have processes and life cycles, so do programs. The program life cycle encompasses the five phases of a program: pre-program planning, program initiation, program formulation, program delivery of benefits, and program closeout. Each phase of the life cycle is unique and can overlap with other phases. Each phase benefits the program by providing a vehicle for enhanced monitoring and control of the program. In practice, the five phases may be slightly different in different environments and should be monitored carefully for variances.

During the preprogram phase, the decision-making committee—whether it be a portfolio, governance, or executive steering team—determines the strategy and the benefits that are needed for the business. To achieve the business benefits, the governing body will determine if a program of work should be created. After the strategic benefits are understood, a plan is created to initiate a program. In this plan, key stakeholders are identified and program objectives and how they align with the business are defined. At this point, initial project approval is given.

In program initiation, the program is further defined. Prior program benefits are analyzed, and a program charter is produced. The program charter describes the program's mission, vision, objectives, benefits, assumptions, and constraints. Many areas can be included in this document.

- *Background.* The background section should describe the current situation or need for new benefits. Here, you want to have the justification for the program clearly explained.

- *Scope.* The scope covers what is included and not included in the program. This is where the future state of the business is described. What is the vision of the future of the business, and how will it be achieved by this program?

- *Results.* What are the results that you plan to achieve with this program?

- *Benefit strategy.* What are the benefits that will be achieved with this program?

- *Concerns, constraints, and assumptions.* It is important to list all concerns, constraints, and assumptions at this point in the program. Later, a program manager will take these items and analyze the risk each presents and determine how to manage these risks.

- *Elements.* Are there any projects or other program elements that must be included for planning purposes for this program?

- *Risks.* What are the high-level risks or issues with this program?

- *Time.* A high-level timescale and major milestones are presented.

- *Resources.* What resources are needed for the program?

- *Budget.* What is the cost plan for the program?

- *Stakeholders.* Who are the key stakeholders?

- *Governance.* How will this program be managed, structured, and controlled? This is a high-level description of how the program will function in later phases.

This document is analyzed and, if approved, signed by the appropriate person(s) to authorize the program to move to the next phase. The program manager and sponsor are assigned in this phase.

Program formulation

When the program reaches the formulation phase, it begins to become officially "set up." In this phase, further definition is given to all the elements of the charter. The scope of the program and its objectives are further defined. The budget and timeline reach another layer of detail. Dependencies are reviewed. The program is further analyzed to see that it continues to align with the mission and strategy of the company. At the end of this phase, a detailed program plan is produced and ready to be executed upon. Also, the program infrastructure is created to support the program work. A program management office is created, and facilities to support the staff are defined. Other necessary tools, techniques, and processes are formulated and/or provided.

Program delivery of benefits

At a high-level, this phase of the program delivers the benefits that it was created to deliver. The projects under the program produce their deliverables, and the program manager manages interproject activities. A number of activates occur during this phase.

- A project governance process is created to monitor and control project activities and deliverables.
- Projects are initiated to meet program deliverables.
- Iterative work continues, where the program plan is managed and updated according to changes in the environment or any other factors that could affect the program.
- *Analyze metrics.* Benefits are identified, measured, analyzed, and reported to the appropriate stakeholders.
- Review and approve change requests when appropriate.
- Communicate with and manage stakeholders.
- Work with the program governance board and other appropriate executive stakeholders to provide them an update on the program status and receive guidance on strategy or direction changes.
- Provide guidance to the project managers throughout the projects.

Throughout this phase of the program, benefits will be realized at different points. The program manager provides vision and guidance during this process.

Program closeout

After all the benefits are realized, the program reaches the fifth and final phase, program closeout. All program work reaches completion. The benefits are transferred to the appropriate operation, and resources are released.

A program may reach the closeout phase before all the benefits are realized due to external factors that prevent its continuation. Regardless of how the program reaches closeout, the same procedures still apply.

- Benefits are reviewed with stakeholders.
- Resources are released. This includes the program organization, team, tools, equipment, and facilities.
- Lessons learned are documented and provide feedback to the organization for improved future program process.
- Documents are stored for history and guidance and analysis for future programs and projects.
- Benefits are transitioned to operations.

As you see in Fig. 2-4, between each phase is a decision point, or gate. At this gate, a determination needs to be made whether to continue with the program and add the appropriate resources to move on to the next phase, if it should be placed on hold, or if it should be killed. If the program ends prematurely at an early phase, it skips successor phases and jumps directly to closeout and follows the necessary steps to disband the program. However, if the program is approved to go to the next phase, the appropriate resources are assigned and processes are followed, as identified previously.

Program management processes

Program management is a process to achieve a company's strategic objectives and benefits by coordinating efforts in a centralized manner. A program is created to strategically manage interdependent activities among projects. Issues are resolved between projects appropriately to best direct efforts to achieve program benefits. There are five program management process groups.

1. *Initiation.* Programs and projects are defined and approved.
2. *Planning.* Plans are developed to build the roadmap to achieve benefits.
3. *Executing.* Plans are enacted and benefits are realized.
4. *Monitoring and controlling.* This happens through all the process groups, where the program is monitored, progress is measured, and variances are identified and managed according to the plan. Corrective action is taken to remain on track to deliver program benefits.
5. *Closing.* Benefits are accepted and the program is formally closed.

Figure 2-4   Program life cycle.

As each of the process groups are discussed, the knowledge area that maps with the process will also be mentioned. Knowledge areas are defined by PMI and are topics or themes that should be understood by all program managers. Knowledge areas are broken down into processes or methods, practices, inputs, tools, techniques, and outputs. Inputs focus on what is contributing to the initial part of the process. The tools and techniques are used to achieve the desired output. The output is what has been produced in that process. PMI has the following list of knowledge areas for program management:

1. Integration management

2. Program scope management

3. Program time management

4. Program communications management

5. Program risk

6. Program procurement management

7. Program financial management

8. Program stakeholder management

9. Program governance

Integration management.   You can compare integration management to a conductor of an orchestra. Each part of the program can be compared to the different parts of an orchestra: woodwinds, brass, strings, and rhythm sections. Each section of a program is planned and often executed in components. However, they all have integration points that need to be planned for, monitored, and controlled. Just like a conductor conducts the orchestra, bringing in and out instruments and maintaining the pacing of the music, the program manager synchronizes the different actions of all the process groups.

- The following are the process activities that occur in this knowledge area. All process activities are described in the next section under the appropriate process group.
  - Initiate program
  - Develop program management plan
  - Develop program infrastructure
  - Direct and manage program execution
  - Manage program resources
  - Monitor and control program performance
  - Manage program issues
  - Close program

Program scope management. The scope management area is responsible for defining, planning, monitoring and controlling, and delivering the results required for the program. The scope can define what is included and excluded from the program of work. It specifies what is required to complete the program.

- The following are the process activities that occur in this knowledge area. All process activities are described in the next section under the appropriate process group.
  - Plan program scope
  - Define program goals and objectives
  - Develop program requirements
  - Develop program architecture
  - Develop program WBS (work breakdown structure)
  - Manage program architecture
  - Manage component interfaces
  - Monitor and control program scope

Program time management. The focus of time management is to develop and control the program schedule. Here, activities are defined, assigned durations, and sequenced appropriately. Resources are assigned to activities and leveled as required.

- The following are the process activities that occur in this knowledge area. All process activities are described in the next section under the appropriate process group.
  - Develop program schedule
  - Monitor and control program schedule

Program communication management. Communication is extremely critical to the success of the program. Communication management involves planning who, what, where, when, and how information is communicated.

- Who needs program information?
- What information is needed by each stakeholder?
- Where should this information be delivered (to stakeholders' electronic inboxes, shared storage site, or hard copies mailed to offices)?
- When should the information be delivered (within 24 hours of a meeting, weekly status updates, monthly progress reports, etc.)?
- How is the information distributed?

Who	What	Where	When	How
Stakeholder 1	Status Updates	E-mailed to the stakeholder and posted on shared server site	Weekly	Program manager to create and send weekly updates
Stakeholder 2	Progress Reports	Posted on the shared server site	Monthly	Program manager creates and posts progress reports
Stakeholder 3	Program Dashboard	Printed and distributed at governance meeting and posted on shared server site	Quarterly	Program manager receives status reports from project manager and rolls up the information into a graphical summary document for the governance meeting
Stakeholder 4	Meeting Minutes	E-mailed to team and posted on shared server site	Within 24 hours after the meeting	Program manager takes notes at the meeting and then creates the meeting minutes

Figure 2-5   Stakeholder communication plan.

A communication matrix can be developed for the communication plan that could look like Fig. 2-5. Other processes areas under communication management include distribute information and report program performance.

- The following are the process activities that occur in this knowledge area. All process activities are described in the next section under the appropriate process group.

  - Plan communications
  - Distribute information
  - Report program performance

Program risk management.   The purpose of program risk management is to identify events that, if they happen, could have an effect on the program, either positive or negative. Program risk events are identified, analyzed, and responded to in the appropriate fashion, as planned for in the risk management plan. Depending on the impact and probability of a risk occurring, the response to the risk could be to ignore, transfer, mitigate, or avoid it.

- *Ignore or accept risk.* You might ignore a risk if the probability and impact is low. Alternatively, you can accept the risk if there is no other option.

- *Transfer risk.* To transfer the risk, you shift the impact of the risk event to a third party. This is a common practice if there is a great financial impact. In practice, you could develop a contract with penalties for late delivery equal to or greater than the cost that would be incurred by the program if the risk happened.

- *Avoid risk.* To avoid a risk, the program plan would be changed to eliminate the possibility of the risk occurring. One way to avoid a risk is to reduce program scope.

- *Mitigate risk.* To mitigate a risk, you reduce its impact or probability. Mitigation of risk can be done in the following ways:

  - Designing a backup or redundant process or system
  - Conducting more tests
  - Developing a less complicated system
  - Planning for secondary resource needs
  - Building a duplicate system at a secondary site
  - Modeling the development on a smaller scale to validate the plan

Figure 2-6 is a risk response matrix. A program manager would plot the identified risks on this chart and then make a determination as to how they will respond to the risk. The chart lists possible risk responses, depending on the impact and probability of the risk. The impact refers to what degree the risk will affect the project.

- Will it have little or low impact on the program?
- Will it have some impact, but not too great of a change to the program?
- Will it have significant impact to the program and be considered a high-impact risk?

The matrix also looks at the likelihood or probability of the risk occurring.

- Is there little or no chance of the risk occurring in the program?
- Is there a good possibility of the risk occurring in the program?
- Is there a strong possibility of the risk occurring in the program?

The risk management plan defines the meaning of low, medium, and high for impact and probability. It is one thing to plan for a risk, but another to plan for and develop the response to the risk. The matrix illustrates possible responses, depending on where the risk falls on the matrix.

- The following are the process activities that occur in this knowledge area. All process activities are described in the next section under the appropriate process group.

Impact of risk occurring

	Low	Medium	High
Low	Ignore	Ignore Avoid Transfer	Ignore Avoid Transfer Mitigate
Medium	Ignore Avoid Transfer	Ignore Avoid Transfer Mitigate	Avoid Transfer Mitigate
High	Ignore Avoid Transfer Mitigate	Avoid Transfer Mitigate	Avoid Transfer Mitigate

Probability of risk occurring

Figure 2-6 Risk response matrix.

- Plan program risk management
- Identify program risks
- Analyze program risks
- Plan program risk responses
- Monitor and control program risk

Program procurement management. This knowledge area focuses on acquiring external products and services that are needed for the program. It concentrates on planning for acquisition needs, developing contracts, selecting vendors, and managing vendor relationships.

- The following are the process activities that occur in this knowledge area. All process activities are described in the next section under the appropriate process group.
  - Plan program procurements
  - Conduct program procurements
  - Administer program procurements
  - Close program procurements

Program financial management.   In financial management, the financial picture for all the projects are integrated into one budget. The costs are monitored and controlled throughout the program.

- The following are the process activities that occur in this knowledge area. All process activities are described in the next section under the appropriate process group.
  - Establish program financial framework
  - Develop program financial plan
  - Estimate program costs
  - Budget program costs
  - Monitor and control program financials

Program stakeholder management.   Stakeholders play a critical role in the success of a program. In stakeholder management, stakeholders are identified, analyzed, and managed according to what is appropriate to their role in the program. Anyone who can positively or negatively affect a program is a stakeholder.

- The following are the process activities that occur in this knowledge area. All process activities are described in the next section under the appropriate process group.
  - Plan program stakeholder management
  - Identify program stakeholders
  - Engage program stakeholders
  - Manage program stakeholder expectations

Program governance management.   The scope and complexity of a program requires that there is a governance process to ensure that decisions are made and that activities performed align to the overall organization's strategy and objectives.

- The following are the process activities that occur in this knowledge area. All process activities are described in the next section under the appropriate process group.

- Plan and establish program governance
- Plan for audits
- Plan program quality
- Approve component initiation
- Provide governance oversight
- Manage program benefits
- Monitor and control program changes
- Approve component transition

Initiation process

In Fig. 2-7, you see the initiate program process. In the preprogram setup, information is gathered and used to initiate a program. The initiating process begins to execute on a strategic initiative developed by the company. The governing body determines the best way to achieve the benefit of the objective. When you initiate a program, it can include organizing current and proposed projects into the program. In addition, approval is needed to move on to the next process. The final output of this process is the program charter, which ties the work that the program is commissioned to perform with the organization's strategy. There are two main functions in the process, which are initiate program and establish program financial framework.

The initiate program function receives information or inputs. The inputs for this work are created in the program management life cycle process of preprogram planning. Initiate program falls under the program integration management knowledge area. To get started in the process, you need:

- *Strategic directive.* This comes from the governing body and outlines what the program is to achieve.

- *Program business case.* This is used to define the business need. It can also be used as a feasibility study or concept development, where it establishes

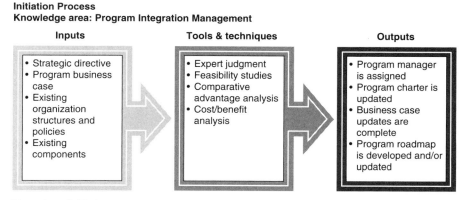

Initiation Process
Knowledge area: Program Integration Management

Inputs	Tools & techniques	Outputs
• Strategic directive • Program business case • Existing organization structures and policies • Existing components	• Expert judgment • Feasibility studies • Comparative advantage analysis • Cost/benefit analysis	• Program manager is assigned • Program charter is updated • Business case updates are complete • Program roadmap is developed and/or updated

Figure 2-7   Initiate program.

estimates for cost, effort, and scope of the program. At the end of the program, the benefits are measured against the business case to determine the program's level of success.

- *Existing organization structures and policies.* The program must have an understanding of the business in which it will be operating.
- *Existing components.* The program may be established to encompass existing or in-flight projects or other business components that it will have to manage and subsequently take on their risks.

The outputs of the initiate program are:

- Program manager is assigned
- Program charter is created
- Business case updates are completed
- Program roadmap is developed

To achieve the required outputs for this process, the following tools and techniques can be used:

- *Expert judgment.* You are relying on an expert's knowledge and experiences to provide advice and guidance for the program. Experts can be internal or external to the organization. The expert can have a functional or technical background. This person could be a consultant, from a professional or technical association, or a specialist from a government agency.
- *Feasibility studies.* These are conducted to determine the technical or economic viability of the program. They can also be used to determine if the program will attain the benefits as planned. A SWOT analysis can help determine the viability of the program.
- *Comparative advantage analysis.* This is used to compare alternatives to the proposed solution. What-if analysis could also be done at this time to determine if the program benefits could be accomplished in other ways.
- *Cost/benefit analysis.* This is used to determine if the program is a good solution to achieve the planned benefit. All program costs are divided by the benefits of the solution. The benefits can be financial and nonfinancial.

The other part of the initiating process is establishing the program financial framework, which falls under the program financial management area. This area defines the financial process for the program and determines if it relates to the financial functions of the business.

The inputs to this process are:

- The program funding source is determined before the program begins.
- Funding goals for the program.

- Funding constraints that will limit the program's funding options.

- Program business case is used to define the business need. It can also be used as a feasibility study or concept development where it establishes estimates for cost, effort, and scope of the program. At the end of the program, the benefits are measured against the business case to determine the program's level of success.

The results or output for the establishing program financial framework are:

- The program financial framework will have appropriate process that interfaces with the business's established processes.

- Program business case updates and financial plan information would be added or updated in the business case.

As seen in Fig. 2-8, to achieve the required outputs for this process, the following tools and techniques can be used:

- *Program financial analysis.* The financial analysis gathers all the cost-related information for the program. How will the program be funded? What factors may affect costs in labor, materials, and availability? The analysis also looks at feature or function/cost tradeoffs. It investigates the best mix of benefits for the cost. Analysis can also include benefit/cost analysis, net present value (NPV), and return on investment (ROI).

- *Payment schedules.* Payment schedules are milestones in the schedule to pay vendors. Using payment schedules, you can forecast cashflow for your program.

- *Funding methods.* There are many sources of funding for a program. The program can be funded internally or externally to the organization. External sources can include investors, government, and lending institutions.

Initiation Process
Knowledge area: Program Financial Management

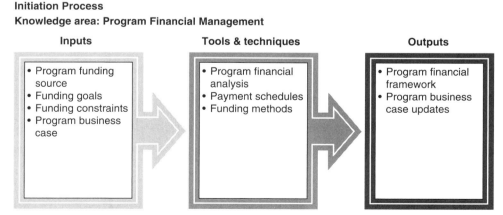

Figure 2-8   Establish program financial framework.

Planning process

The foundation of the program provides a platform for the program's success. Further definition is provided to earlier documents, and the next steps in formalizing the program are taken. Every aspect of the program is defined and planned for in the program planning process. Questions that are answered in this process include:

- How will the program be organized?
- How will the governance structure be handled?
- What will be needed for the program audits?
- What are the scope, costs, and risks of the program?
- Who are all the program stakeholders, and how do they need to be managed?
- What is the schedule and major milestones of the program?
- What is the program's work breakdown structure?
- How will procurement be handled?
- How will communications be handled for the project?
- What are the interdependencies between projects?
- What are the external factors that could affect the program?

The program planning process is iterative, and the plans are updated throughout the program until closeout. Due to the nature of a program, this can be a lengthy process, as updates to the plan will need to be assessed as additions are made to it. For example, updates need to be administered when a project or component closes, risk responses are executed, or outside factors that influence the program occur.

Under the knowledge area of program scope management, there are five parts to be accomplished in the planning phase:

1. Plan program scope
2. Define program goals and objectives
3. Develop program requirements
4. Develop program architecture
5. Develop program work breakdown structure (PWBS)

Plan program scope.   When you plan the program scope, as you can see in Fig. 2-9, you are tasked with developing a detailed description of the major deliverables, program objectives, assumptions, constraints, and concerns. This document provides a basis for making decisions about the program. The scope should also distinguish between what is included and not included in the program. When you plan the program scope, you need to start with some basic information.

Planning Process
Knowledge area: Program Scope Management

Inputs	Tools & techniques	Outputs

Inputs	Tools & techniques	Outputs
• Program business case • Program charter	• Expert judgment • Program management information systems	• Program scope statement updates • Program scope management plan updates

Figure 2-9   Plan program scope.

- The program business case is used to define the business need. It can also be used as a feasibility study or for concept development, where it establishes estimates for cost, effort, and scope of the program. At the end of the program, the benefits are measured against the business case to determine the program's level of success.

- The program charter was completed in the initiating process.

At the conclusion of this work, you will have a program scope statement and a program scope management plan.

To achieve the required outputs for this process, the following tools and techniques can be used:

- *Expert judgment.* You are relying on an expert's knowledge and experiences to provide advice and guidance for the program. Experts can be internal or external to the organization. This person can have a functional or technical background. The expert could be a consultant, from a professional or technical association, or a specialist from a government agency.

- *Program management information systems.* This system is a central server that houses your program documents. Access to the system is determined by rights management protocols. The system may also have tools incorporated within it to help with forecasting, planning, and analyzing your program information.

Define program goals and objectives.   From the scope statement, we move to defining the program goals and objectives. In this part, as seen in Fig. 2-10, we take the program scope statement and management plan and then create updates to the scope statement and create the benefits realization plan. A

Planning Process
Knowledge area: Program Scope Management

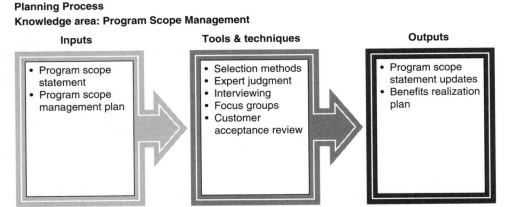

Figure 2-10   Define program goals and objectives.

program is created to achieve benefits. The benefits realization plan describes the program's goals, objectives, and benefits and how they will be accomplished.

All the program benefits are listed and described in this plan. This plan also maps the program outcome to the benefits listed within it. The benefits realization plan includes information on cost and expected return on investment, which proves the feasibility of the program and helps acquire funding. Later, cost, ROI, balanced scorecard, key performance indicators, and value realization tools are used for metrics to determine if the execution of this plan is on track. This plan will also be affected by the organization's environment and external factors, which should also be included in this document. Roles and responsibilities and a transition plan are also described. This plan is updated throughout the program's life cycle iteratively. As the program progresses, increasing granularity and refined details are incorporated into the plan.

Information-gathering techniques are used to assemble the information that is encompassed in the plan. Program benefits are identified and aligned with the organization's strategic objectives. In addition, a determination is made as to how the benefits will be realized.

- *Selection methods.* These determine how benefits will be selected for inclusion in the program.

- *Expert judgment.* Using the advice of experts in the field to determine what benefits should be included in the plan.

- *Interviewing.* Talking to internal and external stakeholders is an effective way to gain valuable information for the program.

- *Focus groups.* A focus group is a synergistic effort where the group talks about the subject at hand and is a great way to understand the needs for the program, solve problems, and find innovative solutions.

■ *Customer acceptance review.* This is an effective method to understand customer needs and requirements, improve deliverables to better meet customer needs, and obtain buy-in from the customer by having them involved in the program process.

Develop program requirements

The develop program requirements process further enhances the requirements for the program. The information feeding this process, as seen in Fig. 2-11, is the program scope statement, program business case, program roadmap, and program change requests. At the conclusion of this process, you have a program requirements document and component requirements document.

To achieve the required outputs for this process, the following tools and techniques can be used:

■ *Requirements gathering.* There are many methods to gather requirements. You can conduct interviews with internal and external stakeholders or experts. You can use focus groups or send surveys or questionnaires to the stakeholders to gather information.

■ *Requirements analysis.* When you analyze the requirement information, you need to make sure all the information is accurate and complete, then begin the process of further clarifying the information. In this process, prioritization of the information should be conducted using the assistance of critical stakeholders.

■ *Design reviews.* The program is reviewed by peers and subject matter experts to make sure that it complies with the appropriate standards and best practices of the organization.

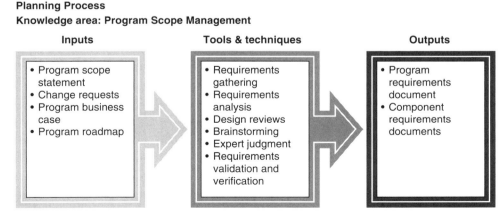

Planning Process
Knowledge area: Program Scope Management

Figure 2-11 Develop program requirements.

- *Brainstorming.* Sessions should be conducted with the appropriate stakeholders who have knowledge of the discussion topic. Brainstorming sessions are a valuable tool to help with requirements identification.

- *Expert judgment.* You are relying on an expert's knowledge and experiences to provide advice and guidance for the program. Experts can be internal or external to the organization. This person can have a functional or technical background. The expert could be a consultant, from a professional or technical association, or a specialist from a government agency.

- *Requirements validations and verification.* Requirements are further validated and verified for accuracy and the level of need for the program.

Develop program architecture

The program architecture focuses on how the program is going to be organized. This document also looks at how each component of the program is going to relate to the others. It focuses on creating policies and procedures on how the program is going to operate. To start this document, as seen in Fig. 2-12, you need the program requirements document. At the end of this process, the program architecture baseline is created. The architecture baseline describes the interfaces of the program, such as their characteristics, deliverables, abilities, and timing. In order to prepare the architecture baseline, you need technical knowledge to develop the optimal solution for the program.

Develop program work breakdown structure (PWBS)

The program work breakdown structure (PWBS), as seen in Fig. 2-13, breaks down the scope information into program deliverables. The PWBS can be

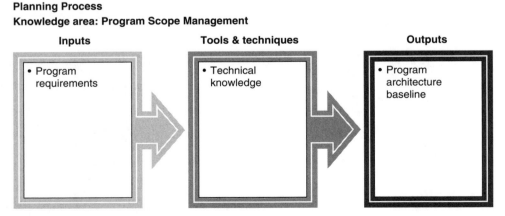

Figure 2-12 Develop program architecture.

Planning Process
Knowledge area: Program Scope Management

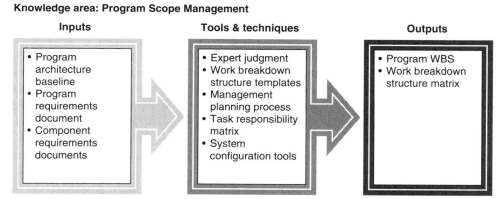

Figure 2-13   Develop program work breakdown structure (PWBS).

displayed in outline or organizational chart format. Each deliverable is numbered, and the numbers can be used in other documents to keep the information organized and aligned throughout all the documents in the plan. Work is broken down into smaller, more manageable segments. The structure of the PWBS requirements is hierarchal and deliverable-oriented. The lowest level of the PWBS is the program packages. These are assigned to project managers to deliver. The program package is where the program ends and the projects begin. It is also the control point for program managers. Anything not included in the PWBS is outside the scope of the program and will not be completed.

The information needed to begin a PWBS includes the program requirements, components documents, and the program architecture baseline. At the conclusion of this phase, you will have developed a program work breakdown structure document and matrix.

To achieve the required outputs for this process, the following tools and techniques can be used:

- *Management planning process.* This process defines how the benefits of the program will be achieved.

- *Task responsibility matrix.* This document defines the roles and responsibilities of each of the team members.

- *Expert judgment.* You are relying on an expert's knowledge and experiences to provide advice and guidance for the program. Experts can be internal or external to the organization. This person can have a functional or technical background. The expert could be a consultant, from a professional or technical association, or a specialist from a government agency.

■ *Work breakdown structure templates.* The WBS follows a specific structure, and the program should use the structure outlined by the organization.

■ *System configuration tools.* These provide a definition of how documentation and version control should be managed to keep the information consistent.

Continuing with the planning of the program, we move from the scope to the program integration management knowledge area. The planning activities in this knowledge area include developing the program management plan and infrastructure.

Develop program management plan

The purpose of the develop program management plan process is to consolidate all the information created in all the other planning activities into one cohesive set of documents. To start this process, as seen in Fig. 2-14, you need the program charter, all existing plans, the best practices information, and program roadmap. The existing plans could include the following:

■ Benefits management plan

■ Scope management plan

■ Communication management plan

■ Staffing management plan

■ Resource management plan

■ Interface management plan

■ Schedule management plan

Planning Process
Knowledge area: Program Integration Management

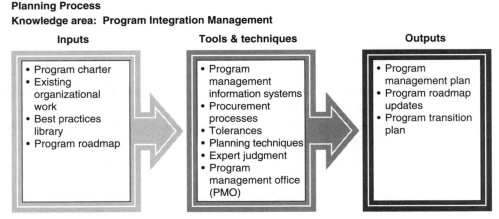

Figure 2-14 Develop program management plan.

- Risk response plan
- Cost management plan
- Contracts management plan
- Procurement management plan
- Quality management plan

Tools and techniques can help facilitate this process. Most organizations have a centralized document storage system where documents can be stored electronically. Information is collected, updated, and distributed from this storage system. User access is granted according to a user rights management plan. This type of storage system allows for version control. Another tool could be an organization's standards plan that describes how information is presented, what documents are required, and how changes are managed. The change control process provides information on:

- How information should be submitted and to whom
- How changes are analyzed and reviewed
- A way to track and review requests and changes
- The approval process, accounting for escalation levels of approvals if needed
- How to deploy approved changes

At the conclusion of the develop program management plan process, the plan reflects the changes and updates. The program roadmap is updated, and the program transition plan is created.

Develop program infrastructure

The program infrastructure is created to provide the program manager a place to define the structure of the program and the technical resources needed to support the work. The information you need, as seen in Fig. 2-15, to begin this process is the program management plan, organizational standards and policies, and the program roadmap.

The program infrastructure describes how the program is going to be supported. Information that should be included in this document is as follows:

- *Program team.* The roles and responsibilities of the program team are defined. The team is responsible for performing the activities of the program.
- *Program office.* This provides administrative support for the program.
- *Program board.* This is the governing body that has the authority to make decisions on important issues concerning the program's budget, scope, or schedule. Issues are sent to this body for guidance. Members of this body include the program director, who is responsible for program policies, and

Planning Process
Knowledge area: Program Integration Management

Inputs	Tools & techniques	Outputs

- Program management plan
- Organizational policies and guidelines
- Best practices library
- Program roadmap

- Expert judgment
- Component analysis
- Review meetings
- Capacity planning

- Core team assignments
- Program resource plan
- Program management processes
- Program infrastructure

Figure 2-15   Develop program infrastructure.

the program executive sponsor, who is responsible for removing barriers to the program's success.

- *Program management office (PMO).* The PMO supports the program board. It defines and manages program policies and procedures. Generally, the PMO oversees many programs.

- *Technical infrastructure.* This includes facilities for the program to operate in and computers and other tools needed by the program members.

The develop program infrastructure phase concludes with the creation of core team assignments, a program resource plan, program management processes, and a program infrastructure plan.

To achieve the required outputs for this process, the following tools and techniques can be used:

- *Expert judgment.* You are relying on an expert's knowledge and experiences to provide advice and guidance for the program. Experts can be internal or external to the organization. This person can have a functional or technical background. The expert could be a consultant, from a professional or technical association, or a specialist from a government agency.

- *Component analysis.* The deliverables for the program are broken down into components, which can be broken down further into benefits packages. Each component could be an activity, project, or subproject. The sum of components provides the benefits planned for by the organization.

- *Review meetings.* These meetings provide the opportunity for the project managers to present the status of their projects to the program manager and other executive members to keep them informed of the progress they are making on their projects.

- *Capacity planning.* Is the effort of managing the scarce resources that are involved in the program. When planning, conflicts, constraints, alternatives, or mitigation practices are activated to provide the optimal capacity of resources for all the components in the program.

Develop program schedule

The planning now turns to the knowledge area of the program time management phase, as seen in Fig. 2-16. The program schedule is developed during this process. Schedules are an iterative process, and there will be updates to the schedule throughout the program. A base schedule is developed. Progressive elaboration, meaning continually improving and adding details to the plan as more information becomes available, thereby creating a more accurate and complete plan, continues throughout the life of the program. The schedule includes component milestones and interdependency with other component tasks. All the program deliverables are listed and then sorted into a logical order. Each deliverable is given a duration, and a resource is assigned to it. Major milestones are defined, which appear in a GANTT chart as zero-duration tasks. To start this process, you need the PWBS, program constraints, program architecture baseline, program charter, contracts, and the program's risk register.

The program schedule shows dependencies among activities within the program. The schedule incorporates resources. Resource leveling can also be done within the schedule. To develop your schedule, you need to know the program components and the order in which they need to be executed, as well as the resources needed for each component.

Planning Process
Knowledge area: Program Time Management

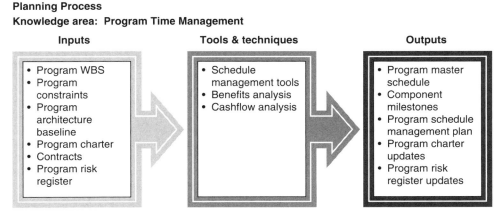

Inputs	Tools & techniques	Outputs
• Program WBS • Program constraints • Program architecture baseline • Program charter • Contracts • Program risk register	• Schedule management tools • Benefits analysis • Cashflow analysis	• Program master schedule • Component milestones • Program schedule management plan • Program charter updates • Program risk register updates

Figure 2-16 Develop program schedule.

The develop program schedule phase ends when the following items are completed:

- *Program master schedule.* The master schedule summarizes the component schedules and dependencies between the components. This schedule allows the program manager to determine when benefits will be delivered.
- *Component milestones.* These identify the component benefit deliverables.
- Program charter updates
- *Program risk register updates.* Updates to the risk register are completed as risk information is discovered as the result of component deliverables or interdependencies.
- *Program schedule management plan.* This plan details how the schedule will be managed and updated in the future. A schedule is a living document and it is updated as information becomes available.

To achieve the required outputs for this process, the following tools and techniques can be used:

- *Schedule management tools.* There are many scheduling tools on the market, and using one is extremely beneficial for a program manager to manage their program schedule.
- *Benefits analysis.* This document identifies benefits and how the program will achieve them. It also looks at how the benefits strategically align with the organization's goals. During the course of the program, as incremental benefits are achieved, adjustments are made to the schedule to ensure delivery of future benefits.
- *Cashflow analysis.* Determines when and how money comes into (revenues) and out of (expenses) the program; the "flow of cash."

Plan communications

As the plan develops, a communications plan needs to be created, as seen in Fig. 2.17. The communication process is included in the program communications management knowledge area. The communication plan needs to include who needs to communicate to whom, what information needs to be conveyed, in what manner (written or verbal), and in what timeframe.

There are a number of inputs to this process:

- Program charter
- Program management plan
- Governance plan
- Program stakeholder management plan
- Organizational communications strategy

Planning Process
Knowledge area: Program Communications Management

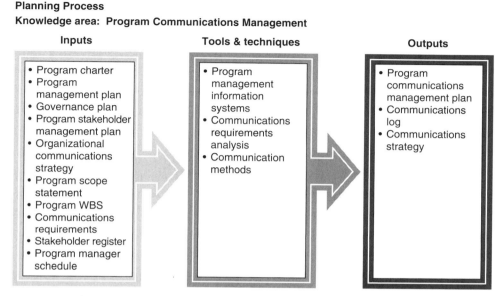

Figure 2-17 Plan communications.

- Program scope statement
- Program WBS
- Communications requirements
- Stakeholder register
- Program master schedule

At the conclusion of this process, the program communications management plan, communications log, and communications strategy are established. The communication plan will outline who will receive what information, in what way the information will be conveyed, and the frequency of the information. If needed, an escalation process is described for urgent matters that cannot be handled by staff. A best practice is to have a glossary of terms so that everyone is "on the same page" and understands the information. The communication log keeps track of all communication. The communication strategy addresses the needs and expectations of the stakeholders as they pertain to communication.

To achieve the required outputs for this process, the following tools and techniques can be used:

- *Program management information systems.* These provide the tools to house data in a central storage system.

- *Communications requirements analysis.* In this analysis, the communication needs of the stakeholders are determined. Who should communicate what information to whom? What are the expectations of the stakeholder with regard to the information they are seeking from the program?
- *Communication methods.* Depending on the complexity and urgency of the information, different methods of communication can be used. More complex information may require a more formal mode of delivery to fully explain it.

Plan program risk management

The plan program risk management process exists in the program risk management knowledge area, as seen in Fig. 2-18. The plan defines how the program will handle risk-related activities. Risk planning is critical to a program's success and provides visibility to the risks in the program. To create this plan, you need:

- Program scope document
- Program management plan
- Program architecture baseline
- Program governance structure
- Resource plan
- Program stakeholder management plan
- Lessons-learned database

Planning Process
Knowledge area: Program Risk Management

Inputs	Tools & techniques	Outputs
• Program scope document • Program management plan • Program architecture baseline • Program governance structure • Resource plan • Program stakeholder management plan • Lessons-learned database	• Planning meetings and analysis • Lessons-learned review	• Program risk management plan

Figure 2-18 Plan program risk management.

At the conclusion of the plan program risk management process, the program risk management plan is established. The plan will be used throughout the program as a guide to managing risks. This plan is a detailed document explaining the approach of managing risk in a program, risk categories are defined, a probability and impact matrix is included, and risk tolerances are defined. The document also explains how risks are going to be tracked, monitored, and reported. The roles and responsibilities of those involved in program risks are specified. Finally, contingency budgeting and timing is added to the overall program for "relief" from a risk occurrence.

To achieve the required outputs for this process, the following tools and techniques can be used:

- *Planning meetings and analysis.* Meetings are conducted to discuss risks. The meeting attendees include appropriate program stakeholders and subject matter experts in the areas of discussion. At the meeting, risk costs and schedule factors are analyzed and added to the program's budget and schedule, where appropriate. Risks are analyzed for the likelihood of occurring or probability and the impact they would have on the program. Depending on the results of the analysis, an appropriate risk response is created.

- *Lessons-learned review.* Reviewing the lessons learned from previous programs and projects can help indicate opportunities for risks in the current project.

Also within the margins of the risk knowledge area are identify program risks, analyze program risks, and plan program risk resources.

Identify Program Risks.   The identify program risk process is where all program risks are identifies and defined, as seen in Fig. 2-19. The appropriate stakeholders and subject matter experts are involved in identifying risks. Risk identification is an iterative process and is done throughout the life of the program. Information you need to start this process includes:

- *Program scope document.* Any program dependency or assumption listed in the scope document should be identified as a risk.

- *Program risk management plan.* The plan identifies the roles and responsibilities of those involved in program risk and the methodologies that should be used to identify risks.

- *Component risk management plan.* The risks that are listed in the component plans of the program and how the program integrates the components should be investigated for program risk elements.

- Program management plan

- *Program governance structure.* The governance structure defines how the program functions with a set of standards and procedures.

Planning Process
Knowledge area: Program Risk Management

Inputs	Tools & techniques	Outputs
• Program scope document • Program risk management plan • Component risk management plans • Program management plan • Program governance structure • Lessons-learned database • Program stakeholder management plan	• Documentation reviews • Information-gathering techniques • Checklist analysis • Assumption analysis • Diagramming techniques • SWOT analysis • Lessons-learned review • Scenario analysis	• Program risk register • Root causes or risk updates

Figure 2-19   Identify program risks.

- Lessons-learned database
- Program stakeholder management plan

At the conclusion of this process, you will have a program risk register and the root causes of risk updates. The risk register lists all the potential risks, with an explanation for each. With each risk there is also a description of a proposed response. The root causes for risks are updated and can be used later for further analysis to discover additional potential risks.

To achieve the required outputs for this process, the following tools and techniques can be used:

- Documentation reviews
- *Information-gathering techniques.* There are a number of information-gathering techniques, such as interviewing, brainstorming, business case analysis, root cause identification, and Delphi technique.
- *Checklist analysis.* Checklists can be built for information gathering.
- *Assumption analysis.* The assumptions of the program are analyzed to check the validity of the understanding.
- *Diagramming techniques.* A number of diagramming techniques can be used, such as affinity diagrams, influence diagrams, cause-and-effect diagrams, and program dependency analysis.
- *SWOT analysis.* In a SWOT analysis, one examines the strengths and weaknesses of the organization and the opportunities and threats of the environment around the organization.

- Lessons-learned review
- *Scenario analysis*. This technique creates a scenario and uses it to determine what possible outcomes could be derived out of it. The scenarios should focus on best-case and worst-case outcomes.

Analyze program risks.   Analyzing program risks takes the identified risks and analyzes them quantitatively and qualitatively according to the probability of them occurring and the impact the risk would have on the program, as seen in Fig. 2-20. Information you need to start this process includes:

- *Program architecture baseline.* There are "typical" risks for each type of program structure.
- *Program risk management plan.* This defines how risks should be analyzed and the roles and responsibilities of those involved in the program risk work.
- *Program risk register.* Lists, describes, and categorizes the potential risks for the program.
- *Program management plan.* Describes how the program should be managed and controlled.
- Lessons-learned database

Planning Process
Knowledge area: Program Risk Management

Inputs	Tools & techniques	Outputs
• Program architecture baseline • Program risk management plan • Program risk register • Program management plan • Lessons-learned database	• Risk data quality assessment • Risk probability and impact assessment • Probability and impact matrix • Risk categorization • Risk urgency assessment • Impact assessments of interdependencies • Data-gathering and representation techniques • Quantitative risk analysis and modeling techniques • Independent reviewers	• Program risk register updates

Figure 2-20 Analyze program risks.

At the conclusion of this process, updates are made to the risk register.

To achieve the required outputs for this process, the following tools and techniques can be used:

- *Risk data quality assessment.* It is important to understand the assumptions for the risk.

- *Risk probability and impact assessment.* The probability and impact of the risks are analyzed based on the characteristics of the risk.

- *Probability and impact matrix.* This is used to plot the risk information.

- *Risk categorization.* Further defines the risk and places it into a category based on its characteristics. Is the risk a skill-set issue, government regulation, or a technology to be developed?

- *Risk urgency assessment.* What is the urgency of the assessment? Is there something pending?

- *Impact assessments of interdependencies.* How do the risks affect the program, organization, or strategic goal of the program?

- *Data-gathering and representation techniques.* There are many data-gathering techniques, including Delphi, causal maps, and brainstorming.

- *Quantitative risk analysis and modeling techniques.* The risk numerically quantified. There are a number of techniques that can quantify a risk. They include sensitivity analysis, simulations, utility theory, and financial methods.

- *Independent reviewers.* Those with previous experience can be consulted to analyze risk.

Plan program risk responses.   You have identified and analyzed risks. The next step in the process is to plan program risk responses, as seen in Fig. 2-21. In the response to the risk, the cost of the risk occurring needs to be assessed in terms of budget, schedule, and the effect to the scope of the program. Information you need to start this process includes:

- *Program risk register.* Lists all the risks, their definitions, priority, and possible responses.

- *Component risk response plans.* Each of the component plans is reviewed and a determination made if it should be included in the program response plan.

- *Program risk management plan.* Includes how risk will be managed in the program and the roles and responsibilities of the risk team. The risk tolerances of the stakeholders are taken into account as well.

At the conclusion of this process, the program register is updated. Contingency plans and reserves are created. Change requests are produced.

Planning Process
Knowledge area: Program Risk Management

Inputs	Tools & techniques	Outputs
• Program risk register • Component risk response plans • Program risk management plan	• Strategies for negative risks or threats • Strategies for positive risks or opportunities • Contingency plan preparation • Risk response action planning	• Program risk register updates • Contingency reserves • Contingency plans • Change requests

Figure 2-21   Plan program risk responses.

To achieve the required outputs for this process, the following tools and techniques can be used:

- Strategies for negative risks or threats:
  - *Ignore or accept risk.* You might ignore a risk if the probability and impact are low. Alternatively, you can accept the risk if there is no other option.
  - *Avoid risk.* To avoid a risk, the program plan would be changed to eliminate the possibility of it occurring. One way to avoid a risk is to reduce program scope.
  - *Mitigate risk.* To mitigate a risk, you reduce its impact or probability.
  - *Transfer risk.* To transfer the risk, you shift the impact of the risk event to a third party. This is a common practice if there is a great financial impact. In practice, you could develop a contract with penalties for late delivery equal to or greater than the cost that would be incurred by the program if the risk happened.
- *Strategies for positive risks or opportunities.* When a risk has a positive outcome, you can accept, enhance, share, or exploit the opportunity.
- *Contingency plan preparation.* A reserve needs to be created in an event of a risk occurring to help prevent a negative effect on the program. The contingency includes money, time, or additional resources to handle the issue.
- *Risk response action planning.* The response selected for the risk needs to be practical and not outweigh the effects of the risk itself.

Plan program procurements

In the knowledge area of program procurement, you have the planning process of plan program procurement, as seen in Fig. 2-22. When you plan for procurements,

Planning Process
Knowledge area: Program Procurement Management

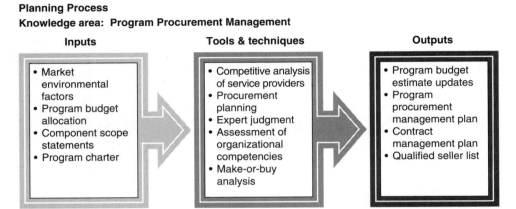

Figure 2-22   Plan program procurements.

you want to create the most effective use of the resources available to you for program purchases. The procurement plan should be reviewed and assessed for any changes and updates. To start this process, you need:

- *Market environmental factors.* What are the conditions that will affect procurement with regard to regulations, laws, contracts, and the economy of the location in which you are doing business, as well as the home office location?

- *Program budget allocation.* The funds for the program are allocated by the program manager, using their best efforts to accommodate the needs of the components while remaining on track to meet the business benefits.

- *Component scope statements.* These are used to understand the procurement needs of each program component.

- Program charter

At the conclusion of this process, you will have completed:

- *Program budget estimate updates.* These are updated based on what is discovered through the analysis.

- *Program procurement management plan.* The procurement plan will have a number of components, including:

  - Identification of required procurements

  - Definition and templates for the contracts that will be used for procurement

  - How the proposal process will function

  - How sourcing decisions will be made

  - List of qualified vendors or service providers

- Contract management plan
- Qualified seller list

To achieve the required outputs for this process, the following tools and techniques can be used:

- *Competitive analysis of service providers.* All the vendors are identified and their products or services are analyzed for the optimal mix of quality versus cost.
- *Procurement planning.* The appropriate stakeholders are involved in procurement planning efforts.
- Expert judgment
- *Assessment of organizational competencies.* An assessment of the organization's competencies must first be captured before outside products or services are considered.
- *Make-or-buy analysis.* For each component of the program, a make-or-buy analysis is done. Is it more economical to produce the component in-house, or is there a better solution external to the organization?

Financial plans

In the knowledge area of financial management, you have three planning processes: develop program financial plan (as seen in Fig. 2-23), estimate program costs, and budget program costs. Due to the nature of a program, which is generally long in duration, funding can come from many different resources. When

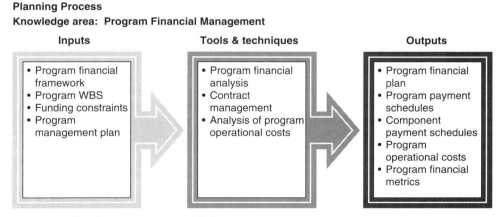

Planning Process
Knowledge area: Program Financial Management

Inputs	Tools & techniques	Outputs
• Program financial framework • Program WBS • Funding constraints • Program management plan	• Program financial analysis • Contract management • Analysis of program operational costs	• Program financial plan • Program payment schedules • Component payment schedules • Program operational costs • Program financial metrics

Figure 2-23   Develop program financial plans.

you develop a program financial plan, you need to keep in mind risk contingency needs, exchange rates, material cost fluctuations, cash flow expectations, and changes in the economy. The information you need to develop a program financial plan is as follows:

- Program financial framework
- Program WBS
- *Funding constraints.* It is rare to receive all the funding required for a program up front. Often, there are milestones in the program where funding is discussed.
- Program management plan

At the conclusion of this process, you will have completed:

- *Program financial plan.* The financial plan will have all the appropriate financial component documents, including payment schedules, financial reporting and metrics, and funding schedules.
- Program payment schedules
- Component payment schedules
- Program operational costs
- *Program financial metrics.* Metrics are compared to earlier metrics to determine the health of the program.

To achieve the required outputs for this process, the following tools and techniques can be used:

- Program financial analysis
- Contract management
- *Analysis of program operational costs.* The operational costs are those internal to the program's organization and need to be taken into account for the purposes of program financials.

Estimate program costs

The program cost estimates (Fig. 2-24) are done in a number of stages. Initially, you may have a rough order-of-magnitude estimate. As the planning further refines the requirements of the program, the estimates also become more refined and closer to reality. The information you need to develop program financial plan is as follows:

- Program architecture baseline
- Contingency reserves
- Program management plan

Planning Process
Knowledge area: Program Financial Management

Inputs	Tools & techniques	Outputs
• Program architecture baseline • Contingency reserves • Program management plan • Program risk register • Contracts	• Total cost of ownership analysis • Architecture/cost tradeoff analysis • Reserve analysis • Estimating techniques • Procurement analysis • Computer cost estimating tools • Expert judgment	• Program cost estimates • Component cost estimates

Figure 2-24   Estimate program costs.

- Program risk register
- Contracts

At the conclusion of this process, you will have completed:

- Program cost estimates
- Component cost estimates

To achieve the required outputs for this process, the following tools and techniques can be used:

- *Total cost of ownership analysis.* This technique determines the total cost for implementing the component and ongoing costs such as maintenance.

- *Architecture/cost tradeoff analysis.* The design of the component is analyzed against the cost and is refined until the best design/cost structure is determined.

- *Reserve analysis.* This determines an appropriate level of reserve for the program in case of external factors that affect the program.

- *Estimating techniques.* A number of estimating techniques can be used to estimate costs.

  - Top down: Estimates for costs are given by executives based on historical information.

  - Bottom-up: Estimates are developed from the individual or team that is closest to the cost acquisition.

  - Comparison: Estimates are created based on a comparison to other program's or components' costs.

- Expert opinion: This technique uses the advice of experts who have had experience in the area in question.

- Parametric: Uses historical information to develop correlations between cost drivers and other parameters. This technique can be used if you know the cost of one tile and the area that the tiles will be laid to get an estimate of the total cost of the tile floor.

- Trend analysis: The estimate looks at the trend of a cost to determine the cost of the current program. Does the cost of lumber go up 25 percent in February every year? If so, the increase is factored into the cost estimate.

- Procurement analysis
- Computer cost estimating tools
- Expert judgment

Budget program costs

The budget is created at this time with cost estimates from the components, as seen in Fig. 2-25. All payment schedules and income information are used to determine the program's cost. This budget creates a foundation for costs to be tracked against to determine how successfully the program is progressing. Budgets are progressively elaborated, meaning that as information becomes available, updates are made. During the life of the program, the budget is continuously being improved upon, with more details being added as more accurate estimates become available. To begin this process, you need the program cost estimates, the architecture baseline, the program management plan, contracts, and component cost estimates.

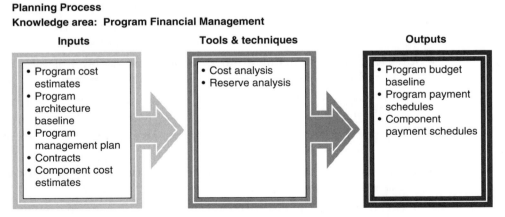

Planning Process
Knowledge area: Program Financial Management

Inputs	Tools & techniques	Outputs
• Program cost estimates • Program architecture baseline • Program management plan • Contracts • Component cost estimates	• Cost analysis • Reserve analysis	• Program budget baseline • Program payment schedules • Component payment schedules

Figure 2-25   Budget program costs.

At the conclusion of this process, you will have completed:

- Program budget baseline
- Program payment schedules
- Component payment schedules

To achieve the required outputs for this process, the following tools and techniques can be used:

- Cost analysis: The costs are analyzed for accuracy and completeness. Items that are analyzed include contract amounts, payment schedule, and funding amounts. When this process is concluded, the program has an understanding of the cashflow of the project.
- Reserve analysis

Plan program stakeholder management

In the program stakeholder management knowledge area, Fig. 2-26, you need to create a program stakeholder management plan. The plan specifies how stakeholders will be identified, analyzed, involved, and managed for the duration of the program. To successfully manage a program, all stakeholders need to be identified and analyzed to determine their level of involvement and communication needs within the program.

The information you need to develop the program stakeholder management plan is as follows:

- *Strategic plan.* This is the organization's strategic plan and it provides valuable information on stakeholders who should be included in the program.

Planning Process
Knowledge area: Program Stakeholder Management

Inputs	Tools & techniques	Outputs
• Strategic plan • Program charter • Program sponsor identification	• Program management information systems • Stakeholder analysis	• Program stakeholder management plan • Component stakeholder management guidelines

Figure 2-26   Plan program stakeholder management.

- Program charter
- Program sponsor identification

At the conclusion of this process, you will have completed:

- *Program stakeholder management plan.* This plan includes all the details on how stakeholders will be identified, analyzed, involved, and managed.
- *Component stakeholder management guideline.* This guideline provides feedback information to all the leaders of the components.

To achieve the required outputs for this process, the following tools and techniques can be used:

- Program management information systems
- *Stakeholder analysis.* A program manager needs a firm understanding of the political culture the program is operating in within the organization. The analysis of stakeholders looks at the degree of influence, attitudes, and communication needs of each of the stakeholders.

Identify program stakeholders

Another planning process for program stakeholder management is identifying program stakeholders, Fig. 2-27. To effectively identify program stakeholders, the task should be undertaken systematically. Both internal and external

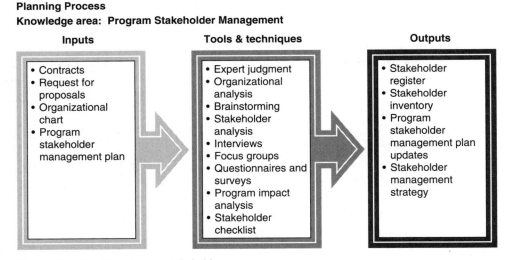

Planning Process
Knowledge area: Program Stakeholder Management

Inputs	Tools & techniques	Outputs
• Contracts • Request for proposals • Organizational chart • Program stakeholder management plan	• Expert judgment • Organizational analysis • Brainstorming • Stakeholder analysis • Interviews • Focus groups • Questionnaires and surveys • Program impact analysis • Stakeholder checklist	• Stakeholder register • Stakeholder inventory • Program stakeholder management plan updates • Stakeholder management strategy

Figure 2-27   Identify program stakeholders.

stakeholders need to be identified. Any person or group that can affect the program either positively or negatively should be identified as a stakeholder.

- *Contracts.* These are mutually binding agreements between two or more parties. Contracts stipulate conditions and obligations for the seller to deliver the specified projects or services and the buyer to pay for in it in the terms defined.

- *Organizational chart.* This graphically displays the relationships among members of a specific group.

- *Requests for proposals (RFP).* This is a procurement document to request proposals from potential sellers for products or services. This document could also specify any terms, conditions, and needs requirements that the buyer has from a potential seller.

- *Program stakeholder management plan.* Specifies how stakeholders will be identified, analyzed, involved, and managed in the program environment. It also describes the process of acquiring and managing stakeholder information.

At the conclusion of this process, you will have completed:

- *Stakeholder register.* The register lists all the stakeholders for the program. This information is used to ensure that stakeholders are communicated with in the correct method, form, and with the appropriate content. Stakeholders can include:

  - Project managers: The individuals working on components of the program.

  - Project team members: Those responsible for completing project activities.

  - Program manager: The manager of the program responsible for the successful delivery of the benefits of the program.

  - Program team members: Those working on program activities.

  - Program sponsor: The sponsor is the champion for the program and is responsible for helping remove "road blocks" that are in the program's way.

  - Customer: The organization or individual who will use or receive the benefits from the program.

  - Program director: The person who will oversee a number of programs and have executive ownership of them.

  - Program management office (PMO): The group responsible for managing program-related processes, templates, and procedures. They provide support for the program teams by handling administrative functions for the program.

  - Program governance board (committee or group)/steering committee. This group is responsible for making strategic decisions, supporting the program, and ensuring that the program delivers the benefits it set out to accomplish.

- Funding organization: The group providing the funds for the program.
- External stakeholder: There can be stakeholders outside the program's organization. This could include vendors/suppliers, potential customers, and government agencies.

- *Stakeholder inventory.* Provides information on stakeholder issues and possible responses, and how stakeholders will be affected by the program.
- *Program stakeholder management plan updates.* The stakeholder plan is updated initiatively throughout the course of the program.
- *Stakeholder management strategy.* The strategy defines how a program manager is going to manage and mitigate areas that affect stakeholders. Mitigation plans can include training and training documentation, and constant communication with the stakeholders to make sure that they are involved in the process and have bought into the program's benefits.

To achieve the required outputs for this process, the following tools and techniques can be used:

- *Expert judgment.* Using the opinions of those who have experience and knowledge in stakeholder management.
- *Organizational analysis.* Analyzing the organization to understand the roles and responsibilities of the members. This should be done for internal and external potential stakeholders.
- *Brainstorming.* A way to creatively gather data that can be used to identify potential stakeholders.
- *Stakeholder analysis.* In this process, you analyze the needs and potential impact or influence a stakeholder can have on the program.
- *Interviews.* These discussions allow for a better understanding of the needs of the stakeholders. Generally, these are open-ended questions to gather data that is important to the stakeholder.
- *Focus groups.* Gather data from a group to better understand the attitudes of stakeholders. In this technique, group members interact with one another and build upon each other's thoughts to achieve a deeper understanding of the needs of the stakeholders.
- *Questionnaires and surveys.* Solicit feedback from stakeholders to better understand their needs and requirements.
- *Program impact analysis.* From the data gathered, an analysis needs to be done to discover how the information provided affects the program. Negative impacts need to be managed, and a response plan needs to be developed.
- *Stakeholder checklist.* This provides a simple, high-level list of the roles and interests of the stakeholders.

Program governance structure

In the knowledge area of program governance, there are three planning processes: plan and establish program governance structure, plan for audits, and plan for program quality.

When you plan and establish the program governance structure, Fig. 2-28, you focus on identifying the goals and defining the roles, responsibilities, and structure of the governance board. The governance process is a proactive process. The purpose of the governance structure is to ensure that the program remains strategically aligned with the organization's goals. The governance board provides support with conflicts, decisions on program changes, and ensures that the program follows the appropriate policies and procedures for the organization and external entities such as government laws and regulations. The governance board includes the following members:

- The program manager is responsible for managing the program and ensuring the deliverables are achieved.

- Project managers and project team members are responsible for delivering the project's deliverables.

- The executive sponsor is responsible for supporting and leading the program manager and the program to success.

- The program management office (PMO) provides administrative support for the program.

A number of items are needed to begin this process. First, you need the strategic directive to make sure that the program is aligned with this mandate. Also,

Planning Process
Knowledge area: Program Governance Management

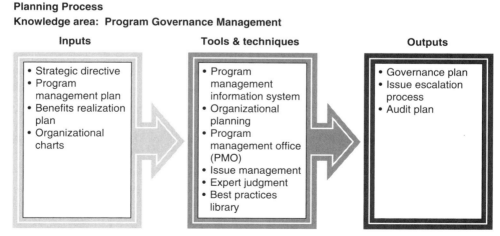

Figure 2-28   Plan and establish program governance structure.

you need the program management plan and benefits realization plan. Finally, you need the organizational charts.

At the conclusion of this process, you will have completed:

- *Governance plan.* Describes the structure, roles and responsibilities, goals, and how the group will govern. Is there a formal gate process where the program checks in to provide a status update?
- *Issue escalation process.* This process addresses how issues will be raised to the governance level.
- *Audit plan.* This should be planned for and have periodically defined intervals.

To achieve the required outputs for this process, the following tools and techniques can be used:

- Program management information system
- Organizational planning
- Program management office (PMO)
- Issue management
- Expert judgment
- Best practices library

Plan for audits

A program should always be prepared for an audit, Fig. 2-29. Audits help determine if what is being delivered is what was originally agreed upon. Audits are a formal examination of the program's financial situation and provide a process

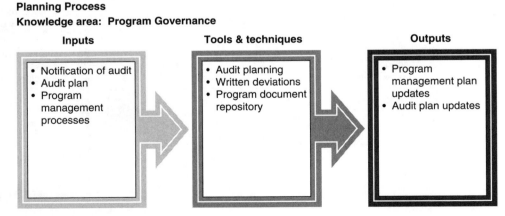

Planning Process
Knowledge area: Program Governance

Inputs	Tools & techniques	Outputs
• Notification of audit • Audit plan • Program management processes	• Audit planning • Written deviations • Program document repository	• Program management plan updates • Audit plan updates

Figure 2-29   Plan for audits.

to examine the program in parts or as a whole to assess the overall progress performance. Audits can be performed at any time, but usually occur during the execution or closeout of a program.

- *Notification of audit*. Audits are usually planned for in advance, and the program manager will be notified prior to the audit so that he or she can assemble the appropriate documentation. There are times when audits have little or no notification.

- *Audit plan*. Part of the program management plan, it includes how the audit are conducted.

- *Program management processes*. These documents provide guidance on the auditor process and the components that will be reviewed. Programs should be managed to ensure successful audit outcomes.

At the conclusion of this process, you will have completed:

- *Program management plan updates*. Updates are made to the program management plan to effectively plan and prepare for audits.

- *Audit plan updates*. After an audit is complete, updates are made to the audit plan to address any issues raised in the audit process.

To achieve the required outputs for this process, the following tools and techniques can be used:

- *Audit planning*. Program managers need to follow the appropriate processes and ensure that all program documentation is in alignment with what is expected.

- *Written deviations*. There are times when the program process should deviate from prescribed processes. When a deviation is necessary, it should be fully documented and approved by the governance board.

- *Program document repository*. All program documentation should reside in a specified location for easy access and retrieval.

Plan for program quality

The plan for program quality is focused on the component level, Fig. 2-30. The components are responsible for ensuring the quality of their deliverables. Planning for quality should be a standardized process that is repeatable in order to build metrics on the quality standards required for the component. To start this process, you need the benefits realization plan, the organizational quality standards, program management plan, and information on external regulations.

At the conclusion of this process, you will have completed:

- *Program quality management plan*. This states the minimal standards, required testing, and planning efforts needed for quality. It also defines the roles and responsibilities for those involved to ensure program quality.

Planning Process
Knowledge area: Program Governance

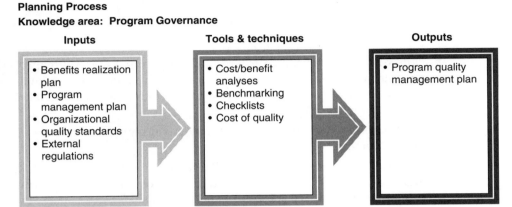

Figure 2-30   Plan program quality.

To achieve the required outputs for this process, the following tools and techniques can be used:

- *Cost/benefit analyses.* There is always a tradeoff between the benefit of a deliverable and the cost that it takes to produce it. The benefit of planning for quality is that there is less rework, less cost, higher productivity, and increased stakeholder satisfaction.

- *Benchmarking.* This is a measure of reference to make a comparison. You are comparing a provided standard to measure the performance of the current system against to determine if it meets the requirements.

- *Checklists.* These help to make sure that requirements are not overlooked.

- *Cost of quality.* This determines the cost required to ensure quality. The types of costs would include appraisal, preventative, and failure costs. These costs can be internal or external and are broken down as such.

Executing process

The executing process carries out all the plans to deliver the benefits of the program. This process follows the plans according to the planned procedures and policies, as outlined in the program management plan and the subsidiary plans. The executing process involves managing quality, costs, and schedules, and making sure that every component is appropriately integrated into the process. Stakeholders receive the information they need, as defined in the stakeholder plan. All government regulations and policies are followed as required.

In the program integration management knowledge area, you have two executing processes: directing and managing program execution and managing program resources.

Executing Process

Knowledge area: Program Integration Management

Inputs	Tools & techniques	Outputs
• Program management plan • Program performance reports • Change requests • Work results • Audit reports • Go/no-go decisions • Program roadmap	• Program management information system • Expert judgment • Program management office (PMO) • Contract management plan • Decision logs • Impact analysis • Tolerances	• Approved change requests • Component initiation requests • Component transition requests • Program issues register • Program roadmap updates

Figure 2-31 Direct and manage program execution.

Direct and manage program execution.   The direct and manage program execution process (Fig. 2-31) provides the benefits intended by the program. It integrates all the components to ensure that deliverables are executed according to plan. As the program is in execution, change requests may be generated. Change requests are handled by the defined program governance process.

- *Program management plan.* The program plan is followed and fully executed at this time.

- *Program performance reports.* Performance reports are produced, documenting the state of the program.

- *Change requests.* These are reviewed for updates.

- *Work results.* These are tracked and managed accordingly.

- *Audit reports.* These are reviewed to ensure that the proper procedures are followed.

- Go/no-go decisions

- *Program roadmap.* This documents when new components are added to execution plan.

At the conclusion of this process, you will have completed:

- Approved change requests

- Component initiation requests

- Component transition requests

- Program issues register

- Program roadmap updates

To achieve the required outputs for this process, the following tools and techniques can be used:

- Program management information system
- Expert judgment
- Program management office (PMO)
- Contract management plan
- Decision logs
- Impact analysis
- Tolerances

Manage program resources.   Program resources need to be effectively managed throughout the life of the program, Fig. 2-32. As part of the execution process, funds, staff, facilities, equipment, and other resources are allocated appropriately. Day-to-day management of resources is a component-level responsibility.

- *Program management plan.* This plan delegates the allocation of program recourses throughout the program's life.
- *Component status reports.* Identifies issues with resources.
- *Resource availability.* Defines when a resource is available to be allocated to a component.
- Program resource plan

At the conclusion of this process, you will have completed:

- *Program resource plan updates.* As new information is acquired, updates are made to the resource plan.

Executing Process
Knowledge area: Program Integration Management

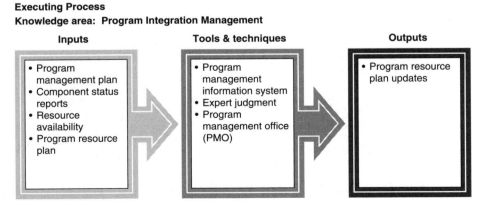

Figure 2-32   Manage program resources.

To achieve the required outputs for this process, the following tools and techniques can be used:

- Program management information system
- Expert judgment
- Program management office (PMO)

In the program scope knowledge area, there are two executing processes: manage program architecture and manage component interfaces.

Manage program architecture

Managing the program architecture is the process for making sure that the component interdependencies are intact. Also, it makes sure that the components are following the rules, regulations, and policies as defined in the plan. Changes may be made to the program architecture if new components are incorporated into the program. As seen in Fig. 2-33, the manage program architecture process includes:

- Program architecture baseline
- Program management plan
- Change requests

At the conclusion of this process, you will have completed:

- *Program architecture baseline updates.* As changes are made, updates are administered to the architecture baseline. This is an iterative process and continues throughout the life of the program.

Executing Process
Knowledge area: Program Scope Management

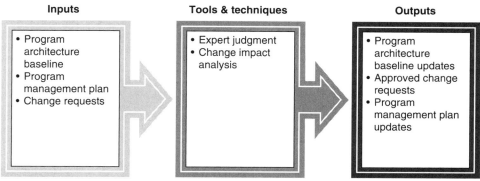

Figure 2-33   Manage program architecture.

- *Approved change requests.* Are incorporated into the architecture of the program.
- *Program management plan updates.* The program plan is updated according to any new information that is approved during this process.

To achieve the required outputs for this process, the following tools and techniques can be used:

- Expert judgment
- *Change impact analysis.* Any change that is considered for the program needs to be analyzed for the impact it will have on the program, other components, and the stakeholders.

Manage component interfaces

Parts of the program scope are operational and project activities, which are considered components. Managing the component interfaces is critical to the success of the program. Component cross-points need to be managed to provide the right support and guidance to ensure success for these intricate connections. As seen in Fig. 2-34, the manage component interface process includes:

- Program architecture baseline
- Program management plan

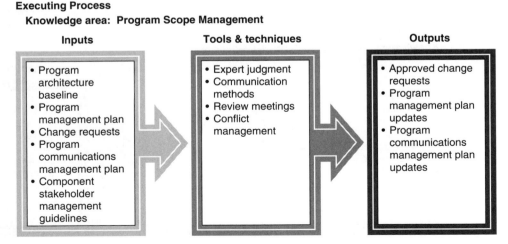

Executing Process
Knowledge area: Program Scope Management

Inputs	Tools & techniques	Outputs
• Program architecture baseline • Program management plan • Change requests • Program communications management plan • Component stakeholder management guidelines	• Expert judgment • Communication methods • Review meetings • Conflict management	• Approved change requests • Program management plan updates • Program communications management plan updates

Figure 2-34 Manage component interfaces.

- Change requests
- Program communications management plan
- Component stakeholder management guideline

At the conclusion of this process, you will have completed:

- Approved change requests
- Program management plan updates
- Program communications management plan updates

To achieve the required outputs for this process, the following tools and techniques can be used:

- Expert judgment
- Communication methods
- Review meetings
- Conflict management

In the program communication management knowledge area, there is one process in the execution process, which is distributing information.

Distribute information.   The distribute information, Fig. 2-35, process provides the required information to the program stakeholders, when needed, in the format

Executing Process
Knowledge area: Program Communications Management

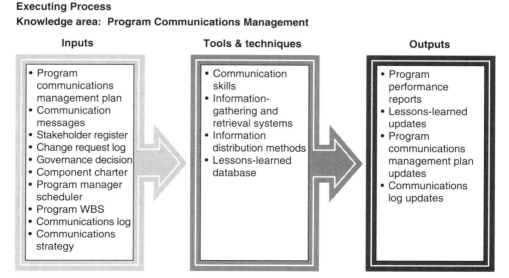

Figure 2-35   Distribute information.

desired, as defined in the stakeholder communication plan. This information can include progress reports, analysis information, change requests, budget information, and updates to external factors.

- *Program communications management plan.* The defined communication plan for the program.
- *Communications messages.* Distributed to stakeholders throughout the life of the program.
- Stakeholder register
- Change request log
- Governance decision
- *Component charter.* Describes the stakeholders and information they require.
- *Program master schedule.* Defines the timing of program communications.
- *Program WBS.* Used to plan communications.
- Communications log
- Communications strategy

At the conclusion of this process, you will have completed:

- *Program performance reports.* The reports that are produced in the program include status reports, memos, presentations, and dashboards.
- Lessons-learned updates
- *Program communications management plan updates.* Updates are made as information becomes available.
- *Communications log updates.* If distribution triggers, level of information, or new stakeholders are added to the program, the communication log is updated.

To achieve the required outputs for this process, the following tools and techniques can be used:

- *Communication skills.* One of the most important skills of the program manager is communication skills. The program manager must be able to translate the information he or she receives into the format that best suits an individual stakeholder.
- *Information gathering and retrieval systems.* Information is gathered throughout the program.
- *Information distribution methods.* There are a number of distribution methods, both formal and informal, including verbal methods (face to face, meetings, video conferencing, and formal presentations) and written methods (e-mails, hard-copy documents, and collaborative software tools).

- *Lessons-learned database.* As information is presented and knowledge is gained, the lessons-learned database is updated. This is an iterative process throughout the life of the program.

In the program procurement management knowledge area, there is one process in the execution process, which is conducting program procurement.

Conduct program procurement.   Conducting program procurements, Fig. 2-36, focuses on obtaining the required resources, products, services, and/or materials needed for the program. This process includes procurement metrics gathering, methods, strategies, and activities.

- *Program assets.* These are specified and a decision made on how they will be obtained. Will the asset be created or produced in-house or is it external to the organization? Does the organization have the skills and expertise to produce the asset?
- *Subcontract procurement plans.* Components of the procurement plan may be subcontracted to a third party that is more capable, efficient, and cost-effective.
- Program procurement management plan
- *Program management plan.* The plan is reviewed for conditions or factors relevant for procurement.

Executing Process
Knowledge area: Program Procurement Management

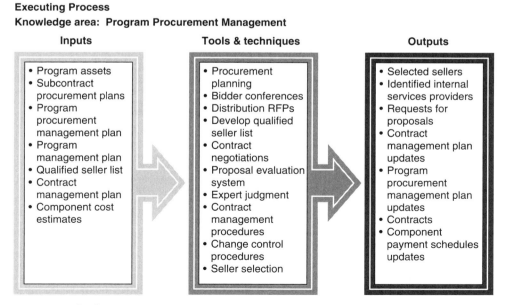

Figure 2-36   Conduct program procurement.

- Qualified seller list
- Contract management plan
- Component cost estimates

At the conclusion of this process, you will have completed:

- *Selected sellers.* The ones who have been determined to best meet the objectives of the buyers.
- *Identified internal services providers.* The providers that can deliver the required goods or services.
- *Request for proposals.* Sent to the qualified sellers.
- *Contract management plan updates.* The plan provides instructions on how to administer the contract process.
- *Program procurement management plan updates.* Updates are made to the program procurement plan, as determined by the decisions made during this process.
- *Contracts.* Final agreements are made between the seller and buyer.
- *Component payment schedules updates.* Updates are made according to decisions made during this process.

To achieve the required outputs for this process, the following tools and techniques can be used:

- Procurement planning
- *Bidder conferences.* Where bidder and buyer meet to discuss the requirements of the organization and answer any questions. This process provides all parties with the same information at the same time.
- *Distribution of requests for proposals (RFPs).* Describes the needs and requirements of the buyer.
- *Develop qualified seller list.* Those on the list who have passed any required buyer stipulations and who will receive the buyer's request for proposal information.
- *Contract negotiations.* Clarifies the agreement terms that binds the seller to the conditions in the document and the buyer to the payment terms as outlined in the contract.
- *Proposal evaluation system.* The criteria for proposal selection is evaluated to determine the best vendor to provide the product or service required by the buyer.
- Expert judgment
- *Contract management procedures.* These are the set procedures as defined by the organization.

- *Change control procedures.* Define how the contract can be modified in the future if certain conditions present themselves.

- *Seller selection.* In this process, the qualified seller list is narrowed to the final candidates.

In the program stakeholder management knowledge area, there is one process in the execution process, which is engaging program stakeholders.

Engage program stakeholders

As seen in Fig. 2-37, engaging program stakeholders is the process to ensure that stakeholders are involved and active in the program as required. Stakeholders' needs, issues, and requirements need to be taken into consideration to determine the appropriate level of involvement.

- *Program charter.* Defines the high-level expectation of the program. It also defines risks and benefits and ways to successfully engage stakeholders.

- *Program stakeholder management plan.* Guidelines on how to interact with the stakeholders.

- *Stakeholder register.* Lists all the stakeholders and their needs, requirements, and issues.

- *Stakeholder inventory.* Specifies all the stakeholders and their roles and responsibilities in the program.

- Stakeholder management strategy

Executing Process
Knowledge area: Program Stakeholder Management

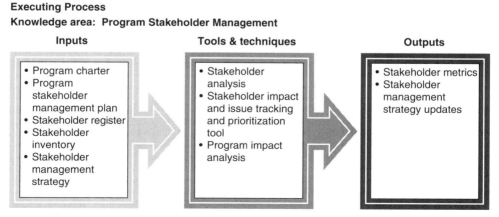

Figure 2-37   Engage program stakeholders.

At the conclusion of this process, you will have completed:

- *Stakeholder metrics.* The level of program participation of the stakeholders is tracked.
- *Stakeholder management strategy updates.* These are meetings to update the stakeholders on the program status and to discuss issues and concerns.

To achieve the required outputs for this process, the following tools and techniques can be used:

- Stakeholder analysis
- *Stakeholder impact and issue tracking and prioritization tool.* All issues and concerns are tracked and prioritized to ensure that they are appropriately addressed.
- *Program impact analysis.* An analysis is done to determine if any issue or decisions have an impact on the program.

In the program governance knowledge area, there is one process in the execution process, which is approving component initiation.

Approve component initiation

Approving component initiation is the process that decides when a component will begin, as seen in Fig. 2-38. This process happens throughout the program, except in the program closeout phase. A number of decision-making factors are needed in order to determine a component should begin. To receive approval for a component to begin, a strong business case needs to be presented, a project manager needs to be assigned, a communication plan needs to be established,

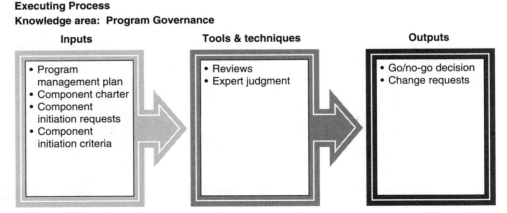

Figure 2-38 Approve component initiation.

and a component tracking procedure needs to be defined. Once a component is approved, it triggers resources to be activated or assigned to produce the required deliverables.

- *Program management plan.* This includes the strategic plan and charter, and is used to guide the approval process.
- *Component charter.* This is required before a component can be approved.
- *Component initiation requests.* This begins the selection process to determine if approval of a component is appropriate.
- *Component initiation criteria.* This is the evaluation information used to help in the decision-making process.

At the conclusion of this process, you will have completed:

- *Go/no-go decision.* Defines whether the component has been approved.
- *Change requests.* Any changes to the component can affect the delivery of the program benefits and needs to be reviewed.

To achieve the required outputs for this process, the following tools and techniques can be used:

- *Reviews.* Review meetings are decision discussion meetings where the component is reviewed and analyzed.
- Expert judgment

Monitoring and Controlling Process Group

In the monitoring and controlling process group, components are examined to ensure that they are progressing in the appropriate manner to achieve the deliverables that they were established to produce. If it is determined that a component is off track, this process proactively takes the steps necessary to coordinate corrective action to redirect the component and achieve the desired results. This process consolidates data on the progress of the components and measures the planned versus actual progress. External factors could cause the need for corrective action to a component, including changes to policies, market changes, environmental conditions, economic fluctuation, and government regulations.

In the program integration management knowledge area, there are two processes in the monitor and control phase: monitoring and controlling program performance and managing program issues.

Monitor and control program performance

The monitor and control program performance process examines the current conditions to make sure that the program follows the approved plan. Monitoring and controlling happens throughout the program life cycle (Fig. 2-39). Information

Monitoring and Controlling Process
Knowledge area: Program Integration Management

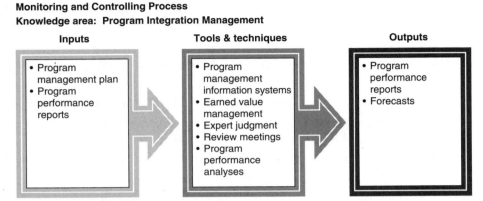

Figure 2-39   Monitor and control program performance.

is collected, analyzed, and distributed to provide trend information to the appropriate stakeholders. Depending on the results of the analysis, corrective action may be taken, which will put the component work in alignment with the benefits delivery of the program. Two areas needed to start this process: program management plan and program performance reports.

At the conclusion of this process, you will have completed:

- *Program performance reports.* These will be a summary of the status of all the components in the program. These reports will have a rundown of progress statements, including what has been accomplished to date, what is planned for in the near future, risks, issues, earned value status, and change requests that are under consideration.

- *Forecasts.* These are updated with new developments. Forecasts are based on current conditions and are predictions or estimates where the program is expected to track.

To achieve the required outputs for this process, the following tools and techniques can be used:

- *Program management information systems.* Software systems that provide tools for documenting and retrieving data.

- *Earned value management (EVM).* Integrates schedules, resources, and scope to objectively measure program performance and progress. Expected results are measured against actual results to determine progress. The trends identified in EVM assist in understanding whether or not the program is on track to deliver its benefits.

- Expert judgment

- Review meetings

- *Program performance analyses.* A number of techniques fall under this category.

  - Trend and probability analysis use program metrics to predict the likelihood of success.

  - Risk analysis examines program risks to ensure that appropriate measures are being taken to respond to the risks effectively.

  - Issue analysis prioritizes all the issues and defines the impact and root cause of the issue and how it should be managed.

  - Gap analysis examines metrics and assesses and identifies any gaps between what is planned and the actual progress in terms of budget, time, or achieving benefits.

Manage program issues

When monitoring and controlling a program, Fig. 2-40, issues will be identified. As issues are identified, they need to be managed and escalated appropriately. An issue can be a concern, an unplanned event, or new information. As issues are identified, they are documented in the issue register and analyzed by the governance board. Managing and controlling issues are similar to managing and controlling risks. The impact of the issue needs to be determined, and a response for the issue needs to be planned and executed. To begin this process, you need the program management plan, audit reports, program risk register, program performance reports, and program issues register.

At the conclusion of this process, you will have completed:

- *Change requests.* Depending on the appropriate issue response, a change request may be generated.

Monitoring and Controlling Process
Knowledge area: Program Integration Management

Inputs	Tools & techniques	Outputs
• Program management plan • Audit reports • Program risk register • Program performance reports • Program issues register	• Issues analysis • Expert judgment	• Change requests • Program issues register updates

Figure 2-40   Manage program issues.

- *Program issues register updates.* The issue register is updated appropriately to reflect the identification, definition, analysis, and response to the risk.

To achieve the required outputs for this process, the following tools and techniques can be used:

- *Issues analysis.* Follows the accepted plan and procedure that has been defined by the program management plan. It is critical to determine the root cause of an issue to make sure that you are addressing it appropriately

- Expert judgment

In the program scope management knowledge area, there is one process in the monitoring and controlling process: monitoring and controlling program scope.

Monitor and control program scope

The program scope is monitored and controlled to ensure that the planned objectives for the program are realized, Fig. 2-41. Programs are large and complex, and have an extended duration. As the program progresses, it is critical to monitor and control scope. Changes to the scope have an enormous impact on the program and the components within it. To change the scope of a program, it will have to go through a formal change control process and receive approval from the governance board. To begin this process, you need the program scope statement, approved change requests, component transition request, governance plan, program architecture baseline, and program management plan.

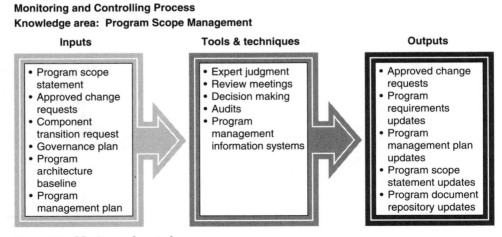

Monitoring and Controlling Process
Knowledge area: Program Scope Management

Inputs	Tools & techniques	Outputs
• Program scope statement • Approved change requests • Component transition request • Governance plan • Program architecture baseline • Program management plan	• Expert judgment • Review meetings • Decision making • Audits • Program management information systems	• Approved change requests • Program requirements updates • Program management plan updates • Program scope statement updates • Program document repository updates

Figure 2-41   Monitor and control program scope.

At the conclusion of this process, you will have completed:

- Approved change requests
- Program requirements updates
- Program management plan updates
- Program scope statement updates
- Program document repository updates

To achieve the required outputs for this process, the following tools and techniques can be used:

- Expert judgment
- Review meetings
- Decision making
- Audits
- Program management information systems

In the program time management knowledge area, there is one process in the monitoring and controlling process: monitoring and controlling program schedule.

Monitor and control program schedule

Throughout the program, the schedule needs to be monitored and controlled to keep it in alignment with planned benefits, Fig. 2-42. This process ensures that components produce deliverables on time. The start and finish time is

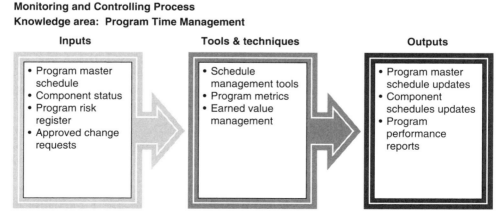

Monitoring and Controlling Process
Knowledge area: Program Time Management

Inputs	Tools & techniques	Outputs
• Program master schedule • Component status • Program risk register • Approved change requests	• Schedule management tools • Program metrics • Earned value management	• Program master schedule updates • Component schedules updates • Program performance reports

Figure 2-42   Monitor and control program schedule.

tracked and compared to planned time estimates. As variances are identified, corrective action is taken. The schedule is analyzed to identify opportunities for acceleration or active compression. To begin this process, you need the program master schedule, component status, program risk register, and approved change requests.

At the conclusion of this process, you will have completed:

- *Program manager schedule updates.* These are made to the schedule according to the information and decisions made concerning the schedule.

- *Component schedules updates.* These are made to the component schedules to align with program changes.

- *Program performance reports.* The timing of activities, components, and deliverables is communicated to the appropriate stakeholders.

To achieve the required outputs for this process, the following tools and techniques can be used:

- *Schedule management tools.* Used to produce program schedules.

- *Program metrics.* Determine the progress of the program. The metrics are, at times, simple graphical illustrations depicting the status of the program. Often, the traffic light analogy is used, with red (issue), yellow (warning), and green (on track) indicators.

- *Earned value management.* Integrates schedule scope and resources to measure the progress of the program.

In the program communication management knowledge area, there is one process in the monitoring and controlling process: reporting program performance.

Report program performance

The performance of the program needs to be documented in a report and distributed to the stakeholders, Fig. 2-43. The report consolidates data to provide stakeholders with accurate, timely, and concise information. The reports include status update, budget actuals and forecast, schedule progress, program metrics, risks, issues, and changes under consideration. To begin this process, you need the program management plan, performance reports, budget baseline, master schedule, go/no-go decisions, variance reports, performance measurements, risk register, approved changes, issues register, benefits realization plan, and forecasts.

At the conclusion of this process, you will have completed:

- *Program performance reports.* These are distributed to the stakeholders as defined in the communication plan.

- Program forecasts are updated.

- *Communications messages.* E-mails, reports, presentations, and voice mails are ways that messages can be communicated.

Monitoring and Controlling Process
Knowledge area: Program Communications Management

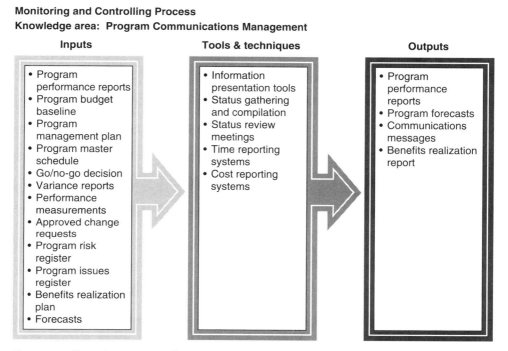

Inputs	Tools & techniques	Outputs
• Program performance reports • Program budget baseline • Program management plan • Program master schedule • Go/no-go decision • Variance reports • Performance measurements • Approved change requests • Program risk register • Program issues register • Benefits realization plan • Forecasts	• Information presentation tools • Status gathering and compilation • Status review meetings • Time reporting systems • Cost reporting systems	• Program performance reports • Program forecasts • Communications messages • Benefits realization report

Figure 2-43   Report program performance.

- *Benefits realization report.* Benefits can be realized throughout the program. The benefits realization report states which benefits are complete and provides the status of the ones in progress.

To achieve the required outputs for this process, the following tools and techniques can be used:

- *Information presentation tools.* Software to help present the performance information.

- *Status gathering and compilation.* Notification, feedback, presentations, reports, and records are gathered and compiled for the performance report.

- *Status review meetings.* Information is exchanged at this meeting. Meetings should be held at regularly scheduled intervals, whether weekly, monthly, or quarterly.

- *Time reporting systems.* Software systems that help with generating schedule reports.

- *Cost reporting systems.* Software systems that aid in generating budget information.

In the program risk management knowledge area, there is one process in the monitoring and controlling process: monitoring and controlling program risks.

Monitor and control program risks

Risks are identified and tracked throughout the program life cycle, Fig. 4-44. Risk triggers are identified. Risks are monitored and controlled, analyzed, and reanalyzed. The probability and impact of the risks are updated as new information presents itself. Risks may change their ratings due to circumstances that may no longer be valid, or due to a greater impact. To begin this process, you need the program architecture baseline, risk management plan, risk register, performance reports, contingency reserves, issue register, and contract information.

At the conclusion of this process, you will have completed:

- *Preventive actions.* These need to be taken to diminish the impact of a risk on a program.

- *Program risk register updates.* Updates are made to the register and new information is presented.

- *Program risk management plan updates.* The plan is updated continually to improve the risk process.

- *Lessons-learned database updates.* These are created as new information is presented.

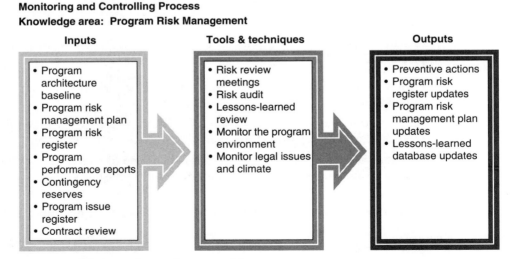

Monitoring and Controlling Process
Knowledge area: Program Risk Management

Inputs	Tools & techniques	Outputs
• Program architecture baseline • Program risk management plan • Program risk register • Program performance reports • Contingency reserves • Program issue register • Contract review	• Risk review meetings • Risk audit • Lessons-learned review • Monitor the program environment • Monitor legal issues and climate	• Preventive actions • Program risk register updates • Program risk management plan updates • Lessons-learned database updates

Figure 2-44 Monitor and control program risks.

To achieve the required outputs for this process, the following tools and techniques can be used:

- *Risk review meetings*. Risks are discussed and analyzed. The appropriate response is assigned to the risk.

- *Risk audit*. Part of the normal program audits, the results may cause the risk process to be updated.

- *Lessons-learned review*. To determine if any procedures should be applied in this type of situation

- Monitor the program environment for changes and areas that need to be addressed.

- Monitor legal issues and climate for changes that could affect the program.

In the program procurement management knowledge area, there is one process in the monitoring and controlling process: administering program procurements.

Administer program procurements

Program procurement agreements are administered according to the conditions documented in the contracts, letters of understanding, or agreement with the supplier, vendor, or service provider, Fig. 2-45. It is also the responsibility of the administrator of procurements to make sure that all agreements follow the

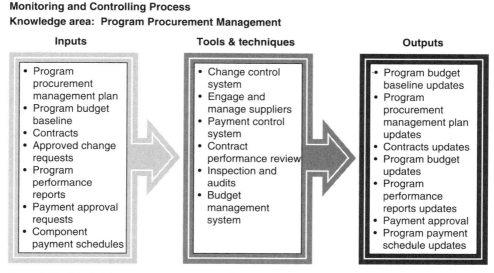

Figure 2-45 Administer program procurements.

appropriate internal policies and procedures, as well as any local government regulations. To begin this process, you need:

- Program procurement management plan
- Program budget baseline
- Contracts
- Approved change requests
- Program performance reports
- Payment approval requests
- Component payment schedules

At the conclusion of this process, you will have completed:

- Program budget baseline updates
- Program procurement management plan updates
- Contracts updates
- Program budget updates
- Program performance reports updates
- Payment approval
- Program payment schedules updates

To achieve the required outputs for this process, the following tools and techniques can be used:

- Change control system
- Engage and manage suppliers
- Payment control system
- Contract performance review
- Inspection and audits
- Budget management system

In the program financial management knowledge area, there is one process in the monitoring and controlling process: monitoring and controlling program financials.

Monitor and control program financials

In order to prevent program cost overruns, the financials for the program have to be vigilantly monitored and controlled, Fig. 2-46. Once the initial funding is granted to the program and expenses are being paid, the monitoring and controlling financial process begins. Metrics are used to determine if the financials

Monitoring and Controlling Process
Knowledge area: Program Financial Management

Inputs	Tools & techniques	Outputs
• Program financial plan • Program management plan • Program budget baseline • Contracts • Change requests	• Cost change management system • Contract cost management • Status reviews • Cost forecasting techniques • Program operational cost analysis • Earned value management	• Contract payments • Component budgets closed • Program budget baseline updates • Approved change requests • Estimate at completion • Program management plan updates • Corrective actions • Program budget closed

Figure 2-46 Monitor and control program financials.

are tracking according to plan. If there are any discrepancies, they are investigated to determine how the program can get back on track.

- Program financial plan
- Program management plan
- Program budget baseline
- Contracts
- *Change requests.* Some change requests may have a financial impact. The governance committee reviews all requests to determine how they will be managed.

At the conclusion of this process, you will have completed:

- *Contract payments.* Payments are made according to the contract conditions.
- *Component budgets closed.* As work is completed for the components of the program, the individual budgets may be closed.
- *Program budget baseline updates.* When changes are approved the corresponding financial baseline is updated.
- *Approved change requests.* The budget is updated according to approval requests and funds are integrated into the budget.
- *Estimate at completion.* The estimate at completion is an iterative process and is updated regularly as changes and new information concerning the budget is presented.

- *Program management plan updates.* Updates are made as needed to the plan.

- *Corrective actions.* When a problem occurs, corrective actions are taken to get the budget back on track.

- *Program budget closed.* As the program reaches completion, the budget is closed and reports sent to the appropriate stakeholders, and they are updated on the closure. Any funds not spent return to the organization.

To achieve the required outputs for this process, the following tools and techniques can be used:

- *Cost change management system.* Determines how financials will be analyzed by looking at planned versus actual financial data and determines how this affects the program.

- Contract cost management

- *Status reviews.* Regularly scheduled review meetings present the financial state to the appropriate stakeholder to determine if any actions need to be taken.

- *Cost forecasting techniques.* Many forecasting techniques can be used to determine how the program is progressing, two of which are estimated at completion and estimates to complete.

- *Program operational cost analysis.* The overall operational costs to run the program have to be monitored and controlled so that they are in alignment with planned costs.

- *Earned value management.* Earned value takes into account the time and cost of the program and determines if the program is progressing as planned.

In the program stakeholder management knowledge area, there is one process in the monitoring and controlling process: managing program stakeholder expectations.

Manage program stakeholder expectations

It is said that perception is reality. Thus, it is essential to manage stakeholder expectations to ensure that they are receiving their information they need and that they are included in the program management process as required, Fig. 2-47. Program managers must to be proactive in managing stakeholders' issues and needs. There will be times when stakeholders have conflicting needs. Program managers must effectively manage the conflicts using negotiating skills. It is also important for stakeholders to participate at the appropriate level to ensure program success.

- *Stakeholder management strategy.* The strategy explains how to keep stakeholders appropriately engaged in the program.

- *Stakeholder register.* The register lists all the stakeholders and the requirements that they have for the program.

Monitoring and Controlling Process
Knowledge area: Program Stakeholder Management

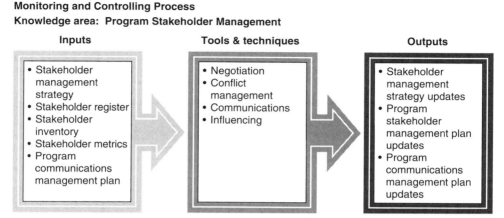

Inputs	Tools & techniques	Outputs
• Stakeholder management strategy • Stakeholder register • Stakeholder inventory • Stakeholder metrics • Program communications management plan	• Negotiation • Conflict management • Communications • Influencing	• Stakeholder management strategy updates • Program stakeholder management plan updates • Program communications management plan updates

Figure 2-47 Manage program stakeholder expectations.

- *Stakeholder inventory*. The inventory lists all the concerns and issues that the stakeholders have had and uses this information to better serve the stakeholders' needs in the future.

- *Stakeholder metrics*. Metrics for stakeholders track their involvement in the program. This information can be used to make sure that stakeholders are participating at the appropriate level.

- Program communications management plan

At the conclusion of this process, you will have completed:

- *Stakeholder management strategy update*s. Updates are made to the stakeholder strategy according to developments made during this process.

- *Program stakeholder management plan update*s. Plans are updated regularly. This is an iterative process throughout the life of the program.

- *Program communications management plan updates*. As the communication needs of the stakeholders change, this plan needs to be updated.

To achieve the required outputs for this process, the following tools and techniques can be used:

- Communication is the single most important tool in stakeholder management. Timely and regularly scheduled communication with the appropriate information is critical in effectively managing stakeholders' expectations.

- Negotiation is a technique used in a program to help reach compromise with the stakeholders. The best solution is always a win-win solution, but this may not always be possible or reasonable.

- *Conflict management.* Conflicts need to be management proactively and escalated appropriately.

- *Influencing.* A technique used in managing stakeholders' expectations that helps shape the understanding of the goals of the program and leads stakeholders to change their attitudes toward decisions and program plans.

In the program governance knowledge area, there are three processes in monitoring and controlling: providing governance oversight, managing program benefits, and monitoring and controlling program changes.

Provide governance oversight

Governance is a system or manner of monitoring and controlling a program, Fig. 2-48. Providing governance oversight for a program ensures that a program follows the determined procedures and policies. Governance is the decision-making body that provides guidance to a program. The governance process is generally run by a board or a committee, which operates according to the program governance structure that was established. The information needed to start this process includes:

- *Governance plan.* This plan provides guidance as to the roles and responsibilities of the governance board members and the activities that they should perform.

- *Gate review decision request.* With a gate review request, the governance board holds a special meeting, or the request is incorporated into one of the regularly scheduled governance meetings for review.

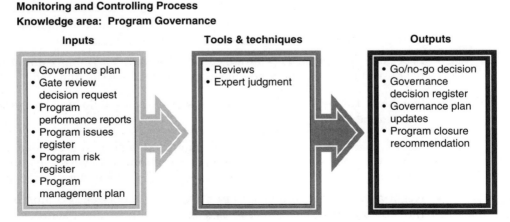

Monitoring and Controlling Process
Knowledge area: Program Governance

Inputs	Tools & techniques	Outputs
• Governance plan • Gate review decision request • Program performance reports • Program issues register • Program risk register • Program management plan	• Reviews • Expert judgment	• Go/no-go decision • Governance decision register • Governance plan updates • Program closure recommendation

Figure 2-48 Provide governance oversight.

- *Program performance reports.* The governance board receives various program reports, such as financial and status reports.
- Program issues register
- Program risk register
- *Program management plan.* This plan defines the benefits that the program is to deliver. The governance group uses this information when making decisions.

At the conclusion of this process, you will have completed:

- *Go/no-go decision.* Decisions are made during review meetings. Pending the results of the decision, the request will receive a go or no-go decision.
- *Governance decision register.* All requests or information put forth to the governance committee is listed in a register, along with an explanation of the item and the result of the decisions and action items that were derived from the decision.
- *Governance plan updates.* The governance structure has built into it a continuous improvement process. Programs last for an extended duration, and as situations arise, new thoughts on how to improve the governance structure are generated and implemented.
- *Program closure recommendation.* The governance board reviews the benefits delivered by the program to determine if the program should be closed.

To achieve the required outputs for this process, the following tools and techniques can be used:

- *Reviews.* Governance review meetings are regularly scheduled meetings with a formal agenda and meeting minutes at the conclusion. During the meeting, decisions are made concerning change requests, the health of the program, and possible improvements to the governance process.
- Expert judgment

Manage program benefits

The status of the program is monitored and controlled throughout the program's life cycle, Fig. 2-49. Appropriate program reports and metrics are created and delivered to the stakeholders according to the communication plan. The reports help stakeholders determine the status and health of the program. If needed, stakeholders can determine from the reports if action should be taken to realign the program to achieve its benefits. Information needed to begin this process includes:

- *Program management plan.* The plan is used as a baseline, and the reports that the stakeholders receive are compared to what was planned to find out whether the program is on track.

Monitoring and Controlling Process
Knowledge area: Program Governance

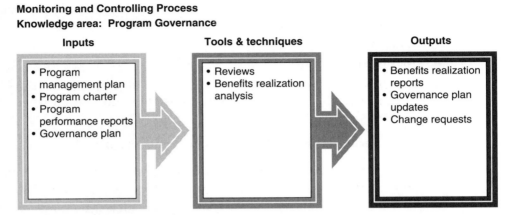

| Inputs | Tools & techniques | Outputs |

- Program management plan
- Program charter
- Program performance reports
- Governance plan

- Reviews
- Benefits realization analysis

- Benefits realization reports
- Governance plan updates
- Change requests

Figure 2-49   Manage program benefits.

- *Program charter.* The charter is reviewed to verify the benefits that are to be delivered by the program.

- *Program performance reports.* Past performance reports are reviewed to track trends or emerging changes to the progress of the program.

- Governance plan

At the conclusion of this process, you will have completed:

- *Benefits realization report.* This report includes metrics that compare planned to actual benefits.

- *Governance plan updates.* Updates are made to the governance plan to improve the plan, structure, process, or roles and responsibilities of the governance members.

- *Change requests.* Change requests may be generated from the review session to help realign the program to best deliver the benefits that it set out to achieve.

To achieve the required outputs for this process, the following tools and techniques can be used:

- *Reviews.* Review meetings are a time when the status of the program is analyzed and the value of the program is reviewed. This could be a gate process to determine whether or not to proceed with the program.

- *Benefits realization analysis.* The benefits of the program and its progress are reviewed against what was planned for the program. A number of areas are addressed in this analysis.

- *Value delivery.* Is this program still delivering the value it was to provide in the fashion it was to provide it?

- *Resource management.* Are the appropriate resources assigned to the program, and are they being utilized appropriately?

- *Performance measurement.* Since programs tend to be long in duration, is it still meeting the required level of performance that was planned?

- *Strategic alignment.* Is the program still strategically aligned with the organization's goals? If there were changes to the strategy, could the program be redirected to meet the new benefit requirements?

- *Risk management.* Are the risks of the program outweighing the benefits that it will return?

Monitor and control program changes

Changes are monitored and controlled to make sure that they align with the organization's strategy and the program's benefit plan, Fig. 2-50. Proposed changes go through the governance process to allow for the appropriate level of decision-making to determine if the requested changes will be rejected, accepted, or modified. The information needed to start the monitoring and controlling program changes process includes:

- *Program management plan.* This is reviewed to make sure that the change is in alignment with the direction in which the program is supposed to be traveling to achieve the benefits that it set out to complete.

- *Change request log.* The log documents all the requested changes, the status of the change, and the final outcome of the request.

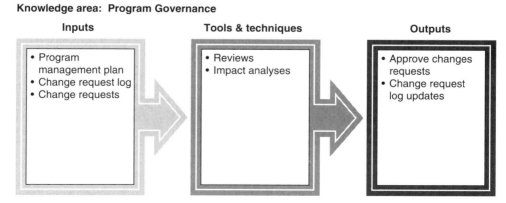

Figure 2-50   Monitor and control program changes.

■ *Change requests.* Changes follow the documented change control process and move through the appropriate channels in order for to be considered a viable candidate for an update to the program.

At the conclusion of this process, you will have completed:

■ *Approved change requests.* These are documented and the appropriate stakeholders notified of the change approval according to the communication plan.

■ *Change requests log updates.* The change request log is updated to serve as a record for the program archive. The log lists all the proposed change requests, the status of the change, and the final outcome of the request.

To achieve the required outputs for this process, the following tools and techniques can be used:

■ *Review meetings.* In the review meeting, the change request is analyzed to verify that it meets the needs of the program and overall organization. A decision made during the meeting could be that further information or clarification of the request is needed, the request is placed on hold, the request doesn't fit into the program and is therefore cancelled, or it is approved and action is taken to set in motion the change.

■ *Impact analyses.* The impact of the change must be analyzed from all aspects of the program, its components, the organization, and external factors. The risk versus benefit of the change is analyzed, and assumptions are reviewed to determine if they are viable.

Closing Process Group

The closing process group formally ensures that all aspects of the program are closed properly according to the plans established for each area. All program activities cease within the closing process, and benefits are formally transferred to the appropriate organization's functional area. The purpose of this process is to confirm that all conditions have been met in the program contracts, all financial responsibilities have been completed, all benefits have been delivered, resources are formally released, and all program information is fully documented and shared with the organization for the lessons-learned growth process.

In the program integration management knowledge area, there is one process in the closing process group: closing the program. Closing the program formally closes all program work and ensures that all program documents are completed and stored.

Close program

When a program is closed, the organization formally accepts the results of the program, Fig. 2-51. Prior to the program closing, all component activities must

Closing Process

Knowledge area: Program Integration Management

Inputs	Tools & techniques	Outputs
• Program transition plan • Program management plan • Program closure recommendation	• Program management information systems • Expert judgment • Contract closure procedure	• Released resources • Final reports • Knowledge transition • Close program

Figure 2-51   Close program.

close first. As each component closes, all their artifacts are transferred to the program to be archived with the final program documents. When the program closes, benefits are transferred to the operation, as required by the program transition plan.

- *Program transition plan.* This plan ensures that benefits are transferred to the organization.

- Program management plan is reviewed to ensure that all the benefits the program set out to complete have been achieved and that all aspects of the program formally come to a close according to the plan.

- *Program closure recommendation.* The program governance committee must sign off on the recommendation to close the program. The program is reviewed for effectiveness in meeting the objectives it set out to accomplish. The governance group must approve the closure recommendation in order for the program to be closed.

At the conclusion of this process, you will have completed:

- *Final reports.* All program plans are updated and final reports for each aspect of the program are created, documenting how it achieved its objectives.

- *Released resources.* Resources are released from the program and transitioned to the areas identified in the program resource plan.

- *Knowledge transition.* The information from the program is transferred to the organization. Knowledge transfer is fully documented, and there may be "training sessions" or meetings to get the receivers up to speed on the information.

- *Closed program.* There a couple of ways a program can come to a close: it can be canceled or end by completing all the objectives it was designed to deliver.

To achieve the required outputs for this process, the following tools and techniques can be used:

- Expert judgment
- Program management information systems
- Contract closure procedure

In the program procurement management knowledge area, there is one process in the closing process group: closing program procurements.

Close program procurements

Program procurement can be closed when all the purchases are complete and contracts are reviewed to ensure that all the deliverables are achieved, Fig. 2-52. Stakeholders need to be notified, including the organization's finance department. The information to start this process includes:

- *Program management plan.* This plan is reviewed to ensure it provides guidance on the proper procedures in closing the procurement process.
- Contracts are analyzed to ensure that all conditions have been met.
- *Program budget.* The budget is reviewed to assess that the funds are available to close the program.

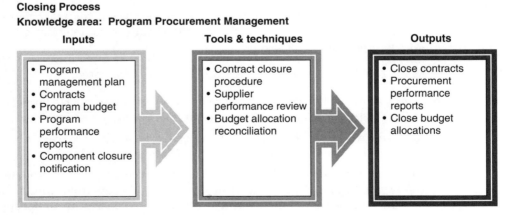

Closing Process
Knowledge area: Program Procurement Management

Inputs	Tools & techniques	Outputs
• Program management plan • Contracts • Program budget • Program performance reports • Component closure notification	• Contract closure procedure • Supplier performance review • Budget allocation reconciliation	• Close contracts • Procurement performance reports • Close budget allocations

Figure 2-52 Close program procurements.

- *Program performance reports*. The reports are reviewed to make sure that all issued documents were properly addressed.

- *Component closure notification*. A notification is sent to all stakeholders, informing them that no additional procurement requests will be taken after the date specified in the document.

At the conclusion of this process, you will have completed:

- *Closed contracts*. Contracts are closed and stored in a secure environment for retrieval if needed at a later date.

- *Procurement performance reports*. The performance results of the procurement process are documented and sent to all appropriate stakeholders.

- *Closed budget allocations*. When all payments are processed, the budget is closed according to the financial plan.

To achieve the required outputs for this process, the following tools and techniques can be used:

- *Contract closure procedure*. Contracts are scrutinized to verify that all the conditions have been met and that there are no residual issues that need to be managed by the program.

- *Supplier performance review*. The performance of the suppliers is reviewed before the procurement processed is closed. This would be the last opportunity to rectify outstanding concerns with the vendor.

- *Budget allocation reconciliation*. The "financial books" for the program are reviewed and reconciled to make sure that all accounts "tick and tie" back to other financial information and processes.

In the program governance knowledge area, there is one process in the closing process group: approving component transition.

Approve component transition

Approving component transition is the final step in the component process, Fig. 2-53. The component is transferred to the operation. Prior to transition, the program manager ensures that the requirements and objectives of the component have been met. Generally, there is a formal review process before transition. The information needed to start this process includes:

- *Component transition request*. A request is made to the governance board to transition the component. The governance board determines if the component met the objectives and requirements it was commissioned to meet. It also has to verify if there were any changes to the program scope that would prevent the component from being transferred to the operations organization of the business.

Closing Process
Knowledge area: Program Governance

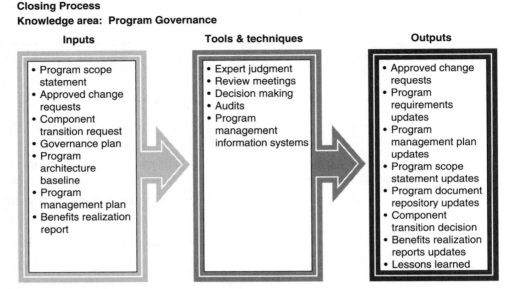

Figure 2-53   Approve component transition.

- *Benefits realization report.* The person or group responsible for the component produces a report on how the component met all the benefits it set out to complete. If the results were not met, the component may not transition and could be terminated.
- Program scope statement
- Approved change requests
- Governance plan
- Program architecture baseline
- Program management plan

At the conclusion of this process, you will have completed:

- *Component transition decision.* The transition decision should be documented according to the program standards and stored in a central server system for archiving purposes.
- *Benefits realization report updates.* This report specifies how the benefits were realized by the component. This report includes all the tools, techniques, and information about the component, benefits it achieved, and possible future benefits that can be realized.
- *Lessons learned.* This is an iterative process that occurs throughout the life of the program. Included information that is documented in lessons learned

are best practices, "words to the wise" information, and experiences gained through the program process.

- *Program management plan updates.* The program plan updates are an iterative process throughout the life of the program. The program log is updated with a description of the decision, when the decision was made, by whom, and why it was made.
- Approved change requests
- Program scope statement updates
- Program document repository updates

To achieve the required outputs for this process, the following tools and techniques can be used:

- *Review meeting.* The review is a gate where the component is evaluated for the successful completion of the activities it was created to accomplish.
- Expert judgment
- Decision making
- Audits
- Program management information systems

The program process has been summarized in Fig. 2-54 to help illustrate what processes or activities happen at what process area.

PMI created program standards so that program managers could all sing out of the same songbook, so to speak. Program managers should take the PMI standards and mold them to best fit their organization's culture. No two organizations are the same; therefore, no two program management processes will be the same. However, from program to program, you will see shades of the same standards presented in this chapter.

Programs were compared to projects and portfolios, each of which carries a great benefit for an organization. The program management office supports the program and the program manager. The program manager needs many different skills to be successful, including communication, technical, people, organizational, time management, project management, political, environmental awareness, program management, strategic vision, and leadership skills.

Program management has a life cycle of preprogram planning, program initiation, program formulation, program delivery of benefits, and program closeout. It also has the following processes: initiating, planning, executing, monitoring and controlling, and closing. Those processes have specific knowledge areas: integration management, program scope management, program time management, program communication management, program risk management, program procurement management, program financial management, program stakeholder management, and program governance management.

Initiation Process	Planning Process	Executing Process	Monitor and Controlling Process	Closing Process
Initiate program	Plan program scope	Direct and manage program execution	Monitor and control program performance	Close program
Establish program financial framework	Define program goals and objectives	Manage program resources	Manage program issues	Close program procurements
	Develop program requirements	Manage program architecture	Monitor and control program scope	Approve component transition
	Develop program architecture	Manage component interfaces	Monitor and control program schedule	
	Develop program WBS	Distribute information	Report program performance	
	Develop program management plan	Conduct program procurements	Monitor and control program risks	
	Develop program infrastructure	Engage program stakeholders	Administer program procurements	
	Develop program schedule	Approve component initiation	Monitor and control program financials	
	Plan communications		Manage program stakeholder expectations	
	Plan program risk management		Provide governance oversight	
	Identify program risks		Manage program benefits	
	Analyze program risks		Monitor and control program changes	
	Plan program risk responses			
	Plan program procurements			
	Develop program financial plan			
	Estimate program costs			
	Budget program costs			
	Plan program stakeholder management			
	Identify program stakeholders			
	Plan and establish program governance structure			
	Plan for audits			
	Plan program quality			

Figure 2-54   Program process.

The benefits of program management are its management of benefits. Program management is a centralized, coordinated management of a program to achieve the organization's strategic benefits and objectives. It involves aligning multiple projects to achieve the program goals and allows for optimized or integrated costs, schedules, and efforts.

The Importance of Program Management Support Systems

Thomas Friedman's *The World Is Flat* confirmed today's movement of information in a seamless stream across the globe, facilitating the almost instantaneous and ubiquitous reach of today's multinational corporation. The remote design and development of programs, projects, and products now require a new business model of strategic collaborating and partnering, and a new dimension of localizing and customizing processes. This requirement touches every aspect of the company, including program and project management, a *retooling* of the infrastructure of the business to provide necessary support in new shapes and forms. The focus of this chapter is on this process of retooling.

Global program management requires global strategies, but often, the key success factor in worldwide program delivery is in support and outsourced activities. Boeing Company outsources whole components of aircraft to various companies in a multitude of nations, integrating components in the assembly process. Sony purchases components from many different suppliers around the world, its selection of suppliers keyed to quality and cost. CNN reaches to any part of the world to gather and report news, working through its complex partnerships with local news systems and sources.

The quality and effectiveness of what we are calling support, or *infrastructure,* systems in this book are the keys to program management success—some say *the* key in future success. This is because program management is a total business-wide effort, not a narrow program concept, and doing business on a global platform changes everything. Programs engage all parts of the business or agency and require support in many planning, operational, marketing, and production functions. Thus, program managers must have a grasp of the total business and its internal and outsourcing potential, and must have strong company-wide and organizational support. Projects can be completed in relative isolation from key business process; programs cannot.

A global program management effort engages all the key business processes and systems. This includes information, administration, contracts and outsourcing, human resources, production and operations, and marketing/distribution.

The program management office (PMO)

To accomplish these support services, a new global program management office (PMO) will be necessary, constructed in a global configuration. PMOs will have local or regional staffs to service program delivery in various worldwide markets.

Figure 2-55 Global program management support model.

The PMO will assist in program planning, program scheduling, cost control, and communications across language and culture barriers.

Figure 2-55 is a model of global program support, which brings into play most of the key factors in a program's success. These are the major success factors in global program management infrastructure and support.

Human resources

Program managers must have a new regard for global talent and building local and regional program workforce expertise in designing and developing programs. This will mean more rather than less virtual teaming and global conferencing. Finding and committing local experts and subject matter experts to program planning and delivery will require a local presence in human resources, from hiring to contracting to engineering, administration, production, and sales and marketing. The program will be localized and decentralized to allow the *center of program gravity* to reflect local conditions and culture. This will mean PMO support on a local scale, guided by a central program strategy.

Information technology

Program management in a global environment cannot operate without worldwide networking and data sourcing. Program charters and other documents will be available on the Web, creating security risks but also opportunities to leverage local information. Program managers will need to promote common methods of program planning and delivery using common tools and techniques, such as a common project management software program accessible to all program stakeholders.

Facilities and supplies

While program management rarely involves permanent facilities and supplies, program planning and delivery usually includes provisions for local business facilities and operational improvements. Programs, by nature, are temporary, but do last for a longer duration than the projects under them. Program managers need support in local sourcing decisions and in real estate and property management issues. If, in the strategic plan for the organization, there is a provision to maintain a local central support area after the program has concluded and transitioned to the organization, a more permanent facility and supply chain may be set up for long-term support in the region.

Customers and new product design

Program design and development typically involve new product lines, new systems, and product concepts. This will require program initiatives to create new product differentiation strategies targeted at local markets around the world. Choices of program priorities will reflect global considerations, and portfolio development processes to create new program categories will use local workforce creativity. Programs involving new products will require new global product design and development processes.

Local regulation

Global companies work in a variety of local cultures, with different public and government regulations and requirements. This requires that program managers have high regard for the role of political and international agencies and trade policies. Associated program risks must be assessed and managed. Local regulation information needs to be monitored and researched on a regular basis to assess any changes and the impact they have on the program. Major concerns regarding changes to local regulations should be escalated to the governance committee in a timely fashion.

Procurement

Procurement focuses on the acquisition of goods and/or services at the best possible price. Procurement decisions need to weigh selection criteria factors of quality, quantity, timing, and cost. Global procurement can deliver a significant competitive advantage if delivery costs are reduced, superior technology resources are optimized, and resources are streamlined and favorably acquired. Global programs will require outsourced partners in designing and developing deliverables and outcomes. Supply chain management options need to be reflected in global program design. Procurement decisions concerning "make or buy" must be housed in an international context.

Communication/translation

If communication is an important success factor in domestic projects, it is certainly a key to successful global program management. Communicating within a global program adds a new dimension of complexity to an already multifaceted situation. You have physical barriers of distance and perceptual barriers of different views on the situation. There are also emotional barriers, where there is fear from free and open communication in a virtual or long-distance team environment. There are cultural barriers, too, where certain behaviors of communication that are accepted within a group are not known or understood by parties that are not local. Different communication styles exist in different territories. Gender communication styles differ in different countries. There can be language barriers, even among those who speak the same language. With language issues, there are, at times, difficulties in understanding expressions, jargons, or even buzzwords. With language differences, you have the added complexity of accuracy in translations. One word could have different meanings in different contexts.

There is also the hurdle of time zones. Time zones could mean that you have a conference call at 4 A.M. or 11 P.M. The definition of a "typical" workday evolves into a "global workday," which is defined according to the terms and conditions of the individual global program teams. Overall, communication and translation issues must be addressed in program planning.

Communication is one of the most important factors in project, program, portfolio, and program management office processes. The goal of any communication is to present the right information, to the right person, in the right format, in the right setting, and at the right time. The program stakeholder matrix would list the stakeholders with their communication needs, delivery methods, and time constraints. It is necessary to present the information in the context that would be most useful for the stakeholder.

Figure 2-56 shows the different qualities of communication message contents that could be presented to the stakeholders. A PMO receives copious amounts of data, which is a collection of true facts. This information is not organized in any particular way. Presenting pure data to a stakeholder would cause a great deal of confusion and frustration. The stakeholder should not be expected to sort through data to find the relevant nuggets.

The PMO should take the data and verify its accuracy. From there, it should assemble and categorize the data to create relevance. This process is not just a sum of the parts, but a logical construction of building blocks to create information. The

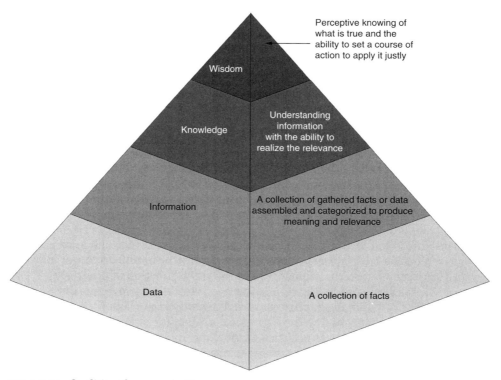

Figure 2-56   Qualities of communication.

information is then analyzed and prioritized before disseminated to the stakeholders. This process creates increased understanding and decreases uncertainty for the stakeholders. The value of the information could be used to confirm expectations, as an input to a decision-making process, or as a trigger for corrective action.

With the information in the hands of the stakeholders, knowledge of the state and condition of the program is accumulated. Knowledge is the understanding of the information, with the ability to realize its relevance. The awareness and understanding of the relevance provides the capacity to act effectively.

Only with the amalgamation of experience, knowledge, and perception is true wisdom realized. Wisdom becomes the perceptive knowing of what is true and the ability to set a course of action to apply it justly. Actions are not taken without good judgment and common sense for the greater good. Reaching wisdom within a program, portfolio, or program management office occurs at a mature process level within the organization.

Demographics/culture

A special regard for local cultures and cultural codes, e.g., local meanings and ways of thinking and making decisions, becomes key in understanding others.

Program managers must understand international *culture codes* (see *The Culture Code* by Clotaire Rapaille) and *what makes local people tick* around the world. The best program managers will be flexible and proactive. They will work to find solutions rather than get frustrated with cultural barriers. They will also be proactive and work to learn and understand. Program managers need to set the tone for the team as to how the group is going to operate and function. Program managers will use their leadership skills to guide and direct their team.

Quality

Since quality is defined differently around the world and customer expectations, needs and requirements vary according to local conditions, program managers will have to be agile and flexible in how they integrate quality into program planning and management. Program goals will be framed in terms of continuous improvement and quality. The program manager works to improve processes incrementally. This is not to say that there may not be breakthrough advancements discovered. Whether the improvements are incremental or produce a larger leap forward, it is critical to keep the momentum in the right direction of progress.

As displayed in Fig. 2-57, the best way to improve quality is to plan for it. After you plan for quality, you need to analyze the benefits with a test case. After success is reached with the test case, the plan is rolled out to all of the groups and becomes a best practice for the organization. Finally, after the rollout of a new process, the PMO processes are assessed to quantify and qualify the benefits that were added to the business. The information from the assessment is then fed back in to the beginning of the process, and planning begins again to identify new opportunities for change. The benefits report is presented to all the stakeholders of the PMO. The quality assurance process is necessary for a PMO because the environment in which the PMO exists is always changing and the quality processes need to advance with the changes.

Economics of scale

An economics of scale occurs when efficiencies in areas of cost, time, or quantity are gained when more is produced. Programs develop improvements and

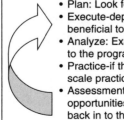

- Plan: Look for and identify opportunities for change
- Execute-deploy changes in "test case" scenarios to validate if the change will be beneficial to the organization
- Analyze: Examine the results of the change to determine if the benefits are valuable to the program process.
- Practice-if the benefits prove to be valuable, the change is then moved into a wide-scale practice.
- Assessment-the practices in the organization are monitored to determine if new opportunities for change and advancement are possible. This information is then fed back in to the planning process.

Figure 2-57 Quality assurance in a PMO.

upgrades in company products and services; thus, program managers will have to understand cost control in program delivery, but also how cost in an international context affects program outcomes and benefits.

Other Considerations

More on outsourcing

The decision to contact or outsource portions of the program management process to other countries is largely a question of cost, efficiency, and effectiveness. This decision is made after several key aspects of the program are "completed" internationally. This means that processes such as product design, manufacturing, distribution, and customer support are opened up for local and regional outsourcing, both to reduce costs and also to gain favor in the local marketing area.

Administrative support

There are copious amounts of logistical work within a PMO. A PMO could be formulated as a support organization. Part of the support that it could offer is to provide administrative support for a portfolio, program and project management teams. The administrative work would encompass clerical assistance for filing and storing program-related data. The PMO could also assist with acquiring materials and support equipment for the program. Other services provided could include acquiring and establishing space and facilities. Additional logistical functions could be performed by the PMO, such as reserving conference room and booking travel arrangements.

More on human resources

Recruitment and outsourcing is performed in local or regional locations in order to find nationals who can support a program in their native countries. This requires language skills and a full appreciation of local workforce characteristics and limitations. Hiring decisions can be difficult within your own culture and expertise, but when you add the global dimension to the decision of selecting candidates for positions that you may or may not have ever met in person, this adds to the complexity of the human resource management of a PMO.

Integration in global and international programs

Integration management focuses on identifying, defining, and coordinating the intersection points of a program. Global and international programs inherently face additional difficulties with integration issues, simply because the normal and typical integration barriers are compounded by political, national, and language issues. Programs to assist developing countries, for instance, are typically managed by the collective of nongovernmental organizations (NGOs), donor country managers, and local team members, who often do not agree on the purposes and goals of the program. Opportunities for fraud, waste, and abuse are again compounded because of the many *players.*

Tools in building an integrated program management system

In building a company or agency system to support program integration, there are 10 areas for process improvement: organization-wide or enterprise-wide program management systems, program portfolio system development, integrated resource management systems, information technology, technical product development (including a stage-gateway review system), interface management system, program portfolio management, program monitoring and corrective action, change control, and program evaluation.

Building an organization-wide global program management system

A global program management system has several key ingredients:

- *Integrated Program Management Culture.* Leaders develop their organizations to accomplish integration through systems and communication. This system involves the development of a culture of defining and capturing work in terms of programs—e.g., all work of the organization outside of recurring production work is considered program work with a customer and deliverables. All training and development and incentive systems are built to encourage that work be accomplished through formal programs and plans and schedules that integrate cost, time, and quality.

- *Generic work breakdown structure.* A generic work breakdown structure is an outline of tasks in sequence but not linked. The purpose of the generic WBS is to integrate the work, which is program-coded to capture costs and task performance history, with the scheduling of any task the company takes on in any program. The generic WBS defines each task in a *data dictionary,* or task definition, that covers what the task expectation is and what its deliverable is for a safety task in a product development program.

- *Scheduling system.* A scheduling system places all work in a program scheduling software—e.g., Microsoft Program or Primavera—assigns resources, and estimates costs in order to control the work. Integration of all the work of the company is accomplished through scheduling, which is seen as a process of *committing resources to work.*

- *Resource assignment.* Resources are assigned to programs and tasks so that the workforce is integrated into the work that is authorized and sponsored by the company. Programs are seen as investments in the business plan; therefore, there is a major impetus to capture the work being performed in a resource assignment system.

- *Task linkages and interdependency*. Programs are consolidated and tasks are linked to stress the interdependency of program work. No piece of work in the company is left unconnected in order to ensure integration. These interdependencies become key management points for the program manager. The interdependencies are actively monitored and controlled throughout the program. Risks at these interdependency points are identified and analyzed, and a planned response is given for each risk.

- *Matrix team structure*. The matrix structure ensures integration because functional departments and program teams are intermingled in every aspect of the company's work, from programs to process development to improvement. Program teams are staffed by functional departments, who are in charge of the quality of the work and the development of technical systems. Program managers manage assigned team members toward program deliverables and earned value.

- *Work authorization system*. In order to ensure that the work that goes on is authorized, and directed by the program manager, all work must go through the defined channels, as planned for by the governance committee and PMO. The way work is approved is through the baseline schedule, which defines the authorized work. As work is completed, tasks are assigned, or changes are approved and the baseline schedule is updated.

- *Knowledge management system*. The PMO is the central warehouse for information. Information is gathered from the business, programs, projects, outside factors, and process. It is then disseminated to the appropriate parties in accordance with their needs and requirements. Finally, information is stored for later retrieval needs.

- *Guidelines for program management plan*. The program management plan is defined in a company policy statement to guide the definition and control of the work. Therefore, the plan must include control points—e.g., stage-gateway reviews—that assure that managers authorize movement from one phase or stage to another. Reporting and monitoring strategies, including the use of earned value to integrate cost, schedule, and quality performance, should be made explicit. Here, the PMO plays a critical role in defining, overseeing, and updating the guidelines for the program management plan.

- The plan should also address accountability, particularly in view of the recent legislative and regulatory requirements of the Sarbanes-Oxley Act. This requirement is in compliance with internal control and accounting standards, and is no longer optional for program managers. In fact, the price of disconnected and inconsistently applied efforts throughout a program and its interfaces, as well as lack of financial tracking systems that provide for audits, could be business-wide. Compliance with Sarbanes-Oxley, therefore, is not a choice but a requirement, and the plan should state standards for estimating costs, tracking the costs, and relating costs to work performed, as well as the integrity of the closeout procedure and invoices to customers for work performed.

- Sarbanes-Oxley standards are accounting practices in the United States. For programs that are based in other countries, additional accounting standards and government polices need to be complied with for the program to be successful. The PMO needs to be aware of the specific requirements to best advise program teams.

Program/portfolio planning and development system

- *Business planning system and strategic objectives.* The integrated company has a business and strategic planning process that produces a statement of strategic objectives to serve as a guide for all planning and budgeting. Such a system helps to shape the program portfolio and assures that the company invests in programs that are integrated with the direction of the business and its ownership. This can also mean that with a change in direction or strategy, company programs may be adjusted accordingly, placed on hold, or cancelled.

- *Decision process.* A defined decision process supports integration because it opens up the decision-making process. If decision points are not addressed for a prolonged time, it can lead to waste and ineffective work. Decision trees are used to assess the commercial value of various decision paths involved in defining the task structure and sequence of approved programs.

- *Budgeting system.* A capital rationing system or some way to allocate company resources in line with the priority of relative strategic objectives is part of integration. Once budgets are identified to carry out business plans, programs are planned and prioritized in the portfolio system, and then costs are estimated. Finally, programs are funded according to their relative merit against business plans and available budget. The program maintains the budget throughout its life cycle and reports the status to the portfolio committee. Budgets are forecasted. When actuals are available, they are compared to the forecasted amounts. The variance between forecast and actuals can pinpoint triggers for risk response plans.

- *Risk management system.* A risk management planning system is needed to identify and assess risks and to generate a risk contingency plan. It is necessary in an integrated program management system. The risk matrix is the format for developing risk information that is used in scheduling and controlling the work.

- *Program definition.* Programs are sets of projects with similarities in processes, products, and customer base. Every company approaches programs differently. It is up to the PMO and the business that it resides in to define what constitutes a program in their organization. Definition of longer-term *product lines* will help to clarify the boundaries of a given program over time.

- *Portfolio pipeline system.* A pipeline of approved programs is maintained so that as funds and resources become available, programs are quickly initiated.

Program plans and schedules are produced for programs in the pipeline so that when authorized they can proceed quickly.

Resource management system

An integrated program management organization must manage resources carefully, simply because there is value in targeting all resources and equipment on the right program work. Waste is compounded in programs with multiple tasks and the effects are felt across the company. A resource pool can be established using Microsoft Project software, or tailored project management software, that records all assignments in order to keep a running view of how people and equipment are being utilized.

- *Workforce planning.* A workforce planning system integrates the hiring and training of personnel with the needs of the program portfolio. In others words, people, equipment, and systems are brought into the company to fill needs that are made explicit in the program resource allocation pool, which reflects both current and planned work.

- *Staffing planning.* A staffing system allocates staff to the priority program needs in order to fully integrate the core competence of the workforce with the priority needs of key programs. Staff is focused on assignments that are visible and reviewed regularly.

- *Financial and accounting control.* Financial and accounting practices are assured in a program management system that captures all program costs, both direct and indirect, and assures internal controls on program costs and equipment inventories.

  - Earned value: Reports on work progress and costs are used to calculate earned value so that the company knows how each program is doing in terms of schedule and cost. EV reports are a beneficial tool in monitoring and controlling a program's progress. It identifies issues in a program. The PMO uses tools like EV to track and monitor the progress of a program and reports the results to executive management.

  - Industry standards: Industrial cost and work standards are used to control the estimated duration of scheduled tasks, e.g., using a trade association to schedule an industry-wide activity on which there are work and industrial standards.

Program information technology system

A program information system that documents all program work in consolidated schedules and resource pools assures that work is staffed, planned, and monitored in a uniform way. This allows a comparison of program progress and supports decisions on where to focus resources.

- *Network system.* All program and program information—e.g., schedules, resource pools, program review and gate review data, and configuration management documents—are kept on a company intranet to allow wide-ranging visibility.

- *Accessibility to key information.* Accessibility to information is controlled and focused on need-to-know criteria. However, customers are regularly informed on program and program progress through Microsoft Program Central web-based reporting systems that allow review of schedules without parent software or a similar software system.

- *Reliability planning.* Reliability planning targets products with failure mode effects assessments and functional hazard assessments, along with risk matrix documents, to consistently design and test reliability of product performance to customer requirements and specifications.

- *Workforce training.* Workforce training is designed to meet program needs as evidenced in work performance feedback reviews and lessons-learned exercises with program teams in closeout. Workforce training can be developed to best serve the needs of the users. The training can include detailed documentation, with screenshots of the system, a flip card with high-level information, live training events with hands-on skills development, and/or online training tools.

Product/service development process

Integrated program management cannot be accomplished without integrated product development processes with strong stage-gate milestones.

- *Cooper stage-gateway process.* Program management is a process of managing time, cost, and quality, but the underlying strength of any program integration process is a strong, phased development process, with clear controls on entry to the next stage. Gate reviews are documented, and generic work breakdown structures and data dictionaries are developed for all product and service development activities. The PMO defines the process and the stages that the programs need to follow. As the maturity PMO of the organization increases, this process will improve over time.

- *Technology support and testing.* Technical support that meets industry standards assures that product integration and testing is verifiable. Designs are tested against specifications, specifications are tested against scope of work, and scope of work is traced to customer requirements and expectations. It is best if tests are aligned to the PWBS (program work breakdown structure) and tasks in the timeline.

Interface management

- *Matrix organization.* Interfaces between functional departments—e.g., accounting, engineering, program management, and testing—are assured through strong interface management. Separate departments and functions are brought together constantly through information and reporting systems and face-to-face review meetings at key gates.

- *Program review meeting formats.* Review meetings are controlled by generic meeting agendas and data and information support from a professional PMO or staff. This way, review information is objective and consistent. Having a consistent format allows stakeholders to easily understand the information. The PMO is responsible for creating the formats that the programs follow for meetings and the documents such as meeting minutes and agendas. Here, you obtain an apples-to-apples comparison of information.

- *Procurement interface.* Due to the importance of contract and outsourced work, contractor personnel and processes are integrated with sponsor company personnel and processes. Common scheduling and reporting systems are designed.

- *Financial, accounting, and internal control interface.* New impetus for strong accounting and accountability reporting now requires that program managers capture costs and related them to work performed and equipment purchased and in inventory.

- *Marketing and sales interface.* Integration of marketing, sales, and program work is accomplished by assigning marketing and sales personnel to program teams. They attend and input to the teams on customer developments and learn what they can and cannot commit to customers and when.

- *HR interface.* The interface with HR is important to integrate personnel and HR policies and procedures with program work and priorities. Performance reviews are left flexible, yet important, in assigning resources to future programs.

Program monitoring and control system

When monitoring and controlling a program, constant communication between stakeholders, company executives, and project teams is required for successful delivery of the program benefits.

- *Program management office (PMO).* Monitoring is based on earned value reporting, and quality is assured by a task planning system that relates percent (%) complete to defined milestones in the baseline schedule.

- *Corrective action / risk management process.* Contingencies and corrective actions are based on remaining work and are forward-oriented. Contingencies are embedded in schedules to assure that should risks occur, contingencies have already been scheduled and budgeted.

- *Escalation system for decisions.* Conflicts and differences within program processes are reviewed regularly by top management to assure that decisions are not delayed.

Change management system

- *Change order system.* All changes to a scope of work are submitted by program team members or the sponsor/customer to assure that changes are reviewed and managed. It is critical that a formal change control system be in place. The PMO monitors the changes to projects and programs to advise as needed on the direction of the change and how it could affect the business or deliverables for the program.

- *Change impact system.* Change impact statements are prepared for all substantial changes, with risk, schedule, cost, and quality impacts specified. Any change to a program has an impact. The impact could be minor or major. The impact system would define what the change is, how it will be executed, who would be involved in the change, the environments that will be affected, and will notify all appropriate stakeholders of the change if it is approved for implementation.

Program evaluation system

- *Document lessons learned.* Closeout includes a lessons-learned meeting with all appropriate program stakeholders present and documentation of outcomes. During this process, the entire program is reviewed from start to finish. The team discusses what went well with the program process. It also reviews areas for improvement for the next program. The PMO is made responsible for assuring that lessons learned are integrated into future programs and stored in a central archive for future retrieval needs. The PMO guides and educates current and future programs on the lessons learned where appropriate.

- *Financial auditing system.* A financial and program audit system is managed to assure accountability and internal control of all assets. This system audits all assets from cradle to grave. Assets due to retire can be added to the portfolio process to be prioritized for replacement among the other investments the portfolio is managing.

Limitations of integrated program management systems

Systems don't integrate programs, people do. Even if the organization is able to design and install compatible systems to help integrate programs, they will not work if the people who manage the work don't use them. Configuration management as an integrating function in product development between design and production cannot be effective if the configuration manager does not see both ends of that spectrum. Program managers who are obsessed with schedules and on-time delivery at any price, and who do not care about costs simply because top management or the customer has not focused on costs, *will not be successful in the business as a whole.* The lesson is this: Individual program success should not be at the expense of the business itself. To assure that this does not happen, company leadership must continuously work toward an integrative vision and process at all levels of the organization—they must daily walk the integrative talk.

Microsoft program support for program management

Microsoft Project is a program integration tool and is particularly useful in integrating tasks, schedules, costs, and quality at the program level. This tool can be used to build schedules, manage resource allocations, and determine if a program is on track.

WBS and task outline

As shown in Fig. 2-58, tasks are outlined to show major summary tasks—e.g., develop program management—and then are broken down into subtasks—e.g., program scope management—that build, or roll up, to the summary task. This is the beginning of integrating tasks so that the level of work breakdown is scaled to the particular needs of the program.

Linkages between tasks are accomplished through identification of predecessors. These linkages integrate tasks *with each other;* thus, this is a second integration function of the software.

Tasks are integrated with resources through the assignment of resources to individual tasks—that is, by highlighting the task and adding resources and percent assignment in the task information box.

Costs are then estimated for each resource after resources are assigned. Microsoft Program allows cost entry in the Resource Sheet view, where personnel resources are listed (see Fig. 2-59).

ID	% Complete	ⓘ	Task Name	Duration	Start	Finish	Predecessors	Resource Group	2001 Q3
32	0%		**Develop Project Management Plan**	**38 days**	**Tue 11/27/07**	**Thu 1/17/08**			
33	0%		Project Scope Management Plan	2 days	Tue 11/27/07	Wed 11/28/07	31		
34	0%		Generic Work Breakdown Structure	2 days	Thu 11/29/07	Fri 11/30/07	33		
35	0%		Schedule Management Plan	2 days	Mon 12/3/07	Tue 12/4/07	34		
36	0%		Cost Management Plan	2 days	Wed 12/5/07	Thu 12/6/07	35		
37	0%		Quality Management Plan	2 days	Fri 12/7/07	Mon 12/10/07	36		
38	0%		Process Improvement Plan	2 days	Tue 12/11/07	Wed 12/12/07	37		
39	0%		Staffing Management Plan	2 days	Thu 12/13/07	Fri 12/14/07	38		
40	0%		Communication Management Plan	2 days	Mon 12/17/07	Tue 12/18/07	39		
41	0%		Risk Management Plan	2 days	Wed 12/19/07	Thu 12/20/07	40		
42	0%		Procurement Management Plan	2 days	Fri 12/21/07	Mon 12/24/07	41		
43	0%		Milestone List	2 days	Tue 12/25/07	Wed 12/26/07	42		
44	0%		Resource Calendar	2 days	Thu 12/27/07	Fri 12/28/07	43		
45	0%		Schedule Baseline	2 days	Mon 12/31/07	Tue 1/1/08	44		
46	0%		Cost Baseline	2 days	Wed 1/2/08	Thu 1/3/08	45		

Figure 2-58 GANTT chart view, entry table.

ID	ⓘ	Task	Resource Name	Type	Material Label	Initials	Group	Max. Units	Std. Rate	Ovt. Rate	Cost/Use	Accrue At	Base Calendar	Code
1			Test Eng	Work		T		100%	$70.00/hr	$0.00/hr	$0.00	Prorated	Standard	
2			ME	Work		M		100%	$70.00/hr	$0.00/hr	$0.00	Prorated	Standard	
3			Bob Smathers	Work		B		100%	$70.00/hr	$0.00/hr	$0.00	Prorated	Standard	
4			Bill Carter	Work		B		100%	$70.00/hr	$0.00/hr	$0.00	Prorated	Standard	
5			John Smoltz	Work		J		100%	$70.00/hr	$0.00/hr	$0.00	Prorated	Standard	
6			Ryan Brookings	Work		R		100%	$70.00/hr	$0.00/hr	$0.00	Prorated	Standard	
7			Ryan Brown	Work		R		100%	$70.00/hr	$0.00/hr	$0.00	Prorated	Standard	
8			Pete Hallings	Work		P		100%	$70.00/hr	$0.00/hr	$0.00	Prorated	Standard	
9	◈		**Bill Dow**	**Work**		B		**100%**	**$70.00/hr**	**$0.00/hr**	**$0.00**	**Prorated**	**Standard**	
10			Ben Gay	Work		B		100%	$70.00/hr	$0.00/hr	$0.00	Prorated	Standard	
11			Bob Harris	Work		B		100%	$70.00/hr	$0.00/hr	$0.00	Prorated	Standard	
12			Bart Starr	Work		B		100%	$70.00/hr	$0.00/hr	$0.00	Prorated	Standard	
13			Dennis Bloom	Work		D		100%	$70.00/hr	$0.00/hr	$0.00	Prorated	Standard	
14			Bud Manaker	Work		B		100%	$70.00/hr	$0.00/hr	$0.00	Prorated	Standard	
15			Bart Werrel	Work		B		100%	$70.00/hr	$0.00/hr	$0.00	Prorated	Standard	
16			Ben Schwartz	Work		B		100%	$70.00/hr	$0.00/hr	$0.00	Prorated	Standard	

Figure 2-59   Resource sheet view.

ID	% Complete	ⓘ	Task Name	Duration	Start	Finish	Predecessors	Resource Group	2001 Q3 Q4
176	0%		**Performance Requirements**	**95 days**	**Tue 6/3/08**	**Tue 10/14/08**	**71**		▽
177	0%	▦	Requirements analysis	15 days	Wed 6/4/08	Tue 6/24/08	71		
178	0%	▦	Data analysis	19 wks	Tue 6/3/08	Tue 10/14/08	71		
179	0%		Simulation Studies	14 days	Wed 6/4/08	Mon 6/23/08	71		
180	0%		**Interface Integration**	**70 days**	**Wed 6/4/08**	**Tue 9/9/08**	**71**		▽
181	0%		Electrical	16.5 days	Wed 6/4/08	Thu 6/26/08	71		
182	0%	▦	Mechanical	5 wks	Wed 6/4/08	Tue 7/8/08	71		
183	0%		Software	4 wks	Wed 6/4/08	Tue 7/1/08	71		
184	0%	▦	Powertrain	*14 wks*	Wed 6/4/08	Tue 9/9/08	71		
185	0%	▦	Wheels	0.6 wks	Wed 6/4/08	Fri 6/6/08	71		
186	0%		**Outsource Controls**	**5 days**	**Wed 6/4/08**	**Tue 6/10/08**	**71**		▯
187	0%	▦	Contracts	0.1 wks	Wed 6/4/08	Wed 6/4/08	71		
188	0%		Supplier negotiations	2.5 days	Wed 6/4/08	Fri 6/6/08	71		
189	0%	▦	Collaboration and partnering	1 wk	Wed 6/4/08	Tue 6/10/08	71		

Figure 2-60   GANTT chart view, entry table.

Then, durations are estimated for each task based on the best guess or standard duration for that task, as shown in Fig. 2-60. This completes the definition of work that is the basis for Microsoft Program assumptions, that is, work = time × resources. Each task is measured, not only in terms of the definition of the work itself in a generic work breakdown structure with deliverables, but also in terms of time and resources—e.g., *5 person days*. Thus, the actual work of the task is defined in terms *of how long it will take, given the resources assigned to do it.*

Fixed costs—e.g., equipment, capital assets—are estimated and entered in a different view and table. Since fixed costs are associated with tasks, not resources, fixed costs are entered into the cost table (see Fig. 2-61).

The integration of risk into the scheduling process is accomplished through the use of the PERT analysis tool, as shown in Fig. 2-62. Here, three alternative durations are estimated: expected, optimistic, and pessimistic. Then weights are

ID	Fixed Cost	Fixed Cost Accrual	Total Cost	Baseline	Variance	Actual	Remaining	2001	
								Q2	Q3
67	**$0.00**	**Prorated**	**$22,400.00**	**$22,400.00**	**$0.00**	**$0.00**	**$22,400.00**		
68	$0.00	Prorated	$2,100.00	$2,100.00	$0.00	$0.00	$2,100.00		
69	$0.00	Prorated	$4,200.00	$4,200.00	$0.00	$0.00	$4,200.00		
70	$0.00	Prorated	$0.00	$0.00	$0.00	$0.00	$0.00		
71	$0.00	Prorated	$4,900.00	$4,900.00	$0.00	$0.00	$4,900.00		
72	$0.00	Prorated	$2,800.00	$2,800.00	$0.00	$0.00	$2,800.00		
73	$0.00	Prorated	$8,400.00	$8,400.00	$0.00	$0.00	$8,400.00		
74	**$0.00**	**Prorated**	**$822,860.00**	**$737,240.00**	**$85,620.00**	**$0.00**	**$822,860.00**		
75	**$0.00**	**Prorated**	**$127,740.00**	**$3,360.00**	**$124,380.00**	**$0.00**	**$127,740.00**		
76	$50,000.00	Prorated	$50,560.00	$560.00	$50,000.00	$0.00	$50,560.00		
77	$0.00	Prorated	$560.00	$1,120.00	($560.00)	$0.00	$560.00		
78	$75,500.00	Prorated	$76,060.00	$1,120.00	$74,940.00	$0.00	$76,060.00		
79	$0.00	Prorated	$560.00	$560.00	$0.00	$0.00	$560.00		
80	**$0.00**	**Prorated**	**$295,840.00**	**$191,520.00**	**$104,320.00**	**$0.00**	**$295,840.00**		

Figure 2-61   GANTT chart, cost table.

ID	% Complete	⊕	Task Name	Duration	Start	Finish	Predecessors	Resource Group	Apr
153	0%	▦	**Safety**	10.8 wks	Wed 6/4/08	Mon 8/18/08	71		
154	0%		Security	14 days	Wed 6/4/08	Mon 6/23/08	71		
155	0%		**Seating**	2 wks	Wed 6/4/08	Tue 6/17/08	71		
156	0%		Visibility	1 day	Wed 6/4/08	Wed 6/4/08	71		
157	0%		Controls	2 days	Wed 6/4/08	Thu 6/5/08	71		
158	0%	▦	Dashboard	3 days	Wed 6/4/08	Fri 6/6/08	71		
159	0%		Trunk	10 days	Wed 6/4/08	Fri 6/27/08	71		
160	0%	▦	Hood	5 days	Wed 6/4/08	Tue 6/10/08	71		
161	0%	▦	Tires	3 days	Wed 6/4/08	Tue 6/10/08	71		
162	0%		Capacity	5 days	Wed 6/4/08	Tue 6/10/08	71		
163	**0%**		**Six Sigma Supplier management**	**86.81 days**	**Wed 6/4/08**	**Thu 10/2/08**	**71**		
164	0%		Key Processes	11.5 days	Wed 6/4/08	Thu 6/19/08	71		
165	0%	▦	**Process Performance Indicators**	8 wks	Wed 6/4/08	Thu 10/2/08	71		
166	0%		**Process improvement strategy**	3 wks	Wed 6/4/08	Tue 6/24/08	71		
167	0%		**Measures**	3 days	Wed 6/4/08	Fri 6/6/08	71		
168	0%		Data collection	3 days	Wed 6/4/08	Fri 6/6/08	71		

Figure 2-62   GANTT chart, PERT chart view.

placed on each scenario by the program manager, depending on risk severity in that task and other information in the risk assessment, and documented in a risk matrix. Then Microsoft Program calculates a new duration based on the weights and durations estimates, as shown.

This function integrates risk into scheduling by giving the program manager a way to reflect risk information and insight from the risk management process directly into the scheduling and planning process.

ID	Task Name	Act. Start	Act. Finish	% Comp.	Act. Dur.	Rem. Dur.	Act. Cost	Act. Work
	Critical: No	**Thu 4/17/97**	**NA**	**17%**	**9.72 wks**	**47.48 wks**	**($95,936.00)**	**849 hrs**
1	Select Architect	Thu 4/17/97	NA	2%	0.77 wks	38.63 wks	$1,040.00	49 hrs
2	Recruit & Train Managers	Thu 4/17/97	Fri 6/27/97	100%	10.28 wks	0 wks	$5,872.00	312 hrs
5	Create Production Design	Tue 6/17/97	Mon 7/14/97	100%	4 wks	0 wks	$3,528.00	164 hrs
6	Building Concept	Thu 5/1/97	Wed 5/14/97	100%	2 wks	0 wks	$4,312.00	88 hrs
10	Permits and Approvals	Tue 8/5/97	NA	12%	0.36 wks	2.64 wks	$1,680.00	24 hrs
12	Plant Personnel Recruiting	NA	NA	0%	0 wks	8 wks	$0.00	0 hrs
13	Equipment Procurement	NA	NA	0%	0 wks	24 wks	$0.00	0 hrs
14	Raw Material Procurement	Tue 7/15/97	NA	80%	6.44 wks	1.56 wks	$2,632.00	132 hrs
16	Product Distribution Plan	NA	NA	0%	0 wks	2 wks	$0.00	0 hrs
17	Landscaping	NA	NA	0%	0 wks	3 wks	$0.00	0 hrs
18	Truck Fleet Procurement	NA	NA	0%	0 wks	8 wks	$0.00	0 hrs
22	Sales/Revenues	Mon 6/2/97	NA	5%	2 wks	36 wks	############	80 hrs
	Critical: Yes	**Thu 4/17/97**	**NA**	**39%**	**########**	**36.11 wks**	**$228,488.03**	**707.92 hrs**
3	Select Real Estate Consultant	Tue 7/1/97	NA	89%	1.79 wks	0.21 wks	$1,824.00	68 hrs

Figure 2-63   Tracking GANTT and tracking table.

Actual costs are entered in Microsoft Program through the tracking GANTT and tracking table (see Fig. 2-63). Here, the program manager can enter actual hours worked or costs incurred during a given period. However, this requires deactivating the automatic calculation of actuals that Microsoft Program figures based on work performed. The best integration can be achieved through an electronic interface between the company timesheet and purchasing system to allow automatic entry of real-time costs into program. This now allows the program manager to see true earned value at any point in the program without entering any actual costs or hours manually.

Earned value is calculated and presented in Fig. 2-64, showing the schedule variance (SV) between the budgeted cost of work performed and the budget cost

ID	Task Name	BCWS	BCWP	ACWP	SV	CV	EAC	BAC	VAC
112	Vertical Integration	$1,120.00	$0.00	$0.00	($1,120.00)	$0.00	$1,120.00	$1,120.00	$0.00
113	International consortia	$0.00	$0.00	$0.00	$0.00	$0.00	$0.00	$0.00	$0.00
114	Supply interfaces	$5,600.00	$0.00	$0.00	($5,600.00)	$0.00	$4,480.00	$5,600.00	($1,120.00)
115	**Configuration Management**	**$2,800.00**	**$0.00**	**$44,880.00**	**($2,800.00)**	**($44,880.00)**	**$53,920.00**	**$2,800.00**	**$51,120.00**
116	Software tailoring	$0.00	$0.00	$44,880.00	$0.00	($44,880.00)	$52,800.00	$0.00	$52,800.00
117	Data Entry	$0.00	$0.00	$0.00	$0.00	$0.00	$560.00	$0.00	$560.00
118	Change management	$0.00	$0.00	$0.00	$0.00	$0.00	$560.00	$0.00	$560.00
119	**Tooling**	**$0.00**	**$0.00**	**$0.00**	**$0.00**	**$0.00**	**$11,200.00**	**$0.00**	**$11,200.00**
120	Pre Manufacturing inspection	$0.00	$0.00	$0.00	$0.00	$0.00	$2,800.00	$0.00	$2,800.00
121	Safety system	$0.00	$0.00	$0.00	$0.00	$0.00	$2,800.00	$0.00	$2,800.00
122	Drawing	$0.00	$0.00	$0.00	$0.00	$0.00	$2,800.00	$0.00	$2,800.00
123	Alignment	$0.00	$0.00	$0.00	$0.00	$0.00	$2,800.00	$0.00	$2,800.00
124	**Electrical Components**	**$4,480.00**	**$0.00**	**$0.00**	**($4,480.00)**	**$0.00**	**$2,240.00**	**$4,480.00**	**($2,240.00)**
125	Component designs	$2,240.00	$0.00	$0.00	($2,240.00)	$0.00	$2,240.00	$2,240.00	$0.00
126	**Chassis Assembly**	**$85,120.00**	**$0.00**	**$0.00**	**($85,120.00)**	**$0.00**	**$168,000.00**	**$85,120.00**	**$82,880.00**
127	Panels	$36,400.00	$0.00	$0.00	($36,400.00)	$0.00	$36,400.00	$36,400.00	$0.00

Figure 2-64   Tracking GANTT and earned value table.

ID	🛈	Resource Name	Details	Apr 12, '98							Apr 19, '98
				S	M	T	W	T	F	S	S
1	⟡	**Facility Specialist**	Work		8.33h	8.33h	8.33h	8.33h	3.42h		
		Recruit & Train Managers	Work								
		Select Real Estate Consultant	Work								
		Building Concept	Work								
		Building Design	Work								
		Site Selection	Work								
		Select General Contractor	Work								
		Building Construction	Work								
		Plant Personnel Recruiting	Work								
		Equipment Installation	Work		3.42h	3.42h	3.42h	3.42h	3.42h		
		Landscaping	Work		4.92h	4.92h	4.92h	4.92h			
	📊	*Select Architect*	Work								
2		Project Manager	Work								
		Recruit & Train Managers	Work								
		Select Real Estate Consultant	Work								
		Select General Contractor	Work								
		Permits and Approvals	Work								

Figure 2-65   Resource Usage view.

of work scheduled at any given point in the program. Cost variance, the variance between the budgeted cost of work performed and the *actual* cost of that work performed, is shown, along with estimate at completion, budget at completion, and variance at completion.

In order to keep track of actual work to be performed by resources assigned to individual or multiple programs, the Resource Usage view (Fig. 2-65) allows the program manager to see each assigned worker on the program and a breakdown of work to be performed based on the plan by task. Note that the facility specialist is overallocated and associated work is shown in red—based on an eight-hour day, the specialist is overallocated by .33 hours during the week shown.

Integrating programs in Microsoft Project

One of the challenges with program management is running several programs at once, orchestrating and monitoring several programs, all with their own program managers. The IMS (integrated master schedule) can be constructed using the Microsoft Program Insert function. Highlighting an open task below a current program, a new program can be inserted to allow for a consolidated master schedule.

In Fig. 2-66, we have integrated two programs: Integrated Transportation System (ITS) and New Plant. This was accomplished using the ITS program as the base,

ID	% Complete	🛈	Task Name	Duration	Start	Finish	Resource Group	2003 Q2
1	0%		**Integrated Transportation System**	**343 days**	**Mon 6/25/07**	**Wed 10/15/08**		
243	0%	📄	**New Plant**	**263.1 days**	**Fri 6/1/07**	**Wed 6/4/08**		

Figure 2-66   Integrated programs.

ID	Resource Name	Details	September					October		
			8/26	9/2	9/9	9/16	9/23	9/30	10/7	10/14
	Integration Gateway 3: Organizational Development	Work								
	Integration Gateway 4: Global Team Composition and Development	Work								
	Integration Gateway 5: Support Systems Audit	Work								
	Business Strategy	Work	40h							
	Portfolio Development	Work		32h						
	Project Selection Criteria	Work		8h	24h					
	Demand studies	Work			16h	24h				
43	Ben Sheets	Work								
	Systems design	Work								
	Hardware	Work								
1	Facility Specialist	Work	12h	10.8h	6h	8.4h	12.85h	8h	4.55h	2.4h
	Select an Architect	Work								
	Select Real Estate Consultant	Work								
	Building Concept	Work								
	Building Design	Work				1.6h	8h	8h	4.55h	1.6h
	Site Selection	Work	12h	9.6h						

Figure 2-67   Resource Usage view of integrated programs.

then going to "Insert," then "Program," and highlighting "New Plant" from another directory and inserting it. This is the beginning of the process of integrating a program of programs in one schedule to allow multiprogram control.

Resources from both programs can now be displayed in the Resource Usage view (Fig. 2-67) to monitor resources committed to both programs and to identify potential conflicts, overallocations, and underallocations of personnel.

Here we can see how staff are allocated to two separate programs: Ben Sheets working in the ITS Program and the Facility Specialist working in the New Plant program. The Resource Usage view allows us to view all resources by task.

The Tasks Usage view allows us to view resources from a tasking view—e.g., each task from each integrated program is shown with all resources assigned to it (Fig. 2-68). Here, we can see that the Prototype Full Integration task,

ID	Task Name	Details	2008					
			Nov	Dec	Jan	Feb	Mar	Apr
	Lour Bentt	Work						0h
240	Chassis Available	Work						0h
	Lour Bentt	Work						0h
241	**Prototype Full Integration**	Work						40h
242	Build	Work						40h
	David Bart	Work						40h
243	**New Plant**	Work	########	869h	863.8h	766.4h	466h	527.2h
1	**HUNTSVILLE PROJECT**	Work	########	869h	863.8h	766.4h	466h	527.2h
2	Select an Architect	Work						
3	Recruit and Train Managers	Work						
4	Select Real Estate Consultant	Work						
5	Pre Production Plan	Work	21.6h					
	Production Specialists	Work						
	Manufacturing Engineering	Work						
	Project Manager	Work	21.6h					
6	Create Production Plans	Work	108.8h	147.2h				

Figure 2-68   Tasks Usage view.

Build, is staffed by David Bart at 40 hours in April, while the Preproduction Plan task in the New Plant program is staffed by the project manager at 21.61 hours in November.

Summing up the Microsoft Project integration function

No software is going to integrate programs unless the managers and program staff involved *want to integrate programs.* As evidenced previously, Microsoft Project has the capacity to document and display schedule and resource information on a multiprogram portfolio and provide for cross-comparisons of tasks, resources, and costs. However, the motivation and incentive to integrate programs must come from the leadership of the organization and be demonstrated daily in program execution decisions. As implied in the new PMI PMBOK standard on integration, program execution is the "proof of the pudding" in integration. It is in the execution that key program decisions are made—e.g., how scarce resources are allocated to competing programs. These decisions can be made in a vacuum by individual program managers entirely in the interest of *their own programs,* or they can be made by those same program managers in the interests of the whole organization and its business goals. This is where company leadership sets the tone; if programs are to be managed for the good of the company as a whole, then each program manager sees himself or herself as part of a collaborative management team, making judgments about company investments *beyond their own interests.* That is the integration challenge.

Tools used to build an integrated program management system

When building a company or agency system to support program integration, there are 10 areas for process improvement: organization-wide or enterprise-wide program management systems, program portfolio system development, integrated resource management systems, information technology, technical product development (including a stage-gateway review system), interface management system, program portfolio management, program monitoring and corrective action, change control, and program evaluation.

Implementing a global PMO

A PMO should be implemented as an integral component of the organization. Unlike a project or program, a PMO doesn't have a specific duration. A PMO is similar to a portfolio, where once correctly established, becomes a mission-critical functional department for the organization. An organization can outsource many functional areas within the business, including manufacturing, accounting, finance, marketing, and human resources. However, the one area that should never be outsourced is the mission, vision, and strategy for the company, which is achieved through the portfolio process and supported by the PMO. The portfolio process allows for the selection of company investments. What the company invests in determines the business's future viability.

Many have heard of failed or troubled PMOs. Failure can be caused by "biting off more than you can chew," by providing services that are not relevant or a priority for the organization, or by the organization misunderstanding the benefits that a PMO provides.

To properly set up a PMO, first, an executive sponsor for the process needs to assemble a team. There are two schools of thought on the team. First, the team could be composed of small group representing key stakeholder interests, along with the selected program director to develop the program. The advantage to this process is that you embed stakeholder acceptance and support in the process by having them intricately involved in it. The disadvantage with this method is that it could take time to get the stakeholders up to speed on the process and the importance of the standard.

The second approach is to have a small group consisting only of PMO members. The advantage of this is that the PMO members would be "experts" on the PMO process and would be able to implement a valuable PMO. The downside to having only PMO members involved, however, is that there may be problems with buy-in from the organization and long-term support for the process. It is also possible that PMO experts might not possess the breadth or depth of understanding of the business to achieve a successful implementation of a PMO that meets the business's needs. To offset stakeholder lack of support, the PMO sponsor has to be an active advocate for the process. Moreover, the PMO needs to continually market itself to the business and present the value they bring to the organization in a timely fashion.

With either method, it is best to have a small team (five to seven people) involved in the creation of the PMO. When there are too many cooks in the kitchen, the soup becomes spoiled. Through the PMO development process, the team would regularly report on its status and gain approval from key stakeholders as the PMO develops.

Building a PMO

When you take on the responsibility to create a new service, it is best to define it as a project and manage it with a standard project management process. Following a project management methodology when building your PMO provides a framework for success. Some believe that you can take a shortcut and build a PMO within a few weeks. Depending on the size and project maturity level of the organization, a PMO could be fast-tracked. However, in most organizations, it is important to build a PMO one step at a time. When you rush into creating a new department for your organization, critical elements can be missed and important bridges might become damaged.

The PMO should be developed using the fundamental steps of initiation, planning, executing, monitoring and controlling, and closing processes. When the PMO is fully developed, the project would then close, transfer to "normal" operations, and the ongoing maintenance of improving the functions of the PMO would continue. As you can see from Fig. 2-69, a PMO development team needs

- Identify needs of the organization
- Analyze and prioritize the needs
- Build a plan of action
- Gain support from key PMO stakeholders and sponsor
- Implement the plan in achievable stages
- Report status to stakeholders to further gain support
- Continually build, develop, and refine the process

Figure 2-69   PMO development and ongoing maintenance process.

to go through a number of steps to successfully build and maintain a PMO. PMO requirements need to be gathered, identified, analyzed, and prioritized.

The business needs to define why a PMO is needed before it determines what it needs to do. The next step would be to build a plan of action on how best to implement the PMO. The plan needs to be socialized with the organization to gain support and confirm the goals of the PMO. After support is gained, the PMO implements the approved plan and continually updates stakeholders on its status. Once the PMO achieves the goals of the plan, the team works to continually improve the process. Over time, the needs of the business may change and grow. After the initial implementation, alterations and/or additions to the process may be required. The PMO team then works with the organization to reassess the needs of the business, builds additional services as required, and refines current process as needed to the PMO framework.

Organizational needs identification

To build a successful PMO, you first need to research the current organization using standard tools and techniques to gather information. It is best to understand the immediate needs of the business from the stakeholders' point of view. Some of the tools that could be used for information gathering include:

- *Information-gathering techniques.* There are a number of information-gathering techniques, such as business case analysis, root cause identification, and the Delphi technique.

- *Brainstorming.* Sessions should be conducted with the appropriate stakeholders who have knowledge of the discussion topic. Brainstorming sessions are a valuable tool to help with requirements identification.

- *Expert judgment.* You are relying on an expert's knowledge and experiences to provide advice and guidance for the program management office. Experts can be internal or external to the organization. This person can have a functional or technical background. The expert could be a consultant, from a professional or technical association, or a specialist from a government agency.

- *Diagramming techniques.* A number of diagramming techniques can be used, such as affinity diagrams, influence diagrams, cause-and-effect diagrams, and program dependency analysis.

- *Organizational analysis.* Analyzing the organization to understand the roles and responsibilities of the members. This should be done for potential internal and external stakeholders. The organizational analysis should also include observing the environment. Much can be learned from observing the organizational processes that may or may not be communicated fully by the stakeholders. Observations should be reviewed with stakeholders.

- *Interviews.* These discussions allow for a better understanding of the needs of the stakeholders. Generally, these are open-ended questions to gather data that is important to the stakeholder.

- *Focus groups.* Gather data from a group to better understand the attitudes of the stakeholders. In this technique, group members interact with one another and build upon each other's thoughts to achieve a deeper understanding of stakeholders' needs.

- *Questionnaires and surveys.* Solicit feedback from stakeholders to better understand their needs and requirements.

The output of this process is a PMO requirements document. The requirements document includes details on the process in which the information was gathered, who was involved in the process, and a list of all the requirements, with a description of each requirement.

Analyze and prioritize

The PMO requirements document then goes through a refining process, where the requirements and their definitions are analyzed and prioritized. A few standard tools can help with this process.

- *Requirements analysis.* When you analyze the requirement information, you need to make sure that all the information is accurate and complete, and begin the process of further clarifying the information. In this process, prioritization of the information should be conducted using critical stakeholders to help in this process.

- *Requirements validations and verification.* Requirements are further validated and verified for accuracy and the level of need for the program.

- *Assumption analysis.* The assumptions of the requirements are analyzed to check the validity of the understanding.

At the conclusion of this process, the PMO requirements document is updated.

Build a plan of action/PMO charter

From the PMO requirements documents, a plan of action is created to implement the PMO. This plan is similar to a program/project plan. One of the documents that is created is the PMO charter. The charter describes the business need for the PMO and who is sponsoring it. It also describes the role and direction of the PMO. A number of questions should be asked to help address the formation of the PMO.

- What are the current issues with the business process that can be addressed with a PMO?
- What are the mission, vision, and strategy of the PMO?
- What services will it offer initially and in the future?
- What services will it not offer?
- How would new services be rolled out to the business?
- What are the roles and responsibilities of the PMO team?
- Who would be the sponsor for the PMO?
- Who are the stakeholders for the PMO?
- How is the PMO going to be funded?
  - Is it going to be a fully allocated service?
- What key performance metrics will be used to measure success?

The same tools and techniques you would use to build a project plan are used for this process.

- *Work breakdown structure templates.* The WBS follows a specific structure, and the PMO should use the structure outlined by the organization.
- *PMO schedule.* The schedule details the tasks involved in the creation of the PMO.
- *Feasibility studies.* These are conducted to determine the technical or economic viability of the PMO. They can also be used to determine if the PMO will attain the benefits as planned. A SWOT analysis can help determine the viability of the PMO.
- *Task responsibility matrix.* This document defines the roles and responsibilities of each of the team members of the PMO and any stakeholders that are needed for building it.
- *Capacity planning.* This is the effort of managing the scarce resources that are involved in a PMO. When planning, conflicts, constraints, alternatives, or mitigation practices are activated to provide the optimal capacity of resources for all the components in the program.
- *PMO risk register.* Creation of the risk register and updates to it are completed as risk information is discovered as the result of component deliverables or interdependencies.
- *Benefits analysis.* Identifies benefits and how the PMO will achieve them. It also looks at how the benefits strategically align with the organization's goals.

- *Cashflow analysis.* Determines when and how money comes into (revenues) and out of (expenses) the PMO: the "flow of cash." PMOs tend to be cost centers. However, it should be determined how the PMO will save the organization money in the long run.

- *Stakeholder register.* Who are your key stakeholders, and what are their needs?

- *Program management information systems.* Provide the tool to house data in a central storage system.

- *Communications requirements analysis.* In this analysis, the communication needs of the stakeholders are determined. Who should communicate what information to whom? What are the expectations of the stakeholders with regard to information they are seeking from the program?

- *Communication methods.* Depending on the complexity and urgency of the information, different methods of communication can be used. More complex information may require a more formal mode of delivery to fully explain it.

At the conclusion of building a PMO action plan, an action plan document is produced and a presentation is created to be shown to executive management. The action plan document would include the purpose and goal of the PMO, the roles and responsibilities of the members, the level, and type of support and services that the PMO would provide initially, and in the future, and how the PMO will be rolled out to the organization.

Gain support

The action plan needs support to be activated! The process of gaining support for the PMO is an iterative process and begins when the PMO concept is created, and it continues as long as the PMO exists. Once a fully developed plan is completed, it is then circulated among key stakeholders for their approval. Refinement to the plan may be needed at this point. Prior to receiving final approval for the PMO, some tools and techniques that can be used.

- *Comparative advantage analysis.* This is used to compare alternatives to the proposed solution. What-if analysis could also be done at this time to determine if the program benefits could be accomplished in other ways.

- *Design reviews.* The PMO structure and functions are reviewed by peers and subject matter experts to make sure that the PMO complies with the appropriate standards and best practices of the organization.

At the conclusion of this phase, the PMO is given final approval to begin with their implementation plan.

Implement the plan

Using the approved PMO action plan, the PMO is rolled out to the organization in achievable stages. The analogy of crawl, walk, run comes into play. If you want to be successful, it is best to begin with a seed, a single process or support structure, and then let it take root. When the PMO becomes assimilated into the organization, the need for services will grow and bloom. Going for the "big bang" to introduce a PMO could lead to a lot of frustration and a struggle for long-term success.

As the plan is implemented, it should follow the communication plan within the overall action plan that describes the methods in which the PMO will be introduced to the organization. A training plan should be presented to train users and stakeholders on the PMO. Throughout the rollout, lessons learned are captured to improve upon the roll out of additional services.

Report status

Reporting on the status is an iterative process and begins when the PMO concept is created and continues throughout its life. The appropriate communication method is presented to the correct stakeholder in a timely fashion. A PMO needs to stay front and center in an organization to best serve its needs. When the PMO takes a back seat, momentum will be lost, and the importance of its services will be diminished and possibly eliminated from the organization. It is critical that the PMO actively present its value to the business.

In addition to status reports on PMO activities, other information that should be presented to the business includes:

- *Performance measurements*. This information clearly outlines the value the PMO brings to the organization.
  - If the PMO provides training, what training was covered? How many students were trained? How many training hours were logged? What was the cost for internal training versus external training?
- *Customer satisfaction survey results*. Surveys should be sent to customers of the PMO and used as a benchmark to show improvement results.
- *Service metrics*. These display turnaround time for issues. How quickly does the PMO respond to requests or problems?

It is important for a PMO to be able to present to the business on a regular basis the value that they bring to the organization.

Continuous improvement process

The PMO is tasked with continually building, developing, and refining the services and support that it provides. The environment in which a PMO exists is always changing and evolving, and the PMO must keep up with new developments. The original implementation plan is reviewed, and the next steps in implementing are reviewed and followed, if appropriate. It is important that the organization's needs be identified early and analyzed for changes and reprioritized as needed. A PMO is a service department. It needs to be flexible and able to change and adapt to the needs of the business.

Value of a PMO

The PMO exists to provide value to the organization. The value of a PMO to an organization is in the areas of knowledge management, standardization, and in-house expertise in program delivery. The PMO is the central warehouse of program and project information. It creates tools, techniques, and templates to provide standardization of information and practices. As the in-house experts, the PMO provides training and guidance to program and project teams. Given that a large percentage of projects and programs fail due to poor program and project management practices, the PMO could prevent failures. They also provide improved reporting by consolidating, analyzing, and prioritizing program and project information. The information is then used to make better decisions and determine the alignment of programs and projects with the strategy of the company. Enhanced communication and collaboration is a residual benefit. The consolidated information could also provide a consolidated risk management strategy. The PMO can focus on workload balance and resource allocation.

The value of the PMO should be calculated in both tangible and intangible ways. Tangible ways include the cost savings inherent in not having to reinvent the wheel for every new program or project, or for supporting internal training. Intangible examples include the improved success of programs that have been supported by the PMO using the tools and templates provided by it. Global PMOs provide support for the impact of cultures and customs that might affect a program. The world is ever-changing, and it is up to the global PMO (GPMO) to stay a step ahead. The GPMO needs to have an optimum communication structure that takes into consideration time, language, and different communication and comprehension requirements.

PMOs can be tasked with:

- Creating and maintaining master program schedules
- Supporting resource decision management and delegation
- Maintaining issue tracking and reporting
- Developing best practices for the organization
- Establishing and enforcing standards and practices
- Producing metrics for the PMO and the projects and program that are supported under its umbrella
- Generating a consolidated dashboard of project and program information; this allows for an apples-to-apples comparison of projects and programs
- Supporting the portfolio process by providing streamlined information and helping with decisions that need to be carried out
- Training the staff and program stakeholders on processes and procedures
- Supporting internal and external policies and procedures (ethical standards, Sarbanes Oxley [SOX], etc.)

Structure of the PMO

The structure of a PMO is based on the specific needs of the organization. Therefore, you will not find carbon copies of PMOs. The size and focus is determined by the needs of the organization and the maturity of the PMO. The PMO can be structured in a centralized or decentralized fashion. The PMO team and framework could be housed with the "home office" building, or the PMO could have a presence in the "home office" and also be dispersed to each location in which a program or project is being executed. To determine the best structure of the PMO, a number of factors need to be evaluated.

- What are the support needs of the program?
- What are the consulting or mentoring services desired?
- What are the necessary coaching and training requirements?
- What is the available funding for a PMO and costs that will be incurred?
- How is program information gathered, analyzed, disseminated, and stored?

Roles and responsibilities of the PMO team

Many people play a part in a PMO. However, the PMO structure will determine the roles and responsibilities of the PMO team. In Fig. 2-70, you see some key positions in a PMO. Depending on the maturity of the program process for a business, these positions may or may not exist.

	Organizational Structure		
Roles	Functional	Matrix	Projectized
Program Director	Nonexistent	Part-time to full-time position	Full-time position
Program Manager	Nonexistent	Part-time to full-time position	Full-time position
Project Manager	Nonexistent or part-time work	Part-time to full-time position	Full-time position
PMO Administrators	Nonexistent	Part-time to full-time position	Full-time position
PMO Team Members	Nonexistent	Part-time to full-time position	Full-time position
PMO Sponsor	Nonexistent	Part-time to full-time position	Full-time position

Figure 2-70   Organizational structure and the roles and responsibilities of the PMO.

The program director is responsible for the day-to-day activities of the PMO. This person manages the staff and liaisons with executive management. Generally, a program manager is responsible for the oversight for one program at a time. The work to manage a single program is extensive and is a full-time job. The program manager would interface primarily with the project managers and the program director.

The project manager within a program may be responsible for managing more than one project within the program, or other projects outside the program. A PMO has an administrative support team to manage the logistics of the PMO activities. The PMO often has team members who report to the program director and are responsible for developing training programs and tools, techniques, and processes for managing programs and projects. They are also responsible for gathering, analyzing, formatting, and distributing communication from the programs and projects under the PMO.

A PMO sponsor encourages and supports the PMO. The sponsor will also help remove barriers to the PMO's success. The PMO program director often reports to the PMO sponsor.

Maturity model of a PMO

As there is a maturity model for project management, there is also one for a PMO. At one end, a PMO would not exist or have no structure to tie the functions together. From there, the PMO begins to solidify and provides basic processes to an organization. This can be considered the first step in a marathon. Next, a PMO begins to offer increased support and offers additional services to the business. The next level provides full PMO practices for all programs and projects. Finally, the PMO reaches the continuous improvement environment, where it fine-tunes the services that it provides to support the business.

A mature GPMO can provide services to an organization to mitigate challenges that it could face in a global environment. A GPMO would have an effective understanding of international challenges to best support the programs. The GPMO socializes lessons learned to provide the most effective tools, techniques, and processes to the frontline of the program world. It supports the localized use of tools and oversees the consistent use of best practices around the globe.

Summary

This chapter covered program management according to PMI, program support systems, and PMO. The standards are a guidelines for organizations to use. The support systems allow for PMO efficacy. The benefits of a GPMO provide increased efficiency in programs and projects. A GPMO provides a centralized database of project and program information. This information is gathered by the GPMO, analyzed, and disseminated appropriately around the world. The

GPMO evaluates risks globally and prioritizes the use of resources. It provides training on the best project and program processes, as well as on cultural, regulatory, and global issues. Collaboration across borders exists in a GPMO world. The GPMO interfaces with the business and fully understands the strategy and direction in which it is headed. It works closely with the portfolio team to ensure alignment with the corporate plans.

Global Portfolio
Management Strategies

"Drive thy business or it will drive thee."
BENJAMIN FRANKLIN
U.S. author, diplomat, inventor,
physicist, politician, and printer
(1706–1790)

Introduction

In Chapter 2, we described the PMI *Standard for Program Management*. We saw that program management is quite different from project management, in that programs focus on broad benefits and business outcomes, are longer-term with regard to timelines, and serve as a management and resource framework for several coordinated project components. In effect, programs can be seen as categories or groupings of business initiatives, focusing a company's resources in target markets globally and generating a number of project deliverables for implementation.

The Global Portfolio

The global portfolio is the potential pool of programs and projects designed to improve the business and/or its products and services globally. Final approval and funding in the budget process then authorizes the programs and projects to be implemented. This chapter explores that portfolio development process and how companies move from planning the "right" programs for a given global target, to funding the program, to implementing its component projects.

Programs as Unifying and Organizational Themes

Developing a global portfolio can get complicated very quickly without some way of categorizing projects across the world. Starting with a global universe for the generation and evaluation of multiple projects, the process can result in a wide variety of project ideas and initiatives without any unifying theme or principle and no way to compare projects. To resolve this situation, a global company needs some way of organizing and categorizing global projects in the portfolio development process. Here is where the concept of program is useful. Programs can be used to frame and align and candidate global projects that have common goals. For instance a program package can be geographic, e.g. the Middle East global program and associated projects in that area, or the European global program. Or a program can be used functionally to organize technology projects into categories of technical projects for comparison, e.g. information systems, utility development and construction, green building maintenance, and training. Managers can be assigned to a given program area during portfolio development to generate and evaluate projects within their program area and to help compare and analyze them within their program and against other programs.

The Tylenol poisoning case in the autumn of 1982 in the Chicago area illustrates what can happen to a global product and how risk events can generate program responses. Tylenol was the leading pain-killer medicine in the United States and a major international product. The poisonings that occurred involved someone lacing Extra-Strength Tylenol with a deadly amount of potassium cyanide. It was discovered that the tampering took place once the product reached the shelves. Seven people died from this tampering, and the perpetrator was never apprehended.

Since there was no precedence for a crisis of this nature, Johnson & Johnson, the makers of Tylenol, had to think fast to prevent any more deaths and to protect its brand long-term, both in the United States and abroad. The company decided to conduct an immediate product recall, first for the entire United States. This recall totaled 31 million bottles of Tylenol at a loss of more than $100 million (Lazare, 2002).* At that time, Tylenol controlled 37 percent of the market, with revenue of about $1.2 million (Mitchell, 1989).[†] After the poisonings, Tylenol's market was reduced to 7 percent (Mitchell, 1989). Johnson & Johnson was praised by the media for its handling of this tragedy. The company then reintroduced Tylenol products with triple-seal tamper-resistant packaging and with heavy price promotions. Because of this swift action, after a few years, Tylenol was able to regain much of its previous market share.

Why is the Tylenol case so important? It is significant because of the *butterfly effect*. The butterfly effect is the concept of sensitivity dependence on initial conditions in chaos theory. It looks at small variations of an initial state, which produce a large variation. Think of a butterfly flapping its wings, which only creates tiny changes in the atmosphere. One butterfly could have a far-reaching ripple effect on succeeding events. For example, in Australia, a butterfly begins to flap its wings preparing to take off. This motion may have a rippling

---

*Lazare, Lewis, *Chicago Sun-Times:* Tylenol crisis triggered brilliant PR response. Accessed on Dec. 4, 2002.

[†]Mitchell, Mark L., Economic Association International, 1989: The impact of external parties on brand name capital: the 1982 Tylenol poisonings and subsequent cases.

effect in the atmosphere that causes a tornado in the state of Kansas in the United States.

The Tylenol case can be considered a *global butterfly*. Though it may not have been the intention, the actions of one criminal, the flapping of the wings of this case, caused a major rippling effect in the pharmaceutical, food, and consumer product industries. These industries were now required to develop tamper-resistant packaging, and they had to invest project resources to quickly develop a cost-effective way to create such packaging. These packaging projects were most likely not considered prior to this event, nor were their funds and resources reserved for this work. Not only did this crisis affect commercial industries, but it also affected government. It became a federal crime in the United States to tamper with products.

Having an effective project portfolio process in place could be the difference between success and many costly and fruitless efforts of project trial and error. In organizations that use project portfolio management effectively, they can successfully realign their projects, people, and partners with their new corporate objectives. As a result, changes in corporate strategy and financial priorities can be quickly implemented—preventing the waste of valuable time, dollars, and resources.

To address the Tylenol case, the company chose to develop a major new program area—safety compliance. The portfolio-to-program-to-project process looks like this: The portfolio generated candidate programs, long-term plans to address a particular global market or support need. A good example of a program generated by the Tylenol case would be the company's major effort to safeguard all company products worldwide and to comply with regulatory and legal requirements in given countries as they surface through major, targeted resource investment. The program goal might be to achieve zero risk in product safety worldwide through innovative technology solutions by 2015. Projects generated in this general program area could include:

1. A project to design packaging with flexible features to address particular requirements

2. A project to develop techniques to build controls into the distribution of products online to avoid risks in a store environment

3. A project to develop a web-based reporting system on product safety violations in various countries

4. A project to train and inform product handlers at every level globally on product's best practices in product safety

A World Without a Program and Project Portfolio and Program Management System

A company without a portfolio process and without a program and project management system will not be successful globally. Development of a program in the portfolio process requires that a global company look at the *big picture.* Without a portfolio process, a company is likely to flounder against global competition. Long-term goals and objectives are not defined through programs with international impacts and benefits. Planning is not focused on the company's global investment strategy, and the workforce and individual program and project teams are not informed on where the company is going and how it will get there in particular, worldwide markets.

Project teams may find it difficult to find the right people to handle deliverables for the project. There might also be delays caused by not having the right resources at the right time. There might be a lot of turnover in the project team due to burnout. Or, there is not one person or group making sure that work is not duplicated in the different departments. There is no overall oversight into the investment strategy. Because of the lack of a project management system, it is likely that no one is making sure that resources are being assigned to the higher-priority projects because there is no context for projects and no way to know which projects are important. In fact, projects are all considered at the same level and at the same level of priority, which is high. On the other hand, project statuses may be changing, often moving from high priority to extremely high priority and then back.

Teams "spin their wheels," putting in a great deal of effort to get projects completed successfully. But because the teams are not organized, no one on the project team is at the senior level and no one on the team necessarily communicates with senior management on a regular basis. Once the project team receives approval for the project concept, the team sets off and creates the project plan with the deliverables that they found necessary for the operation for their customers at the ground level.

There is intense competition with internal project teams for limited resources, like testing or training areas. Usually, the project manager with the best relationship with the resource will win first dibs on the resource, which might not align with corporate strategic plans for the resources. Projects may eventually come to completion without meeting the business's strategic need.

The bottom line here is that a company without a portfolio process and without a project management system to back it up does not see the value in building a portfolio aligned with business objectives, generating program goals and actionable projects to meet short-term and long-term global goals.

Portfolio Development

There is an analogy between an individual's financial portfolio and a business's program and project portfolio. Let's explore the nature of an individual investment portfolio. The goal in a financial portfolio is to have the appropriate assemblage of investments. A diversification strategy is used to limit risk in a portfolio. By spreading investments into several different assets, certain types of risks can be mitigated. A financial portfolio may contain stocks, bonds, and real estate, among other investments. Even within one type of investment, there can be diversification. As an example, with stocks, you can invest domestically, internationally, in large or small companies, and in numerous combinations. When you have a financial investment portfolio, you would typically conduct periodic analysis to determine the value and success of your investment decisions. Are the decisions still aligned with the goals of the portfolio? Have market conditions changed, requiring a rebalancing of the collection of assets?

Financial portfolio management involves deciding what investments to make and maintain in the portfolio, given the goals of the portfolio owner and the changing economic environment. The investment strategy is not limited to just what to invest in, but also how much, when to purchase, and what to divest. These decisions are weighted by how much opportunity you have to invest versus the availability of these investments and, of course, the risks associated with the investment. The asset selection process involves analyzing performance measurements, such as expected return on the investment, against the risk associated with the return and comparing that to other combinations of opportunities. The goals of the portfolio are also a major consideration. Some investors may be risk-adverse, while others can be risk-neutral or risk seekers. The risk level of the portfolio can be measured using benchmarks, tools, and techniques. Project portfolio management has many of the same attributes as a financial portfolio, and as we will discuss throughout this chapter, project portfolios are often confronted by as many hurdles as financial portfolios.

In building the portfolio, a company faces uncertainty and takes on major risks because of the long-term nature of expensive global programs and the difficulty of anticipating complex forces affecting programs in foreign countries. Investing in the wrong technology project could waste a great deal of time, money, and resources and potentially ruin a business. Furthermore, the global environment, competition, and technology changes are success factors that require the company to make educated decisions rapidly. Rapid decision-making is more important now than in any other decade because of rapid time-to-market cycles for new products.

In this chapter, we are going to dive into the concept of portfolios and how they are a critical business necessity. Portfolios oversee the company's investment in programs and projects. We will go into detail concerning programs and projects in the next chapters. We will also look at why portfolios are a critical business function for global strategies.

What Is Portfolio Management?

Harvey A. Levine once characterized project portfolio management as:

> A set of business practices that brings the world of projects into tight integration with other business operations. It brings projects into harmony with the strategies, resources, and executive oversight of the enterprise and provides the structure and processes for project portfolio governance.

What is missing in this classic definition is the concept of a *program, a* relatively new concept in the traditional project management community. The term program suggests an organizational framework for individual but related projects, each contributing to a particular program goal. Programs provide an organizational, support, and resource context for integrating individual projects and transforming them into broad program benefits over time. Programs provide the classic "big picture," capturing all the business resources—both programs and supporting and administrative activities, such as procurement and human resources—required to achieve a broad business goal. Programs serve to link individual projects to particular global business objectives.

A portfolio aligns business objectives with project investments. At the same time, a portfolio determines how the project investments should be best managed. Is it best for the project investment to be managed as a single project or under a program umbrella? The program umbrella would include dependent or congruent projects and business processes to reach the organizational results as determined by the portfolio. Global projects are usually managed by a program manager within a program framework because they are typically more complex and generate more risks. That program manager functions at a relatively high level in the business, between top management and individual project managers, and can mobilize a wide variety of resources to support effective program delivery on a worldwide scale.

When we look at the concept of a project portfolio, it is essential to understand what a well-run portfolio process brings to a company. The portfolio process is the glue that binds business operations to investment strategy. This process takes the strategy we discussed in Chapter 1 and evolves the strategy to make it a reality. This process can be thought of as two sides of the same coin. The decisions made in your portfolio help define the future state of your business, where the decision you make for your business helps define your project portfolio.

In essence, a project portfolio is similar to an investment portfolio, where your investments are grouped together to meet your goals. Your business objectives help guide your portfolio decision process. While the selection of investments is part of a portfolio, another critical element is managing scarce resources. A resource can be a person, place, or thing that is in need of a corporate project or program investment. There are many types of resources. You can look at resources as the people needed to work on the project and the location where the project needs to be staged, as well as the tools needed to run the projects and raw materials needed to build a product. Resources are also the funds to allocate materials and pay for labor. Depending on what industry you are in, you would produce a different list of resources.

Most resource lists would include staff, computers, and location, but then there are divergent paths for industry-specific tools. Regardless of the resource list, priorities must be set within the portfolio to delegate resources to projects and programs that best meet the needs of the business. No matter how many resources there are, there never seems to be enough to meet all the needs and desires of the project investment load. Therefore, critical decisions need to be made to ensure the best mix of projects and programs to maximize value for the company.

The project portfolio is your link from your corporate strategy to your investment in projects and programs. To best manage a portfolio, you need to have a fully functional project portfolio management structure. Simply put, project portfolio management is the oversight, monitor, and control of the projects or programs grouped together in the portfolio. The administration of a portfolio also includes ensuring that the correct mix of investments is initiated, grouped, funded, and managed.

Why Are Portfolios Needed?

Not all companies have defined processes for capturing, analyzing, and evaluating project proposals. Often, projects are initiated by senior executives, who have their own agendas for improving their division of the business. There are also overly optimistic leaders who believe that every project is needed and needed immediately. The optimism causes a chain reaction, and capacity issues ensue. Low-priority or even unnecessary projects consume resources that could be used by projects that would produce a higher yield. Ineffective capacity and resource allocations become a formula for misfortune.

Companies are seeking investments that return value. Portfolios help you gain control over your investments. In a portfolio process, the team is charged with aligning business strategy with company investments. Part of this process involves balancing riskier investments with more moderate ones. Another critical component of the portfolio process is to monitor and control projects that are currently underway and make sure that they remain on the right track. At times, projects need to be accelerated, realigned, stalled, or even terminated.

Having a process is one part of the equation; the other part is execution. You could have the best process in the world, but if you fall short on the execution, you will never make it to the goal line. Companies that execute their portfolio processes well receive amazing results. They are able to maximize the value of their investments. They focus on higher-priority projects and programs while minimizing the overall risk to the company. They are able to reduce redundancy and kill projects that do not add value to the business. Overall, these companies produce a process that focuses on the greater good of the business, not individual departments.

There is not one single formula to achieving success with portfolio management. Many models and methodologies can be chosen within the vast array of companies, vendors, and software packages that offer portfolio management solutions. Even with all the possibilities, however, there are obstacles to implementing them well. There are ways to build a solid foundation using portfolios that will get you on the right track.

Developing a Portfolio Process

If the company does not currently have a portfolio process, or is looking to improve on its current one, it is important to lead this effort like a project and create the objective of the project plan as "developing and/or improving the portfolio process." A key person should be involved, like a project manager, who leads the effort, as well as key stakeholders from all aspects of the business. The objective for this team would be to design, develop, and implement the project portfolio management process for the organization. Many considerations have to be taken into account.

You would want to start at the top and get a clear picture of the company's mission, vision, goals, objectives, and direction. Is the company focused on growth or on maintaining its core business?

There needs to be support and commitment from the executives of the company. Build buy-in from the departments—work with key stakeholders from this group to further develop the portfolio. The objectives and direction are then divided into manageable groups or categories within the portfolio. Then, projects in-progress or proposed are analyzed. How do these projects match up with the portfolio categories? The projects need to be validated against the organization's objectives. Do these projects meet the company's strategic objectives? If so, they are placed in the appropriate category. If not, further analysis needs to be done to determine whether to maintain this project or kill it. A closer look into each category is needed to prioritize the projects with the buckets ("prioritization" is discussed later in this chapter). Determine the portfolio steering group. Usually, this group is supported by your project management office (PMO) and has the following responsibilities:

- A project master schedule is created to manage the process.

- A database of all projects is maintained.

- Determine the right mix of projects.

- Compare financial needs with availability.

- A watermark for resources might be created.

- Assign resources to the portfolio for project and program work.

- Report on variances to the planned program or project work.

Oversee the projects and programs within the portfolio with established project management process developed by the PMO. Information is rolled up from projects and programs to a portfolio dashboard high-level view of the status. Metrics are built on how projects will be evaluated, and a common framework is created to operate within the business.

Regularly do a health check of the portfolio to make sure that it is still aligned with the business objects. If business objectives change, the portfolio is realigned as needed by maintaining current projects, adding new projects/programs, and suspending or canceling projects/programs that no longer meet the business' objectives.

A corporate culture needs to be developed to support the portfolio process in its development. This would include buying in to how decisions are going to be made, how projects and programs are going to be categorized and prioritized, and understanding that if a project doesn't fit within these parameters, it will be eliminated from the portfolio and lose all resources. Part of this culture involves adding responsibilities to key stakeholders to participate in a steering or governance group.

The governance group is responsible for approving new projects, participating in portfolio projects, clarifying and validating company's strategic objectives, creating and maintaining priorities, communicating portfolio objectives, assigning resources to projects and programs, recognizing areas for improvement, and helping to implement improvements.

The portfolio process is a dynamic one, where new projects are identified, analyzed, prioritized, and possibly selected for inclusion to the portfolio. Current projects may be reviewed within the context of any change in strategic objectives, environment, market, technology, or proposed projects to determine if they should continue to be a viable participant in the portfolio. Projects may be accelerated, decelerated, put on hold, priorities changed, or killed. Resources are evaluated and allocated according to the priorities created by the portfolio. It is critical for the long-term success of a company to build a methodology for managing projects within a portfolio. The methodology needs to be straightforward and simple to maintain. It also needs to be powered with the appropriate processes to make the right decisions for the company. It is best to have a single location for proposed projects to be entered into the system. The proposed project concepts should have enough information with them to allow for decisions to be made by the governance process to delegate resources for further investigation. When the governance members receive a request, they will evaluate the proposal and analyze it against the priorities it has set in place. At this point, a number of outcomes can occur.

The proposed project/program concept:

- Is approved and added to the portfolio and given the resources needed to run the project or program.

- May need further clarification. The governance committee may assign additional resource to the proposal team so that they can add details to their proposal.

- Is put on hold for the time being and may be reevaluated again at a later date when resources are available, or when the priority of the concept is later reevaluated and given a higher rating.

- Is killed because it doesn't align with the company's strategic objectives.

Then all projects are prioritized for funding decisions. Which projects can the business finance this year, and can broad programs encompassing several projects be fully funded? The funding decision is typically made in the company budgeting process.

If the project is approved and funded, the project team should begin to use the templates and standards created by the company's PMO. The information created by the project team is periodically rolled up to the portfolio and delivered to the governance committee for review. The review helps the portfolio management team monitor and control projects and programs. The governance committee ensures that the assumptions made in the approved project proposal are still valid. It looks for any changes to company priorities. Risks are evaluated to see how well the project is weathering the storms it faces and how it affects the company. The governance committee also looks at key metrics to make sure that the project is in alignment with the company's priorities. Finally, the entire process is occasionally reevaluated to create a continuous improvement environment to make sure that the projects and programs are in line with the business objectives and meeting them in a timely fashion.

Portfolio Process

When you implement a portfolio process, it is important to follow some guidelines to ensure a smooth transition.

Phased-in process

Just as it would be difficult to eat an entire pie at once, it would also be difficult to implement an entire process at once. There are pitfalls in implementing unfamiliar and complex processes in an organization. A phased-in approach offers the advantage of providing incremental improvements with a lower degree of risk.

The portion of the process that should be implemented first should be determined by business requirements. Some companies may start with how they approve new projects; others may look at building prioritization standards to measure current projects and help mitigate capacity and resource limitation issues. Other areas could be phased in as lessons are learned and the need of the portfolio process matures.

Standards

When starting any new process, it is essential to have formal policies, procedures, templates, and tools for everyone to use. These standards help people adhere to the process. It also provides a way to monitor and control the process by presenting information in a way to allow for an apples-to-apples comparison.

To successfully implement the standards of the process, it is essential to have a communication plan, training program, and documentation.

Continuous improvement

As each phase of the process is implemented, the business is able to achieve small victories with low cost and risk. This incremental approach allows for the business to learn from the last process and improve the next one. When the entire portfolio process is implemented, there needs to be a constant feedback loop to change and enhance the process as the business needs grow and develop.

The work does not stop after a portfolio process is created. A portfolio becomes a corporate functional area just like any other area in the company, such as finance or marketing. The portfolio must be maintained and improved upon to survive. It must be given resources to manage the process to maintain high levels of service to the business.

Portfolio process management

There are six key steps in creating and managing your portfolio. Each step is critical to the success of the portfolio.

1. Inventory
2. Analyze
3. Prioritize
4. Optimize
5. Mobilize
6. Improve and enhance

Inventory. The inventory is a comprehensive list of all programs and projects at any stage of the project life cycle. Assessing the inventory is the first task in gaining control over your projects and programs. This can be a painstaking process, but it's worth the effort. An inventory is the foundation for aligning projects and programs with business objectives, getting control of resource allocations, and managing your business investment.

Each project and program should have enough descriptive information so that it can be analyzed and compared. There should be a project inventory template that is filled out by the project manager, project team, program manager, or key stakeholder. TA system should be established whereby the project or program can be categorized. An identification number could be assigned to projects and programs to help prevent issues with project or program names changing over time or project/program names being confused with other similarly named projects/programs. The identification number can be sequential or be embedded with information. Some information that could be included in the identification number are the year the project or program was established, the department number that correlates to the project or program, and the category of business the project or program is for, among other indicators. The information that may be captured in the template can include:

- Project/program identification number
- Project/program name
- Business case for initiating the project/program
- Brief description of the project's/program's scope
- Strategic business objectives that the project/program meets
- Key deliverables
- Estimated duration
- Estimated cost
- Resources needed
- Risk assessment
- If the information is from a program, it should also include the details concerning the projects under its supervision

This template information could be compiled on a database or assembled on a spreadsheet for analysis and comparison. The assembled inventory will have all the project's/program's attributes, priorities, budgets, and risks in a single place. The outcome of this process creates an up-to-date collection of information for each project so it is visible to the portfolio. As projects or programs change status, the inventory information needs to be updated to maintain an accurate listing of information.

Analyze.   In the identification process, the inventory of projects and programs are assembled. The next step in the process is to analyze the information. When you analyze, you examine projects, in part to gain a better understanding of them. You evaluate all aspects of the projects and programs and scrutinize, evaluate, question, explore, and probe. Many questions are explored during this phase, for example:

- Is the project/program still meeting the objectives that it was established to meet?
- Have the risks exceeded the planned rewards?
- Is the project/program still aligned with the business strategy?
- Are costs getting out of control?

- Has the project/program added too many features and created a gold-plated project/program that minimizes the margins of planned value?
- Is the resource level allocated to this project/program effective?
- Is this project/program dependent on other projects/programs? Or are other projects/progroms dependent on this project/program?
- Are there project/program conflicts?
- Are there redundant projects/programs offering similar capabilities?

Projects and programs are mapped to objectives and verified that they still fall into the same categories or groups to which they were originally assigned. The findings of the analysis may be presented in reports, charts, and/or graphs. Here is where the portfolio team examines all the projects and weeds out problematic projects/programs that have low business value. The projects/programs that are determined to be viable and have a strong business case move on to the next step in the process.

Prioritize. When you prioritize, you rank projects/programs according to importance or urgency for the business. Priorities may shift from year to year. The analyzed project/program information is compared to established criteria to prioritize the projects and programs. They are then reviewed to develop a balanced list of projects/programs. Balance is found if projects/programs maintain the appropriate level of risk and reward versus cost and resource use for the business.

Projects/programs may be given high, moderate, or low priority. Projects/programs ranked high will receive more support than lower-ranking projects/programs. The beauty of portfolio management is that it allows you the ability to fund the projects/programs that provide you with the most business value.

Projects can be grouped in a number of ways. For example, the groupings could be financial.

Group 1: Projects under $50,000

Group 2: Projects $50,000–$250,000

Group 3: Projects $250,000–$500,000

Group 4: Projects $500,000–$1,000,000

Group 5: Projects $1,000,000+

For each of these groups, you can have a different set of criteria or evaluations to determine if the project makes good business sense. The portfolio management team is tasked with distributing funds and resources appropriately to projects. Using grouping and criteria appropriately and effectively will allow for the best mix of projects to be selected.

There are many ways to categorize initiatives besides financial. Some of the ways you can categorize projects/programs are based on:

- Risk to the business
- Possible success rate/reward
- Technology
- Internal competency
- Business value/need

You could also look at the categories strategically and have them align with your business objectives.

Strategic Buckets

Initiative	Representative of the Types of Projects in This Area
Plant/Office Space—Physical Building	Improvements to the physical workspace New ergonomic furniture HVAC improvements Power improvements, including back-up generators Move to a new office space or expansion space
Sustaining—Replacement and Renewal	Maintenance and utility projects Equipment replacements needed to maintain current business functionality Replace outdated software/hardware systems
Enhancing/Growth	Upgrades to current business functionality to improve efficiency and precision of the operation Upgrades to back-office support and manufacturing software Customer-requested upgrades to services and product offerings
Strategic/Innovative	Investment in new product lines Research and development to create a new product or service Develop new markets Build a global presence

Each category could have different criteria for prioritization and different amounts of funds and resources that the company is willing to allocate to it. Using strategic buckets listed in the table, you can divide funding in the following way:

30%   Plant/Office Space—Physical Building

30%   Sustaining—Replacement and Renewal

25%   Enhancing/Growth

15%   Strategic/Innovative

The allocations to these strategic buckets can fluctuate as the needs of the business or changes in the market environment occur. The portfolio team and process has to be agile to maneuver the transformations that might occur.

The criteria for evaluation could be weighted according to priority, which would give you a weighted score. Following the scoring, the projects are then ranked. A watermark can be drawn on the rankings at a certain business value. For example, the watermark could be $10 million. When the project total reaches that point, the projects above that line are approved and continued. The projects below that point will be placed on hold, moved to next year's project list, or killed.

The prioritization process is the primary reason portfolio management is so effective. Here, leaders from all parts of the company come together. When this team is together, communication improves. This gives them a better understanding of how the projects will affect them and let them have a say in what gets executed.

Criterion.   A criterion is an accepted standard used to make a decision or judgment about a project. A criterion for a project can be a measure, a standard, or a condition. You can measure the project against other projects in the portfolio to see how it measures up to them. There can be set standards that state the level of quality or excellence that needs to be attained by the project or program. There can also be conditions that must be met or exist for something else to happen to make the project or program valid. A project might be dependent on another project that isn't complete, so there is little reason to start the project at this time.

Criteria are created jointly with representatives from all the business units that make up the portfolio team. They are based on internal goals and/or predefined threshold metrics. The criteria for projects can be looked at in a number of different ways. Some examples include:

- Strategic value
- Ease of implementation
- Market need/value
- Financial benefit
  - Return on investment (ROI)
  - Internal rate on return (IRR)
- Cost
- Resource impact/capacity

Figure 3-1 shows projects being ranked against the criteria. It is important to qualify the rankings so that your team has an understanding of how to rank the projects against the criteria. You can always add and change criteria as needed to meet business objectives. Also, weights can be added to the criteria to show the level of importance.

Another priority model could look like this:

Strategy/Objective	Proposal Criteria	Proposal Scoring Anchors
Optimize the use of resources	Does it reduce costs and/or cycle time?	4 = reduces cost and cycle time 2 = reduces cost or cycle time
Business need	What is the immediacy of need?	5 = urgent 3 = pressing need 1 = not urgent
Increased effectiveness and accessibility	Does it enable/improve the ability of the customer and/or company to do what they need to do?	5 = customer and company 3 = either one 1 = neither one
Seamless and interoperable technology	Does it easily integrate with our standards and/or products?	5 = supports standards and is relatively easy to implement 3 = generally supports standards and can be implemented 1 = barely supports standards and requires major modifications to implement
Reaches/supports our customer base	How broad? How varied? How many customers?	5 = extends beyond the customer base 4 = can apply to all current customers 3 = supports many current customers 2 = supports a few current customers 1 = supports one customer
Appropriate technical risk	Is this moving us toward evidenced best practices?	5 = evidence of benefit and transferability 3 = evidence of benefit but no evidence of transferability 1 = No evidence of benefit

Optimize. After prioritization is complete, further assessment still needs to be done. In prioritization, you looked at individual projects and programs and how they weighed against given criteria. The next critical step is to appraise the portfolio as a whole.

When the portfolio team looks to optimize the portfolio to make it more effective and efficient, they need to take a step back from the components and visualize the portfolio holistically. Here, we look at the system as a whole rather than simply each project separately.

- What does the entire portfolio provide?
- Does the portfolio balance our resources as far as capacity versus demand?
- Does the portfolio balance our current needs with our future needs?
- Does it provide maximum value to the business?
- Are critical resources allocated correctly?

Project	Strategic value for the company	Complexity of project	Financial benefit for the organization	Cost	Internal resource impact	Priority of the project	Notes
Replacement of key system (Project 1.1)	3	3	4	2	4	3.2	Average-impact project, but needed to improve current operational processes
New development project (Project 2.1)	5	1	5	2	4	3.4	High-impact project, but with much risk surrounding it
Maintenance project (Project 3.1)	1	5	1	5	2	2.8	Low-impact project, but needed to support key functionality

Legend			
Strategic value	5 = very important 1 = low importance	Cost	5 = low cost to the project 1 = high cost to the project
Complexity of project	5 = very easy 1 = very complex	Resource impact (people, technology, equipment, etc.)	5 = low impact 1 = high impact
Financial benefit for the organization	5 = potential for great financial benefits for the organization 1 = little or no financial benefits		

The higher the score, the higher the project's priority

Figure 3-1   Project ranking model.

- Are projects and resources aligned with the company's objectives?
- Are there any redundancies?
- Are there projects or programs that we should accelerate, place on hold, maintain, or cancel?

It is crucial when you optimize that you test out or model alternative project portfolios to make sure that you have the best mix of projects and programs and that they provide the company with the highest return on investment. When you model a portfolio, you can simulate the mix of approved and must-do projects and select the ones that optimally fit the budget and resource limitations. Here, you can also test out the unselected or out-of-capacity projects and put them back in the mix to see how they could better serve the portfolio. In your simulation or model, you should ensure that must-do projects are approved and that resources are allocated appropriately for them. The outcome of this process is to have an optimal project portfolio.

Mobilize.   When the inventory, analysis, prioritization, and optimization phases are complete, it is then time to set in motion the decisions made for the portfolio. When you mobilize, you prepare your resources for action. The portfolio is unveiled to the troops. A communication plan should be established regarding how this information is rolled out to the workforce. Changes to the portfolio can be published, with explanations on how the projects and programs align with the company's strategic objectives. Resources are then deployed to the appropriate projects or programs at the planned level.

Projects or programs placed on hold have their resources redeployed. Cancelled projects or programs have contracts terminated and resources migrated to other assignments.

All projects and programs have their schedules and budgets revised with any updates or changes planned for in the portfolio. New projects are put into motion.

Portfolios have to be monitored and controlled. Here is where the appropriate individuals alert the portfolio team if critical schedule, financial, or resource thresholds are exceeded, or if project or program conflicts arise. Performance reports on approved or in-progress projects or programs are rolled up to the portfolio team for review and analysis. Metrics are analyzed to ensure that the project/program mix in the portfolio is meeting established criteria. This then feeds back into the inventory of projects and programs with updated information.

Improve and enhance. Periodically, the entire process (inventory, analyze, prioritize, optimize, mobilize) needs to be reviewed for possible improvements. Enhancements can always be made to any process. These enhancements can be incremental or evolutionary, depending on the needs of the system. The process system review should include all key stakeholders. This gives you an opportunity to demonstrate the value of the portfolio process and ensure commitment from sponsors and stakeholders. The portfolio needs to be actively managed to be successful.

Project portfolio benefits

An organization that successfully uses a project portfolio system will see many benefits. Let us face facts. You cannot change what you do not know. The portfolio provides a process for capturing project and program information into a usable format for management decisions. Having a dashboard to review the strengths and challenges of the projects and programs helps executives make better decisions. It allows for a rolled-up view of all projects and programs, with the ability to dive into any one section for further analysis. You can compare variances and build a benchmark for projects and program success. You could evaluate the value earned by the portfolio elements. With this big-picture view, you can analyze the risk in the portfolio and diversify it. Having a project portfolio process that is established and that everyone buys in to can help eliminate internal political players pushing their agendas for resource approvals, and it provides a basis for the system to focus only on what is best for the business. You can analyze the return on investment and manage resource allocations.

Projects and programs that are not meeting expectations and that are no longer aligned with business objectives can be divested. Are any of your projects becoming bridges to nowhere? You could make this kind of assessment by looking at the bigger picture of the project or program in relation to the overall portfolio and business direction.

A business gains the ability to:

- Have oversight on project/program schedule and budget variance
- Measure metrics on return on investment
- Increase resource utilization and reduce headcount
- Understand project and program risk factors and adjust the portfolio as needed
- Cancel projects and reallocate resources

The project portfolio management process is, in a sense, in balance with business objectives. It will provide the ability to select projects based on data, metrics, costs, budgets, and other objective criteria. If a business continues to plan at the individual initiative level, the tactical and reactive nature will remain. An effective portfolio allows for flexibility to adjust with changing objectives and environmental factors, and allows for the mitigation of risk. With a well-executed portfolio, you can prevent the company from putting all its eggs in one basket. You can diversify risk and balance investments that meet corporate objectives. You don't want to spend your entire portfolio on one area. Feeding your cash cow benefits you today, but doesn't provide for tomorrow. On the other hand, being overly aggressive and focusing only on new product development could hurt your core business.

A well-run portfolio can provide the following value items for a business:

- Alignment with business strategy by focusing on doing the right work. Approving the right projects to meet your business objectives. If the objectives change, you can realign the project mix.

- Standard workflows help with consistency in information and understanding in the process, as well as buy-in/support for the process. Helps get things through the process faster. Helps with rolling up information with consistent terminology and metric usage.

- Leverage resources (people, money, assets): Manage them across the business and use the right resources correctly.

- Global teams can be more productive. They share and reuse information, work products, and templates so you can do the work in the correct way. Building upon previous knowledge, the business grows and matures, and doesn't have to reinvent the wheel with each new project.

- Enhance visibility and control organizational transparency so that you can identify and solve problems early. Focus on value versus features and functions. Set up formal checkpoints in the system (Cooper calls these "gates"). At these checkpoints, an evaluation is done to make sure the project or program is still on track. Is it still meeting the project deliverables that it is suppose to meet? Are risks in check?

How Portfolios, Programs, and Project Management Relate

In order to be successful in the projects and programs that your business invests in, there needs to be a best practice to manage it. This practice includes the fundamentals of integrating top-down and bottom-up processes. The more mature the process is, the more successful the company will be in achieving the vision it has for its future. At first, this process may feel like you are building a complex tapestry of multicolored threads woven intricately together and feel too overwhelming to consider, or to even know where to start. However, if you focus first on the core principles and then build from there, you will be able to achieve your goals. Three main elements can help you make business objectives a reality: project portfolios, programs, and projects. A company that can effectively administer these elements can be quite successful in meeting their objectives.

Portfolio	Portfolio focuses on meeting strategic objectives. The portfolio is a collection of projects or programs that are grouped together to facilitate achieving the company's strategic business objectives. The projects or programs of the portfolio may not necessarily be interdependent or directly related, but they are grouped together to achieve business objectives. Through this process, portfolios manage the company's valuable resources.
Program	Program work is focused on results. A program is group of related projects managed in a coordinated way to obtain benefits and control not available with managing them individually. Programs include an element of ongoing work, and may include elements or related work outside the scope of discrete projects in the program.
Project	Projects focus on successfully producing deliverables. A project is a temporary activity taken on to create a unique product, service, or result.

A common saying is that project management is doing projects right and portfolio management is choosing the right projects to do. Program management is doing the right projects right!

Project portfolio management is different from both programs and projects in that it is a continuous process. Both projects and programs have a life cycle, with a beginning, middle, and end. Once a portfolio management process is established, it becomes another functional area in your company, similar to human resources, marketing, finance, and operations. These functional groups are established and continue until the company ends. You might wonder what portfolio management needs to do on a continuous basis.

Well, portfolio management involves the continual process of selecting and managing the optimum set of initiatives to deliver maximum business value to the business. This sounds straightforward, but it takes effort in practice. You can look at portfolio management as the middleware between strategic planning (discussed in Chapter 1) and project delivery. Portfolios provide a cohesive view to track and align the priorities as set by the portfolio process. This process helps the business get its arms around the investment in projects that they are making. They get a clear picture on where their investments are and can periodically reevaluate decisions.

Global project portfolios

Some companies, by their very nature, are global companies; others are nurtured into going global to allow the business to expand into new marketplaces. For example, a computer or technology-based company can go global with minimal effort, but furniture companies have greater barriers to hurdle to become global. Furniture companies have concerns with the high cost of shipping, tariffs, and local tastes. Having a well-formulated project portfolio process for your company can make all the difference. It can allow you into new markets and minimize the risk and provide for better decision-making. It is critical that the company have a strong handle on how they are investing their money and using their resources. There are often inherent integration issues with regard to politics, economics, regulations, and language. These issues can be complex; thus, it is critical to have control over the business, which is most successfully done within a portfolio.

Programs in a portfolio world

Programs are critical elements in a portfolio. A program is created by the portfolio team when the complexity of the work is significant and added oversight is required. The program management efforts add a higher level of integration management to ensure that project investments are delivering the results that the company is striving to receive. The program effort provides a link between the portfolio objectives and project deliverables to secure business results.

Goals of a project portfolio

The objective of a portfolio is to allow the organization to be focused, responsive, and agile. The ability to pick the right projects and make the right investments is critical for a business' success. Every portfolio has three main goals:

- Strategic alignment and direction
- Portfolio value maximization
- Balance risk versus rewards

Each goal is a critical element in achieving portfolio success, and goals can be achieved through a number of methods. Methods should be used in conjunction with each other and not in isolation. Each method produces valuable information, but also exhibits limitations, which should be taken in the context of other tools to help reach a goal.

Goal 1—strategic alignment and direction.   Strategic alignment and direction focuses on the oversight of projects and programs in the portfolio, making sure that they align with the business' strategy. The direction includes communicating the strategic goals of the portfolio to the company and setting project priorities based on the company's needs and visions.

Value maximization focuses on creating the right mix of projects to maximize the value of the portfolio at a given resource allocation. There is a focus on breaking down the spending so that the higher-priority projects have the resources assigned to them first. The breakdown of spending across projects is tied to the priorities of the strategy. The portfolio focuses on its investments and tracks the performance. It also oversees or provides governance over the decisions made. It focuses on controlling and containing costs and efficiently leveraging resources when possible. It also looks to balance project and program efforts them from strategic to tactical.

Building strategy in a portfolio focuses on aligning investments with business strategy. Here, a company might have strategic buckets and priorities with those buckets. There are two basic methods of incorporating strategy into a portfolio. There is the bottom-up method, which focuses on selection tools, and then there is the top-down method, which categorizes items into strategic buckets. It is essential to understand where your products and services are relative to the market to be able to make decisions regarding your portfolio.

While the buckets may be similar for companies, the spending priorities will be different among competitors and in different industries. The strategy must be measured in terms of the company's goal, which is articulated as part of your strategy. Strategy and resource allocation must be intimately connected (like putting your money where your mouth is)—your strategy becomes real when you start spending money. If you don't allocate resources to your strategy, is it nothing but words on a paper.

You want to activate projects that are aligned with the business strategy and that contribute to achieving the goals and objectives set out in the strategy resource allocation across business areas, markets, and project and program types to truly reflect the desired strategic direction of the business. The mission, vision, and strategy of the business must be put into operation in terms of where the business spends money and which development projects and programs it undertakes. If the strategy is to grow, then the projects and programs the company selects will lead to growth rather than simply defending the status quo.

This can be done with top-down approach, where you have a vision, goals, and strategies and then you develop a product roadmap and strategic buckets. This approach starts with a business strategy and attempts to translate it into the right set of projects and their timing.

The bottom-up approach allows for proposals or ideas to bubble up from anywhere in the organization. These projects must be screened, and the good ones allowed to rise to the surface and be funded.

Goal 2—portfolio value maximization.   A number of methods can help maximize the value of a portfolio. Basically, you should look at a project as an investment decision. In this approach, you look at various financial models to help make the decision: net present value (NPV), expected commercial value (ECV), payback period, break-even analysis, return on investment, discounted cash flow (DCF), and internal rate of return (IRR), as well as financial ratios such as productivity index. However, financial tools have their weaknesses, mostly due to a lack of reliable information to support the financial assumptions.

Other tools can be used to measure the benefits of a project or program, such as checklists and scoring models. These tools rely primarily on subjective inputs. You can also use maps, like bubble diagrams, which plot reward against the probability of success. We have to understand that no single process gives the "right answer." Hybrid approaches permit a more tailored approach to portfolio selection.

The goal is to allocate resources so as to maximize the value of the portfolio's return on investment. We will look at each of the following areas individually:

- Net present value
- Expected commercial value
- Balance score card
- Productivity index ratio
- Scoring models

Net present value.   Net present value (NPV) looks at the value of the investment. It focuses on taking the present value of a time series of cash flows. The present value is calculated by estimating the future payment (or series of future payments or revenues from a project) and calculating the present worth of those future payments using a discount to reflect the time. As the saying goes, a dollar today is worth less than a dollar in the future, if that dollar is invested at a given interest rate. I can invest that dollar I have now, and in 10 years it might be worth two dollars. So a two-dollar revenue in 10 years is only worth one dollar today. A two-dollar revenue in five years might only be worth 80 cents today, depending on the interest rate. To correct for all the out-year revenues of a project, they are all discounted to present value. This is a common method for using the time value of money to appraise a long-term project. It looks at shortfalls in cashflow in present terms. The project with the higher NPV should be undertaken. However, other factors have to be reviewed before a final decision should be made, including the risk projects and programs present. The problem with NPV is that it relies only on financial analysis and the validity of the data.

If...	It means...	Then...
NPV > 0	The investment would add value to the firm	The project may be accepted
NPV < 0	The investment would subtract value from the firm	The project may be rejected (There are some projects that are required to be preformed to maintain the operation that have little or no financial value to the business)
NPV = 0	The investment would neither gain nor lose value for the firm	We should be indifferent in the decision whether to accept or reject the project. This project adds no monetary value. The decision should be based on other criteria, e.g., strategic positioning or other factors not explicitly included in the calculation.

With every financial decision, there is an opportunity cost or an economic opportunity loss. An opportunity cost is the value of a project that you omitted from your portfolio of projects and programs. The opportunity cost analysis is an important part of a company's decision-making process. There is a choice being made between desirable yet mutually exclusive results. Opportunity costs are not limited to just financial, but also to any resources that would be assigned to a project or program. Resources that can have an opportunity cost include time, facility space, raw materials, and people.

NPV prioritization.   Once you determine the NPV for all your projects, you need to rank them from highest to lowest. You then drop any projects with negative NPV. This creates a kind of project priority list. You allocate resources to projects starting with the highest NPV. This continues until all the resources are exhausted. There are concerns with using NPV; however, that should be understood. First, this method assumes that financial projections are accurate. It doesn't incorporate risk or resource constraints. It also assumes all or nothing, whereas in some projects or programs you can take an incremental approach. NPV is only one tool, and only using a combination of tools do you yield true portfolio value maximization.

Though NPV has its shortfalls, it does have a number of attractive features. For example, it gets the project team to focus on long-term project deliverables and program results. It also looks at ROI and payback period, and can use those as benchmarks to determine the success of the project or program if selected.

Balanced score card.   A balanced score card approach determines if the portfolio is aligned with the objectives of the company. The focus is on finances, operations, marketing, and risk. The scorecard is a dashboard, or a comprehensive view of the portfolio, and provides the portfolio team with insight on how they are performing in each objective.

Expected commercial value (ECV).   The expected commercial value method seeks to maximize the commercial value of the portfolio, subject to certain budget constraints, but also introduces the notion of risks and probabilities. You can look at the ECV and determine if you want to continue with the project. Some refer to the decision as the "go" or "kill" decision. At this point, a project can continue on its path, be placed on hold, or be killed.

There is a downside to this method. It is dependent on extensive financial and quantitative data and does not consider whether the portfolio has the right balance between high- and low-risk projects, or across the right markets, technologies, or strategic buckets. Finally, it only considers a single financial criterion for maximization.

ECV seeks to maximize the expected value or expected commercial worth of the portfolio within budget constraints.

Attractive features:

- Recognition that the go/kill decision process is an incremental one

- All monetary amounts are discounted to today, which appropriately penalizes projects that are years away from launch

- Deals with the issue of constrained resources

Major weaknesses:

- Dependent on extensive financial and other quantitative data

- Ignores whether the portfolio has the right balance between high- and low-risk projects, or across markets and technologies

Productivity index. The productivity index takes the limitations of NPV and recognizes that resources are limited. There may be projects that have wonderful NPV that you just can't pass up. However, high NPV could also mean that they consume a great number of resources, or even have needs beyond the capabilities of your current resource model. The goal is to get the largest return for the smallest investment. The productivity index is created by dividing NPV by the resource amount and then re-ranking the projects you have.

Scoring model. The scoring model, sometimes called the weighted scoring model of project selection, was discussed earlier in this chapter.

Goal 3—ways to achieve balance. Often, portfolios are weighed down with too many small project tweaks, modifications, updates, and fixes to sustain the growth of the company. A lack of vision when balancing a project portfolio is a risk to the company. The need for diversity, however, is going to vary widely from company to company. Balance the vision and needs of the company with the availability of resources. Visual charts are the most popular way to display balance in portfolios.

Risk-reward bubble diagrams. One type of visual chart is the risk-reward bubble diagram. One axis measures the reward to the company; the other is a probability of success. Too heavy an emphasis on financial analysis can do serious damage, particularly in the early stages of a project. The size of each bubble shows the annual resources spent on each project (dollars per year; it could also be people or work-months allocated to the project). This can show risk versus reward, with circles or bubbles representing individual projects or clusters of projects.

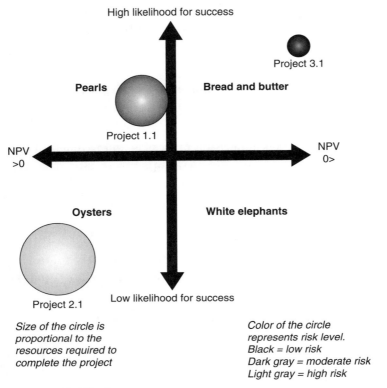

Figure 3-2 Bubble diagram.

Their size usually connotes some important third metric, such as resource requirements. Figure 3-2 provides a sample diagram.

The four quadrants are:

1. *Pearls (upper left).* These are the potential star products, projects with a high likelihood of success that are expected to yield very high rewards. Most firms wish they had more of these.

2. *Oysters.* These are the longshot or highly speculative projects with high-expected payoff, but with low likelihood of technical success. They are the projects or programs where technical breakthroughs will pave the way for solid payoffs.

3. *Bread and butter.* These are small, simple projects with a high likelihood of success, but low reward. They include extensions, modifications, and updating projects. Most companies have too many of these.

4. *White elephants.* These are the low-success and low-reward projects. Every business has a few white elephants, which are difficult-to-kill projects or programs that began life as good prospects but over time became less attractive.

Further information can be added to the diagram to provide clarification on the project status. The size of the circle can represent the amount of resources needed to complete the project. The color of the circle can represent the risk level of the project. The projects listed in Fig. 3-2 are from the chart shown in Fig. 3-1 earlier in the chapter.

Other factors can be measured in these diagrams, such as technical feasibility or market segments. Pie charts can be used to show spending allocation across project types or a breakdown of resources by product lines and markets. Diagrams can be in the form of trees that break down different dimensions. Successful portfolio management incorporates a well-defined and communicated process, as well as tools that add structure and efficiency.

Portfolio management tools can add structure to the portfolio process by encouraging the use of templates and methodologies and providing more efficient tracking and analysis. Individual point solutions (e.g., time management, resource management, skills inventory, etc.) do not provide insight into project feasibility and can keep bad projects from being discontinued. There can be an overload with too many maps, and they don't all have all the answers. These maps are designed to provide information and to help visualize balance. They are not the sole means by which a decision should be made.

Portfolios incorporate both projects and programs to achieve strategic objectives. Projects are tactical and provide deliverables to the organization. Programs can be used to frame and align and candidate global projects that have common goals.

4

Global Program
Management Strategies

Introduction

Chapter 3 explored the portfolio management process and described how companies generate, evaluate, choose, and fund programs and projects. We saw that alignment with business strategy is an important part of that process in order to assure that the programs generated and chosen help to achieve the business's strategies and plans. The portfolio process assures that the *right* programs and projects are initiated.

In this chapter we will address how "going global" changes a business's program and project management strategies, using two case studies. We will explore how different countries, ethnicities, cultures, values, and languages can affect how program managers handle their projects, and how political and regulatory differences can present real risks to global program managers.

Global program managers and their companies face striking changes in the face of global economic growth and development, but the *fundamentals* of global project management remain the same: align business and program strategies to adapt to changing economic, social, and political factors, and tailor program and service delivery to local conditions and local customers and values in targeted markets. Furthermore, the design, planning, and management of a program in tomorrow's global economy must take into account the changes going on in the financial and market worlds. Severe risks are associated with national economies going through major contraction and restructuring, with governments intervening in business investment and finances. How does a program manager approach this kind of global challenge, and how does classic program and project management change as a result? What opportunities are opened up in this kind of this new global dynamic?

In this shifting global milieu, programs that succeed will use one of the following strategies:

Strategy 1. Design programs that can generate early impacts on generating jobs and local development. In other words, how can your program help local economies to succeed—as a program benefit—and meet core program goals as well?

Strategy 2. Frame program benefit goals that align with political, economic, and business growth realities—e.g., how can a program manager align program goals with local objectives?

Strategy 3. Target programs and resources on technology goals that align with applicable foreign government objectives and recovery programs—e.g. how can local interests in technology development be integrated with program delivery?

Strategy 4. Target technologies that stimulate local development, manufacturing, and education goals—e.g. how can programs be designed to produce short-term technology benefits that have social and economic benefits?

Strategy 5. Frame programs as recovery initiatives—e.g. align programs with local attempts to recover from economic downturns.

Strategy 6. Stick to the knitting; deliver core products and services, develop new products that enhance current successful products, and develop user training competencies.

Western Technologies: A Typical Global Corporation and a Framework for Linking Program Management to Business Strategy

Many companies now operate in a global context, simply because they can and must do so to grow and prosper. As an example, a major global producer of manufacturing and automated control systems, let's call it Western Technologies, Inc., might manage its presence in 150 or more countries. Western's strategy is to adapt its products and services to the standards and policies of each country. Western tries to *root* itself in the key target localities so that its people become part of the fabric of the local markets and can understand and interpret customer needs. Western's talent is distributed based on shared skill sets to serve local needs and to adopt best practices. It stresses its contribution to reducing costs, creating jobs, and serving local community and national goals. Western shapes its programs to mobilize all of its resources in a given country or region; in other words, Western thinks of its *programs* as vehicles for core business delivery. Its program managers are its account executives.

In the portfolio process, Western linked its strategy objectives and programs and projects as follows:

Strategy 1. Design programs that can generate early impacts on generating jobs and local development.

Program 1. Market new manufacturing companies in developing countries with new technologies that serve local needs; program goal is to successfully market 10 new production/manufacturing systems in Africa and Asia by 2012.

- *Project 1*. Partner with Kenya manufacturing council to develop manufacturing demonstration project.
- *Project 2*. Partner with four other African nations to develop a demonstration.
- *Project 3*. Develop sales video and supporting materials for marketing production demonstrations.

Program 2. Encourage training of customers and development of local talent and workforce proficiency in measuring and improving manufacturing and operational efficiency, creating the need for more jobs in manufacturing, consistent with local political and economic realities.

- *Project 1*. Develop training materials and video.
- *Project 2*. Deliver training to developing-country clients.
- *Project 3*. Install new production system in one African nation.
- *Project 4*. Produce video showing successful installation of new production system.

Strategy 1/Program Benefit. Successfully improve local economic conditions in targeted countries and regions while marketing new technology for production/manufacturing. Establish local competencies in manufacturing development as program outcome.

Strategy 2. Frame program benefit goals that align with political, economic, and business growth realities.

Program 1. Change Western's culture and value system to encourage innovation in production system applications in Africa.

- *Project 1*. Survey the current culture for indicators of agency culture and readiness for the program.
- *Project 2*. Use survey results to develop an organizational development package to equip Western to manage the new program.
- *Project 3*. Design and deliver the new training project.

Program 2. Align Western's programs with targeted regions and their local economies.

- *Project 1*. Investigate targeted regions and develop a national economic profile.
- *Project 2*. Develop a training project for Western employees.

Strategy 2/Program Benefit. Establish long-term partnerships with local political jurisdictions through technology-transfer programs and ensure long-term marketing opportunity in those regions.

Strategy 3. Target programs with economic and technology goals that align with applicable foreign government objectives and recovery programs around green technology.

Program 1. Develop a series of new products and services in targeted areas for green technology initiatives.

- *Project 1.* Survey market for green technology in target area.
- *Project 2.* Develop new products for marketing green technology.

Program 2. Develop a longer-term approach to focusing green technology on current buildings to house new manufacturing and automation systems.

- *Project 1.* Develop new strategy and product line for equipping current buildings with green technology in ventilation systems.
- *Project 2.* Develop new competence level in Western for the maintenance of green technology.

Strategy 3/Program Benefit. Develop green technology programs to produce immediate benefits while developing longer-term marketing opportunities for manufacturing clients.

Strategy 4. Target technologies that stimulate local development, manufacturing, and marketing competencies.

Program 1. Develop partnerships with local workforce training and developmental nongovernmental organizations (NGOs) in targeted regions to provide green technology materials and instructors for new production systems.

- *Project 1.* Develop a joint training project with Kenya NGOs.
- *Project 2.* Deliver training to all Kenya NGOs by 2012.

Program 2. Design and frame new product concepts to monitor and link to local company information networks.

- *Project 1.* Develop new production assembly line methodology for automobile manufacturing using green technologies.
- *Project 2.* Implement new assembly line methodology in five new plants by 2013.

Strategy 4/Program Benefit. Build local and regional production/manufacturing capacity in foreign nations while developing a market for new products and services.

Strategy 5. Frame programs as recovery initiatives.

Program 1. Develop technology improvements in association with local NGOs and political jurisdictions to generate jobs and economic recovery, as well as local self-sufficiency.

- *Project 1.* Identify the root causes of economic downturn in local target regions and how manufacturing capacity can address those causes.
- *Project 2.* Develop a technology-transfer project that specifically addresses local workforce and unemployment conditions.

Program 2. Design new product development projects that meet technology needs in local jurisdictions and align with local political goals.

- *Project 1.* Develop a new product development project focused on green technology production systems.
- *Project 2.* Implement a new product development project locally.

Strategy 5/Program Benefit. Align with local political goals and develop partnerships with government entities and leading corporations in order to establish the basis for long-term marketing.

Strategy 6. Develop strategic business process enhancement projects around current product lines to reduce costs of development and marketing.

Program 1. Develop a core competency in strategic manufacturing program management as a business process and have program managers adapt this competency to local conditions.

- *Project 1.* Develop a business process training program for Western's global program managers.
- *Project 2.* Implement a training program and certify managers to run global programs.

Program 2. Partner with other companies and consortiums of companies in related fields doing business globally, e.g., manufacturing start-ups, information technology, automated production, Six Sigma implementation, and benchmark industry leaders in manufacturing management to benefit local target regions.

- *Project 1.* Capture Western's core competencies in a business process training manual.
- *Project 2.* Transform the manual into an online and virtual training and development program for local entrepreneurs in targeted countries.

Strategy 6/Program Benefit. Help local companies and governments in targeted regions develop alignment between public goals and private enterprise development in manufacturing, further green goals and objectives, and set up the market for long-term sales.

How does Western manage this process?

Western's programs are integrated with its core products and services and focused on broad program benefits, e.g., programs are initiatives to focus and tailor current products and projects to local market conditions and are tailored to local needs. A Western *program* is characterized by the alignment of its competencies and offerings to foreign conditions, needs, and opportunities, defined by Western-trained professionals who live and work in those countries. Western is keenly aware of cultural and social differences and how they play into successful global program management.

Cultural integration. Alignment with local cultures is important for Western. Using recent research on cultural differences between China and western countries as a framework, Western trains its global managers to align programs with local values. Here is an example. It is well known that Chinese enterprises work within a set of values based on teamwork, and project and program management are quite different from in the West. Here is a matrix of these differences from a research study reported in the *Project Management Journal*, September 2007, by university professors Xiaojin Wang and Lanfeng Liu, entitled "Cultural Barriers to the Use of Western Project Management in Chinese Enterprises: Some Empirical Evidence from Yunnan Province" (p. 64):

Western PM Values and Beliefs	Chinese PM Values and Beliefs
Integration management	Doctrine of the mean (find common ground)
Encouraging disagreements to be surfaced	Requiring people to be less confrontational and less direct
Requiring people to be direct and open	Using compromising and smoothing strategies
Regarding confrontation strategy as the best way of solving conflicts	Strong avoidance of uncertainty
Horizontal management	Strong hierarchy
Small power distance	Large power distance
Cross-functional communication and cooperation	Superior-subordinate vertical work relationships
Influencing and coordinating abilities is important	Line authority and control important
Team considerations	Family consciousness
Short-term orientation of relationships	Long-term orientation of relationships
Work contributions oriented to evaluation of people	Guanxi (personal relationships) evaluation of people
Project team of high diversity	Family members are homogeneous
Everyone plays important roles	Only elite play important roles
Task orientation	Boss orientation
Completing the task	Making the boss happy
Viewing people based on their work performance	Viewing people based on their hierarchical position

Western encourages its local account and product managers, as well as service consultants and sales personnel, to understand these kinds of cultural differences in every country in which they operate and to adapt program initiatives, teamwork, and marketing to local values. Western encourages its people to get to the work and task issues quickly and to focus local clients on technical solutions rather than on developing long-standing teams and partnerships where such teamwork may be inconsistent with prevailing Chinese standards.

Western's approach to program management

Western's global programs are managed using what we would call a loose-tight strategy—it's loose in that it allows maximum discretion in local decisions to allow its program managers on site to make decisions in the context of local conditions—e.g., "if in Rome, do what the Romans do"—but is tight on company values and strategic objectives for continuous improvement, profitability, simplifying processes, and product and service quality. Its work in product development is keyed to broad political, social, and economic trends; thus, its focus on supporting energy efficiency and keying on local, current system-specific improvements in light of the global downturn in new building construction. Western's approach illustrates the importance of a strong, centrally driven product and service mix, combined with a distributed management system that reflects local conditions and dynamics, and tailors its products to site-specific needs.

Simplifying program delivery. Western works to simplify program delivery in countries where the process can become overwhelmed with issues and data. The aim is to simplify and focus all of Western's products and services, and to encourage collaboration across company functions and product lines. Its program portfolio is integrated and its program documents and schedules are transparent, laid out as a baseline on a network for its management and customers to see. Local sourcing and manufacturing is a key to integrating the company into the fabric of the target market. Western emphasizes that it is one global company, and simplifies its support functions—HR, marketing and sales, and financing—to strengthen its supply chain, with stronger emphasis on building longer-term supplier linkages and supporting program and account managers.

Western designs and implements its programs through its strategic planning function and through direct reports and line functions. Programs are not run by separate, offline project teams; everyone is *in the program.* In other words, program management is integrated into the fabric of the company's day-to-day operations in a given country. In effect, Western exemplifies the concept that program management has become a core line activity in global operations rather than simply an R&D or narrow operational function.

Supplying OEMs. Western has targeted original equipment manufacturers (OEMs) with its programs, providing supplier services that are:

- Coordinated across multiple business units, regardless of language and cultural issues, e.g., Western has its *global act* together through integrated program teams
- Collaborative in arranging global sales agreements involving several countries
- Oriented to establishing long-term partnerships in global purchasing, logistics, and processes
- Offering a single interface to coordinate account activities worldwide

- Customized to local needs
- Informed on how to design a manufacturing system that is in compliance with international and local standards
- Planned so that program risks are identified early and addressed in planning and implementation
- Focused on training clients and customers in negotiation skills and other team-based competencies such as performance monitoring

The Eastern Case: Another Global Project Company

Eastern is a manufacturer of aluminum worldwide. This case covers Eastern's approach to developing programs that meet both its own goals and broader program goals addressing global issues. In sum, here is how Eastern handled each one of them.

1. Design programs that can generate early impacts on generating jobs.
   - All of its contingencies involving IT systems were sought out locally in target countries so that local companies could compete for the business.
   - Eastern localized its program teams to be closer to target markets for aluminum, thus generating a need for supporting businesses, e.g., facilities, communication, and logistics.
2. Frame program benefit goals that align with economic and business growth.
   - Eastern developed partnerships with several companies worldwide that shared Eastern's interest in bringing the price of aluminum back up to market levels, thus creating a consortium to spur more profitability in all aluminum-producing companies worldwide.
   - Eastern promoted its new production system initiatives worldwide as global recovering programs that aligned with several western countries' goals of reducing energy consumption and encouraging increases in gross national product.
3. Target programs on economic and technology goals that align with applicable foreign government objectives and recovery programs.
   - Eastern aligned its robotic production team with several foreign governments interested in developing their local R&D capacity, thus helping to stimulate local innovation and business development.
   - Eastern aligned its raw material development programs to benefit countries that had both raw material reserves and national interest in using raw materials to improve energy efficiency and create new jobs around energy independence.

4. Target technologies that stimulate local development, manufacturing, and marketing competencies.

 ▪ Eastern worked with selected national governments to set up marketing teams with local industries to allow an interchange of marketing strategies and information systems with locals.

 ▪ Eastern worked with local safety experts and academic institutions in target countries to develop joint projects to improve aluminum manufacturing safety and efficiency.

5. Frame programs as recovery initiatives.

 ▪ Eastern actually defined its power-saving program as a global recovery initiative that would benefit several national economies.

 ▪ Eastern worked with several international union organizations to educate and inform leadership in ways to improve worker efficiency and productivity while upgrading old aluminum production facilities.

6. Stick to the knitting

 ▪ Eastern stuck to its core competency, the manufacturing of aluminum, but focused on new products to align with new global dynamics.

 ▪ Eastern worked on continuous improvement in its major risk area, e.g., cost of power.

As a project manager, you have the added responsibility to work within the parameters of the program. It is a good practice to work closely with the program manager. In a global environment, you may be located in a different country, speak another language as your first language, and have different customs and beliefs than the program manager; however, it is important to essentially eliminate the "space" between the two of you. The elimination of space can also be thought of as managing up. You may never be able to physically remove the distance, or even the time barriers, that keep you from the program manager; however, you can lower these hurdles by being flexible, taking the time to fully understand what is required by you for the program, and communicating in a timely manner with the program manager.

Background

Programs are complex, relatively long-term, and focused on broad outcomes when compared to projects, which tend to focus on a single product and/or output. It follows that programs that involve many projects and support activities are organizational; they tend to be embedded into the organization, and many become program units in themselves. Programs evolve out of business strategies.

This chapter uses the case approach to address how a real company, termed the Eastern Company for purposes of protecting its propriety, handles program development in its business planning and operations processes. Eastern is a global manufacturer and distributor of aluminum products. Global program management at Eastern involves a staged process of moving from broad strategic planning, to program and project development, to evaluations and assessment, to change and agility. The reason Eastern is a good case is that while it is both a production and marketing company, it is also a global company subject to all the risks and opportunities faced in today's world.

The Eastern company faces major competition and challenges from a tightening global aluminum market and from foreign manufacturers, who regularly "dump" aluminum into western markets at very low prices. Thus, there is a continuous risk in the business from forces beyond the company's control. To address the risks inherent in its future, Eastern prepared a risk-based strategic plan and program.

Eastern manufactures aluminum products from raw materials acquired worldwide, which makes it a truly global company. Its products are produced and distributed in several countries, subject to many different forces and conditions, and it has presence all over the world.

Risk-based strategy and program development

Eastern completed a strengths, weaknesses, opportunities, and threats (SWOT) analysis and identified eight risks and eight strategic goals, as well as program areas to address them.

Risk 1. Required electric power will not be available at an affordable price. Since Eastern's manufacturing plants use large amounts of electricity to convert raw materials into aluminum products, the availability of cheap electric power is essential to its success.

> **Strategy 1.** Secure economically priced power to reduce the risk of power shortage.
>
> > **Program 1A.** Develop alternative, plant-specific electric power generation.
> > **Program 1B.** Develop a global database and reporting system for electricity pricing information.
> > **Program 1C.** Design a new aluminum production system with minimal electric demand.

Risk 2. Cost increase in aluminum manufacturing will increase faster than margin. Raw material costs are rising globally at a faster rate than revenues, thus cutting into margins for the company worldwide.

> **Strategy 2.** Secure other resources at reasonable costs to offset the risk of cost escalation.
>
> > **Program 2A.** Develop a global supply chain program to reduce resource costs.
> > **Program 2B.** Design a global database and reporting system on resource availability and costs.

Risk 3. Customers will not be satisfied with Eastern's products.

Strategy 3. Cultivate customer awareness and promote customer satisfaction to avoid customer satisfaction risk.

Program 3A. Develop worldwide market research program aimed at OEM customers.

Program 3B. Develop new product development scheme driven by global market research.

Risk 4. Eastern's working environment will prove to be unsafe, and the company will experience substantial loss of workforce and financing as a result.

Strategy 4. Create a safe working environment to control the risk of worker injury and associated costs.

Program 4A. Develop a safety training and research program for employees.

Risk 5. The Eastern workforce will not grow with the technology available for continuous improvement.

Strategy 5. Build a responsible and knowledgeable workforce to avoid the risk of workforce instability.

Program 5A. Develop a comprehensive training and development program and continually improve it using blended learning tools, e.g., onsite and online.

Program 5B. Develop an employee recognition program based on performance.

Risk 6. Eastern will not act to improve manufacturing technology in time to keep ahead of competitors.

Strategy 6. Improve technology and plant equipment to produce products more efficiently to control productivity risk.

Program 6A. Develop comprehensive robotic production concept.

Program 6B. Develop new plant performance information system.

Risk 7. Pollution from Eastern facilities will lead to noncompliance with government environmental requirements.

Strategy 7. Improve Eastern's impact on the environment to avoid the cost of pollution and noncompliance.

Program 7A. Identify gaps between requirements to meet environmental controls and actual performance by plant worldwide.

Risk 8. Increasing waste in the manufacturing process and workforce will lead to uncontrolled costs.

Strategy 8. Reduce waste and non-value-added costs to control the risk of wasted effort.

Program 8A. Implement a Six Sigma program in each plant worldwide.
Program 8B. Conduct a global benchmarking study to identify best practices.

Eastern recognized the need to take direct action to sustain its ability to successfully compete on a continual basis in the global aluminum marketplace. The assumption was that despite that fact that Eastern employees, in general, were dedicated to providing the highest-quality products and services to the customer at a competitive price, and to providing a positive return for the owners' investment, they were heavily unionized. The company was committed to the principle that "we will not be able to step up to those challenges unless our employees—and the union—can see where we are going and why, and have the opportunity to 'buy in.'" It is through this strategic plan and its communication that management saw that they could accomplish alignment and reduction of their considerable risk exposure.

The strategic plan was being communicated continuously throughout the plant through special meetings and focus groups to ensure that all employees understood it and could relate it to their work. Employees were encouraged to document actions they or their teams were taking to accomplish or support particular initiatives. This process would continue as the plan was updated annually; and policies, procedures, and the organizational structure were realigned to accomplish the plan. This strategic plan was developed by the directors of Eastern, with support from area managers.

Commitment and partnership.   Eastern management and United Steelworkers of America stated directly that they were committed to this mission for the organization. It was clearly recognized that by working together to accomplish this mission, the interests of all participants would be served. All management and employees were to benefit from long-term job security, job enrichment, and the monetary rewards that result from a successful business that is able to manage its risks. Eastern's stakeholders and owners would benefit from the product recognition and profitability gained by producing superior goods and services. Eastern's customers would benefit from the high quality and service levels delivered to them. Finally, the community would enjoy a stable revenue base from the success of Eastern and from the skills and services individual employees offered.

Stakeholder relations.   The company stated that Eastern's stakeholders were people, organizations, or groups of people who have a vested interest in the success of the company. The major stakeholders were identified as follows:

- Employees, who seek continued employment and income, quality of work life, and opportunities to learn and develop (their perceived risk was related to job security and lack of growth and development) and marketability

- Customers, who seek quality products at low cost and reliable delivery. Their perceived risk was related to product price, quality, and timing, but mostly price. Cost was a major issue as competitors dumped quality aluminum at lower prices.

- Owners, who seek return on investment and continued viability. Their risk was grounded in stock value.

- Regulators, who seek compliance with laws and regulations. Their risk was in noncompliance with regulations and the cost of enforcement and litigation.

- The community, who seeks contributions through taxes and services, with minimal environmental impact. Their risk exposure was in losing the industry tax base but having to pay pollution and environmental control costs.

- Suppliers, who seek to meet Eastern's requirements and continue business with the company. Their perceived risk lies in their inability to meet Eastern's contract requirements and having to share more of the risk in contracted work than they can handle.

To illustrate the documentation of a risk-based strategy, the following document contains an executive summary, situation analysis, and a detailed description of eight key strategies. The situation analysis provides a framework for the strategies, including mission and goals, management direction, SWOT analysis, and linkage to the parent company strategic plan. The eight strategies are supported by specific initiatives and a system to measure their achievement.

This strategic plan for Eastern Aluminum Company covered a five-year period, from 1996 to 2000, and will help guide the company and its employees into the twenty-first century. As the general long-term pathway to growth and profitability, the plan presents the company's approach to achieving Eastern's central strategic goal: to compete successfully on a continuing basis in the world aluminum marketplace. The plan served a wide variety of purposes, including support for ownership decisions; support for budgeting and resource allocation; guidance for management and employee planning, training, and education; and support for long-term capital investment planning. A major element of the strategic planning process that produced this document is the communication of the plan and its underlying vision, assumptions, and values to employees.

Eight strategies and key measurements

Within the overall framework of the basic strategic goal to compete successfully in the world aluminum marketplace, and consistent with the parent company's strategic objectives, eight key strategies were at the heart of this strategic plan:

1. *Secure economically priced power*. Eastern would find ways to lower its power costs through a variety of strategies, including building stronger partnerships with power companies and state and local governments, and through exploration of independent options for generating less expensive power.

2. *Secure other resources at reasonable costs.* As the cost of materials rises, Eastern planned to find low-cost sources for raw materials, as well as explore approaches for using lower-graded materials. Eastern would take the initiative to ensure effective partnerships are built with quality suppliers.

3. *Cultivate customer awareness and promote customer satisfaction.* Eastern would work to educate employees about customers and their requirements and to promote closer ties with customers. Greater appreciation of customers would give employees more incentive for addressing future customer requirements and connecting their daily work more clearly with the "value chain" to the customer.

4. *Create a safe working environment.* Eastern was working to improve its safety record through enforcement of safety and health rules and regulations. Employees would be better educated and trained to understand the safety implications of their work. Safety compliance would be considered a major performance standard for all employees.

5. *Build a responsible and knowledgeable workforce.* Facing a major workforce turnover in the next five years, Eastern placed special emphasis on strategies to build a more responsible and skilled workforce, to improve the partnership with the United Steel Workers, to improve performance and productivity, to lower labor costs, and to find better ways to work together through teamwork. They recognized that if this strategy—grounded in the commitment to building a team-based organization—is not accomplished, Eastern could not thrive and grow, even if the other strategies were accomplished.

6. *Improve technology and plant equipment to produce products more efficiently.* Eastern prides itself on its leadership in technology and technical innovation, and plans to continue this industry leadership. The company was managing several capital improvement projects to make major breakthroughs in productivity and quality. Eastern felt it was demonstrating to its customers and its employees through these improvements that major investments are being made in the plant to meet the challenges of the future global marketplace.

7. *Improve Eastern's impact on the environment.* Through strict compliance with federal, state, and local environmental standards, Eastern would continue to respond to and anticipate environmental impacts and address them. Special emphasis was being made to meet new clean-air requirements.

8. *Reduce waste and non-value-added costs.* Eastern continued to pursue quality and process improvement initiatives to eliminate unnecessary costs due to accidents, rework and scrap, outdated positions and job requirements, and equipment damage. Employees would continue to be trained and educated in process improvement and reengineering to streamline the way work is accomplished.

Overview

Eastern had already been turned around from a high-cost swing plant, with a confrontational labor atmosphere, to a much more competitive operation, practicing effective and efficient management and supervision, worker empowerment, and self-directed team concepts. However, there was a new urgency to ensure that all employees understood that the plant would grow only "by permission" from future customers, and only if it continuously improved its productivity, quality, and internal cohesion and teamwork across departments. The following sections discuss how Eastern was positioned to compete in the future.

Strengths, weaknesses, opportunities, and threats

The following discussion covers Eastern's strengths, weaknesses, opportunities, and threats.

Strengths. Eastern had made a concentrated effort to retain its competitive position in the marketplace through technology. Its major strength is its ability to produce quality products continuously, focus on technology and capital improvement, and keep wages and salaries relatively high for its employees while controlling costs. Capital improvements and improved management and team practices have made it possible to achieve record premium production in the recent past. Eastern continued to demonstrate its leadership in technological improvements and plant capital investment.

Eastern had experienced the longest run at full capacity in its history, remaining one of the few North American smelters not curtailed due to the recent metal surplus caused by the flow of aluminum into the world market from the Commonwealth of Independent States. Eastern continued to show resilience and responsiveness in the face of changing market conditions.

Eastern was working hard to empower its workforce and improve their knowledge, skills, and responsiveness. They sought to align their incentive and reward programs, partnership practices with hourly employees, performance appraisal systems, and quality and process improvement initiatives with key long-term strategies. Eastern faced the future turnover of the workforce with a strong commitment to use the opportunities that changes bring to build a leaner, more integrated, and more productive plant team.

At the heart of its strength was Eastern's traditional core competency to choose, operate, and improve process technology effectively; to produce a variety of difficult-to-produce premium products; and to understand and meet customer needs. Whatever initiatives Eastern undertook, it knew it had to continuously improve these drivers of success.

Weaknesses. Eastern's products (primary, slab, billet, tee, and foundry pig) were priced by the worldwide commodities market. The high quality and excellent service of these products would ensure a positive customer relationship, but Eastern could not control the selling price of the finished product.

Eastern knew that it was a high-cost plant compared to other producers, primarily because of the age of the facility and technology, and because of high wages, salaries, and fringe benefit levels. Because Eastern had little or no control over the market price, the cost of producing aluminum became a key determining factor in remaining globally competitive. In fact, 75 percent of all aluminum in the world was being produced at a cost lower than Eastern.

Eastern faced major challenges in turning over its workforce and creating a more energetic and knowledgeable workforce team; past practices had not always inspired employees to align themselves with the plant's best interests and commit themselves to continuous improvement through teams.

Eastern needed to improve its ability to learn and document its successes—in short, to become a "learning" organization. Past practices had not always taken advantage of what the organization has already learned through the years.

Opportunities. Demand for aluminum was continuing to rise; supplies of aluminum had increased each year, with primary aluminum products now sold on a worldwide basis.

Eastern had the opportunity to position itself at the midpoint on the world cost scale, the point at which 50 percent of world production costs would be higher than Eastern's. In achieving this position, Eastern could take advantage of its high-quality products and services and its improving productivity.

A reduction of 4 cents per pound by 1999 would have placed Eastern in that competitive position, keeping in mind that other aluminum plants are also attempting to reduce their costs.

Eastern's major opportunity was to improve its process efficiency and productivity through a combination of technology and capital improvement, building a more efficient workforce, and reducing labor costs. The 4 cent/pound cost reduction could be achieved by:

1. Conversion of potlines (production lines) to a new "point feed" technology, already underway

2. Reduction of man-hours per ton by 15 percent from 1996 to 2000

3. Reduction of non-value-added costs wherever possible through process improvements, total quality management, ISO 9000 certification, and other quality initiatives

Eastern has a major opportunity to improve its human resource practices and programs as the plant transitions its workforce in the coming five years, both through better training and development of supervisory and hourly employees, and through better, more effective, assessment and hiring practices.

Threats and risks. If Eastern did not continually reduce costs, their position would worsen because:

1. New plants with lower costs would open

2. Existing competitive plants would reduce cost and improve their cost position

3. Other plants with higher costs would close, worsening Eastern's position

The most critical of these risks was the possibility that power costs would continue to rise beyond Eastern's capacity to absorb them. This scenario represented the most significant threat to Eastern's continued growth and had to be avoided. In addition to power costs, the long-term cost of coal could be another important threat to Eastern's growth, as well as unanticipated environmental regulations, particularly from the federal Clean Air Act.

In addition, although Eastern had made major progress in building a more team-based culture, the process could not be slowed by resistance to change and failure to be clear about new roles and functions. Therefore, one source of threat and risk was clearly from within: the threat of slow deterioration of the momentum of teamwork and process improvement already underway. Such a step backwards could always happen as a result of neglect and a lack of trust and respect in the organization.

Eastern's strategic plan

Eastern's strategic plan was an integrated set of strategies, initiatives, and measures supporting an overall goal of competitiveness. Figure 4-1 presents a graphic depiction of the company's eight key strategies. Each strategy was seen as serving the central goal of world competitiveness, but each strategy was also inextricably tied to the others, indicating a strong interdependency of all plan elements. If any one strategy and risk reduction plan was not accomplished, overall achievement of the goal suffers.

The plan describes plant strategies, initiatives, and measures of success. Initiatives are programs and projects now underway or planned to help accomplish a particular strategy. Measures are indicators of progress, and will be used to monitor achievement of the eight key strategies.

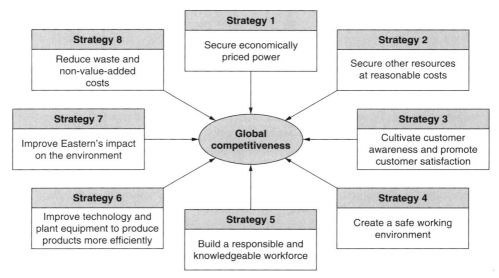

Figure 4-1   Eight key strategies.

Underlying elements of the risk-based strategic plan

Five major elements formed the basis for this risk-based strategic plan: mission, commitment and partnership, driving forces, core competencies, and stakeholder relations. They are discussed in the following sections.

Mission.   Eastern's mission was to be the most cost-effective producer of the highest-quality primary aluminum products, shipped on time to its customers, with optimum utilization of resources. They placed special emphasis on employees and their role in defining the company's mission, and on good community relations. Eastern recognized that accomplishing their mission involved a never-ending journey of continuous improvement.

Commitment and partnership.   Eastern management and United Steelworkers of America indicated that they were committed to this mission for the organization. It was clearly recognized that by working together to accomplish this mission, the interests of all participants were best served. All management and employees would benefit from long-term job security, job enrichment, and the monetary rewards that resulted from a successful business. Eastern's stakeholders and owners would benefit from the product recognition and profitability gained by producing superior goods and services. Eastern's customers would benefit from the high quality and service levels delivered to them. Finally, the community would enjoy a stable revenue base from the success of Eastern and from the skills and services individual employees can offer.

Driving force: production capability.   An underlying element in this strategic plan was the single most important driver of company's success: its capability to convert resources effectively into products through highly organized and managed production processes. Their value added for the future will continue to be their capacity to produce products continuously.

Core competencies and risk contingencies.   Three core competencies separated Eastern from its competitors:

1. Its capacity to effectively choose, operate, and improve process technology: The company's ability to keep up with changes in technology was rooted in its ability to anticipate technology risk stemming from out-of-date technology.

2. Its capacity to produce a variety of difficult premium products: Eastern's ability to change its production systems quickly was rooted in its ability to anticipate the risks of change in product requirements and plan for them.

3. Its capacity to understand and service customer needs: The company's capacity to understand its customers and especially to anticipate and manage customer services helps reduce the risk of failed customer service expectations.

Eastern would strive to maintain and build on these core competencies.

Measures of effectiveness

Eastern identified eight key strategies to carry out its central strategic goal of global competitiveness. Each strategy was carried out through several initiatives and was monitored by the measures shown. They are presented and discussed in the following sections.

Strategy 1.   Secure economically priced power.

The cost of power was a major factor in Eastern's strategic plan. In its partnership with the community, power companies, and state and local government, Eastern would develop support for its efforts to continue to compete in the world aluminum marketplace. Eastern planned to negotiate lower power costs and to explore independent options for generating less expensive power.

Initiatives	Measures	Risks

Address power pricing issues by:
  Maintaining relationships with The public service commission, people's council, local and state government
  Investigate power wheeling sources and benefits
  Develop alternative power sources, including self-generation
  Increase community support for reducing Eastern's power costs
  Eliminate power modulation
	Reduced power costs by 2-4 mills per kilowatt hour (approximately $6-12 million/year)
	Favorable public response and concern for Eastern's power-pricing issues
		That relationships with stakeholder agencies would deteriorate
		That power wheeling sources (independent sources of power created by deregulation) would not provide lower prices
		That self generation of power would fail either from technology problems or cost
		That the community would not support Eastern
		That power modulation—the practice of energy providers to reduce power—could not be anticipated

Power costs became the key factor in maintaining competitiveness because of impending major increases in costs from new sources. Eastern management was working closely with utilities, government officials, the community, and other power sources to ensure that it could achieve independence in power generation should that be necessary. Power wheeling sources and benefits were being pursued, as well as self-generation options.

While this issue is beyond the scope of any one employee, special attention was being given to communicating the power cost issue to all employees so that they could understand the urgency of the situation and help Eastern achieve power independence.

Strategy 2. Secure other resources at reasonable costs.

The cost of materials continued to rise, and had the potential to erase savings created by increased productivity and reduced power costs. Eastern planned to find low-cost sources for raw materials, as well as explore approaches for using lower-graded materials. The company would continue to manage human resource costs through employee attrition and retirement.

Initiatives	Measures	Risks
Obtain raw materials, such as petroleum coke, pitches, alumina, and hardeners Secure high-quality supplies from the most economical sources	Maintain or decrease current raw material costs	That raw materials would not be available on a just-in-time basis
Manage human resource (labor) costs through attrition and retirement	Reduce man-hours per ton by 15% by the year 2000 Contribute to overall efficiency and productivity	That human resource costs would inflate and attrition goals would not be achieved
Explore innovative approaches to using lower-graded materials, such as calciner fines and lower grades of petroleum coke	Maintain or decrease current raw material costs	That lower-graded materials would not be acquired

Key raw materials (alumina, aluminum fluoride, petroleum coke, and liquid pitch) were purchased for all parent company smelters by the same parent office. These costs were rising to a point such that Eastern's overall cost effectiveness was threatened.

This issue challenged the company's capacity to find and use lower-graded materials, such as calciner fines and lower grades of petroleum coke. Eastern would continue to acquire both raw materials and supplies from the most efficient sources, while assuring quality. This involved forming partnerships with suppliers to limit the number of such sources, which would accomplish two objectives: holding down costs and minimizing purchasing and warehousing requirements.

As a major cost element, labor costs had to be controlled while productivity was enhanced through capital improvements and better management, team, and individual performance. Reduction of man-hours per ton by 15 percent by the year 2000 was a major measure of success in reducing risk exposure.

Strategy 3. Cultivate customer awareness and promote customer satisfaction.

Eastern continued to provide consistent and high-quality products and services to end users and customers. The company would work to ensure that all employees were aware of customers and their needs. Emphasis on the customer would encourage the development of new products and services, and help Eastern establish a larger market niche. Eastern would look to external stakeholders to verify gains made in customer awareness and customer satisfaction.

Initiatives	Measures	Risks
Enhance employee awareness about customer and final product satisfaction	Third-party assessment of employees' customer awareness Recognition through accreditation and quality audits (American Association for Laboratory Accreditation, etc.)	That employees were not able to connect their success with company success in end product quality
Selectively diversify products and services to support market expansion	Capacity to change products Number of customer assists through the Metal Quality Group	That its product mix could not be diversified
Support parent company strategy market services to make customers aware of Eastern's capabilities	Alumax Inventory Management System (AIMS) data Customer team visit comments Customer satisfaction data	That the parent company strategy was not consistent with Eastern's strategic plan and core competence
Focus on individual customer demands in metallurgy, product chemistry, packaging, and delivery requirements through process improvement Develop a long-term cast house plan and monitoring systems	International Standards Organization (ISO) 9000 registration ISO (International Standards Organization) accreditation Monitor customer claims and contacts about technology services and products Review customer satisfaction survey results Improved product turnaround indicators	That Eastern's process improvement efforts were not successful because of personnel and union disincentives
Set up cross-functional teams to increase awareness of internal customers	Internal customer satisfaction surveys Extent to which internal customer requirements are met	That cross-functional teams would not work because of internal conflicts and role definitions

Eastern had to establish a market niche in high-quality premium products to remain a viable company and to successfully compete. To meet this demand, Eastern had to work closely with its ownership to identify future customer needs. Eastern would continue to work with parent company marketing teams in the areas of initial order processing, customer team visits, and customer surveys.

Eastern would also make it easier for customers to deal with the plant. Increased use of bar code systems and electronic data interchange was planned, establishing a "seamless" electronic relationship with prospective customers. More attention would be paid to promoting laboratory and metallography capabilities.

The continuing move to quality worldwide was having its impact on the company. More customer inquiries, e.g., from the automotive industry, were expected regarding quality standards. This development prompted efforts to maintain registration and refine documentation to both ISO 9000 and American Association of Laboratory

Accreditation, and for attaining QS 9000 and 14000 certification as well. Increased cycle time was becoming a major customer expectation, generating internal plans to develop systems to measure order entry, production scheduling, and shipping performance.

Finally, because many employees did not have a direct relationship with customers and customer needs, the company was undertaking a program to enhance employee appreciation of customer needs. This program included use of a third-party organization to monitor employees' understanding of these issues.

Strategy 4. Create a safe working environment.

Eastern had significantly reduced accidents in recent years, and needed the support of employees and management to continue these safety efforts. In addition to developing and implementing state-of-the-art safety procedures and guidelines, the company needed to enforce safety and health rules and regulations *consistently*.

Initiatives	Measures	Risks

Eliminate safety and health hazards by:
  Upgrading engineering standards, safety features, and ergonomics consistently
  Promoting employee awareness
  Updating joint safety and health committee guidelines
  Enforcing rules and regulations
  Increasing team and employee accountability for safety and health
	Decrease accident incident rate
	Stay within accident and safety scorecard budget
	Improve safety severity ratio index
	Rate of completion of items on safety list and audits
	Improvements in efficiency and job performance
	Improved plant safety performance record
	Increased safety gainsharing payout
		That safety initiatives would not be accepted and implemented by employees and managers
		That safety guidelines and regulations would shift substantially

Eastern recognized its responsibility and accountability for the safety and health of each employee and for the preservation of property and equipment. The company would continue to incorporate safeguards and procedures into the design and operation of all facilities, which would minimize risks of personal injury and loss of property and equipment. Management was responsible and accountable for the safety and safe work conduct of all employees. Employees were equally responsible and accountable for safe practices, as well as for assisting in the ongoing safety program by reporting unsafe practices, procedures, or conditions when they were observed.

As indicated in the initiatives under this strategy, Eastern was giving special priority to upgrading engineering standards to reflect safety requirements and criteria. In some cases, this could have meant added cost and time constraints on planned capital projects, an expense well worth the investment in a safer working environment.

Strategy 5. Build a responsible and knowledgeable workforce.

By increasing the skills and abilities of individuals, teams, and supervisors and empowering them, Eastern would be able to increase productivity, reduce operating costs, solve personnel problems, and increase teamwork across the entire plant. Initiatives in support of this priority included training and developmental opportunities in support of self-directed work teams.

Initiatives	Measures	Risks

Develop or continue:
  Empowered, self-directed work team development (decision-making and responsibility)
  New performance appraisal system for salaried employees
  Development planning
  Knowledge and skills training for bargaining unit employees
  Supervisory development program
  Strategic plan communication process
  Conversion to parent salary structure
  New bargaining unit job classification (stemming from the labor contract)
  HR strategic plan
	Better communication and coordination within team members and between supervisors and teams
	Innovative, timely, and sound employee and team decision-making
	Better use of tools, equipment, and raw materials
	Employees will be prepared to assume new responsibilities as a result of developmental exposure
	Enhanced partnership agreement
	Increase in ideas and solutions from employees
	Reduce man-hours per ton
		That self-directed teams would not work in the unionized work setting
		That employees would not act on incentives to train and develop new skills
		That the strategic communication plan is not effective in improving employee support of company goals

Offer developmental opportunities to sustain employee education and growth through: Mentoring Inside training Outside technical managerial training Opportunities to manage	Successful development planning Track progress through training records Enhanced employee performance	That the plant could not implement mentoring and training initiatives because of the company culture

Strategy 5 held the key to successful achieving the other strategies—the building of a workforce and organization that (1) was aligned with the strategic direction of Eastern; (2) was structured, capable, and motivated to improve performance; and (3) worked together across departments to provide a "seamless" process of production and quality.

In building a flatter, more streamlined workforce, the company's strategy in the past had been to press for a reduced workforce and more teams and teamwork. As a result, many teams had been generated and trained to take responsibility to solve problems and make the decisions necessary to keep their process operating at peak efficiency. Supervisory and hourly positions were reduced, and roles and functions were changed.

Now in the spirit of building the total Eastern organization, the company's strategic emphasis would go beyond reduced workforce and generation of teams. The strategy would be focused on organizational effectiveness: building the whole organization through a stronger linkage and alignment between management, supervisors, and bargaining unit employees. The opportunity before company management was to build new supervisory roles and functions into a new team-based organization, requiring the development of leadership skills, better business and productivity management and monitoring skills, and more support for technical supervision and cross-department process improvement.

Support services, such as human resources management, were to help lead the effort. Organizational barriers to effective supervision would be identified and eliminated. Organizational and training initiatives were underway to help supervisors function as the guiding force for day-to-day operations.

Development tools would include business and productivity management, process improvement, facilitating and mentoring opportunities, inside training, outside development (technical and managerial), and management opportunities within the organization. To focus on incentives, Eastern would review its performance appraisal and gainsharing structure to ensure that they were aligned with this strategic plan, and would make improvements when called for.

To ensure effective communication, quarterly plant communication meetings would continue, and more information would be provided to employees online, especially in the area of human resources.

Strategy 6. Improve technology and plant equipment to produce products more efficiently.

The company was managing several capital improvement projects to upgrade the condition of equipment and work processes at the plant. The company needed to continue these improvements while employing sound capital project management skills. Eastern would work to speed up completion of these capital projects and to keep them within budget and quality requirements.

Initiatives	Measures	Risks

Complete capital program and budgets each year
Conduct major maintenance projects and overhauls
	Completion of capital improvements, including:
	  Conversion of potlines to point feed technology
	  Substation life extension
	  Cast house continuous homogenizing furnace
	  Rod shop anode cleaner
	  Ladle shop ladle cleaner
	  Bake oven rebuild
	  Potline capacity expansion
	  Rebuilt remelt furnace
	  Developed stack filter systems for metal treatment
	  Facilities expansion
	  Completed stamper upgrades for billet and slab
		That capital budgets would not be completed
		That maintenance projects would not be completed for a variety of reasons

Conduct research to ensure that Eastern adapts or incorporates improved or emerging technologies	Completion of R&D projects within budget	That necessary research on emerging technologies is not conducted
Develop a stronger capital project management system (CPARs) through training and other developmental assignments	Improved capacity to complete projects on time within budget and schedule	That the CPARs system is not made operational

As evidenced in this partial list of capital improvements, Eastern was heavily engaged in upgrading its technological and equipment base in order to maintain its leadership and core competency. The company was a front-runner in keeping pace with required capital improvements to aging plant infrastructure. Improvements are underway in the production lines, carbon plant, cast house, substation, and laboratory, and in general plant functions, such as emission and noise control and information system management.

The focus for this strategic plan was the completion of capital projects within budget, schedule, and technical requirements. This meant developing a stronger capital project management system and employing more effective project management practices.

Strategy 7. Improve Eastern's impact on the environment.

The company would continue to monitor its impact on the local environment. These efforts would be directed toward reducing environmental degradation and pollution.

Initiatives	Measures	Risks
Comply with federal, state, and local environmental regulations by: Providing proactive assistance to regulators Educating employees about regulatory requirements Promptly reporting noncompliance and correcting any violations Filing Title V air permit applications	Eliminate incidents of noncompliance Monitor response time for identifying and fixing violations	That new regulations would be enacted that Eastern could not respond to
Participate in voluntary activities on environment, safety, and health issues, such as EPA Greenlights, reducing greenhouse gases and noise nuisance reduction	Eliminate environmental, safety, or health complaints about the plant or its operations	That voluntary efforts would not improve community relations
Encourage environmentally sound industrial and agricultural growth	Partnerships with state and local agencies	That local growth objectives and dynamics would change substantially
Continue farm production	Farm production and maintenance of safe environmental practices	That the company's efforts at farm production around the fringe of the plant would be unsuccessful

This strategy addressed the company's environmental and community relations practices. Eastern would continue to stay ahead of environmental requirements through two basic approaches: (1) being proactive in assisting regulators at all levels in developing sound and cost-effective regulations that both implement environmental legislation and meet the needs of community and the business; and (2) planning and implementing capital improvements and operating measures to comply with environmental requirements, attempting at the same time to ensure that such improvements also contribute to overall plant productivity.

Costs of compliance would increase as well in the administrative areas of record-keeping, reporting, training, planning, and monitoring; and in the acquisition of necessary monitoring equipment, creating the need to streamline these systems. Eastern would continue to develop the capacity to prevent pollution through technology improvements and through a multimedia approach that addresses losses of material to air, storm water runoff, and solid or liquid waste streams.

Strategy 8.   Reduce waste and non-value-added costs.

Eastern continued to experience waste and non-value-added costs, such as safety and property costs related to accidents, rework and scrap, and equipment damage. Process improvement and problem-solving teams would continue to focus on reducing these costs.

Initiatives	Measures	Risks
Involve quality teams in identifying and resolving quality problems in key production processes	Amount of rework and scrap by department on a monthly basis Stay within approved budget guidelines for rework and scrap costs	That the company's quality teams would not be successful in resolving quality issues
Minimize equipment damage by educating employees, monitoring equipment use, and enforcing rules for properly using equipment	Review monthly maintenance to ensure departmental accountability for responsible equipment use Stay within approved budget guidelines for equipment expenses	That equipment damage rates would continue
Eliminate duplication of effort in administrative processes Process improvement/ reengineering Encourage employee use of best-practice techniques	Benchmark other processes Monitor process costs	That administrative redundancy and increase costs of operation would continue
Improve inventory management of supplies and equipment (includes maintenance, production, and raw material in-process)	Reduce inventory by at least 5%	That inventory management initiatives would not be successful because of internal plant or supplier performance limitations
Minimize waste generation and increase recycling	Waste product reductions	That increasing rates of waste production would continue

This strategy, reducing waste, was in concert with Strategy 5: to build a knowledgeable and productive workforce. Both strategies were required to improve overall productivity. Strategy 8 was key to improving the overall productivity of Eastern by eliminating waste and unnecessary work, for example, by reducing the cost of poor quality through process improvement and ISO and QS 9000 and 14000 documentation.

The company's quality and process improvement efforts started on the production floor, where quality was built in through consistent practices and extensive use of statistical process control methods. Eastern was committed to being quality-driven, not cost-driven; thus, the quickest route to elimination of waste and non-value-added costs was "doing it right the first time." They looked to this strategy to be a major factor in lowering operating expenses by 4 cents per pound.

The quality teams would continue to identify and resolve quality problems in key production processes; a new focus would be placed on administrative and support processes to ensure that they were under review in the context of process improvement as well.

Communicating strategy and risk

The company prepared a communication program to promote the company strategy and to explain the risks inherent in the business and the local plant setting. The following presents the structure of that plan.

Key Strategic Goal: Improve Eastern's capability to compete on a continuing basis in the world aluminum marketplace.

Explanation of strategies. See Fig. 4-2 for a graphic of the agency eight part strategy.

Strategy 1. Secure economically priced power.

The cost of power is a major factor in Eastern's strategic plan. In its partnership with the community, power companies, and state and local governments, Eastern will develop support for its efforts to continue to compete in the world aluminum marketplace. The company plans to negotiate lower power costs and to explore independent options for generating less expensive power.

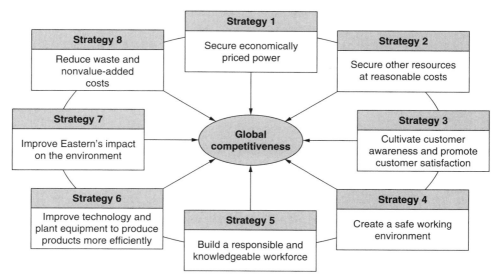

Figure 4-2 Eight-part strategy.

Strategy 2. Secure other resources at reasonable costs.

The cost of materials continues to rise, and has the potential to erase savings created by increased productivity and reduced power costs. The company plans to find low-cost sources for raw materials and to explore approaches for using lower-graded materials. Eastern will also continue to manage human resource costs through attrition and retirement.

Strategy 3. Cultivate customer awareness and promote customer satisfaction.

Eastern continues to provide consistent and high-quality products and services to end users and customers. Eastern will work to ensure that all employees are aware of customers and their needs. Emphasis on the customer may encourage the development of new products and services, and help the company establish a market niche. The company will look to external stakeholders to verify gains made in employee customer awareness and customer satisfaction.

Strategy 4. Create a safe working environment.

Eastern has significantly reduced accidents in recent years, and needs the support of employees and management to continue these safety efforts. In addition to developing and implementing state-of-the-art safety procedures and guidelines, Eastern needs to enforce safety and health rules and regulations *consistently*.

Strategy 5. Build a responsible and knowledgeable workforce.

By increasing the skills and abilities of individuals, teams, and supervisors and empowering them, Eastern will be able to increase productivity, reduce operating costs, solve problems, and increase teamwork across the entire plant. Initiatives in support of this priority include training and developmental opportunities in support of self-directed work teams.

Strategy 6. Improve technology and plant equipment to produce products more efficiently and economically.

Eastern is managing several capital improvement projects to upgrade the condition of equipment and work processes at the plant. Eastern needs to continue these improvements while also employing sound capital project management skills. The company will work to speed up completion of these capital projects and to keep them within budget and quality requirements.

Strategy 7. Improve Eastern's impact on the environment.

Through voluntary efforts and in compliance with state environmental standards, the company will continue to monitor its impact on the local environment. These efforts by the improvement and problem-solving teams will continue to focus on reducing these costs and impact on the environment.

Strategy 8. Reduce waste and non-value-added costs.

Eastern continues to experience waste and non-value-added costs, such as safety and health costs related to accidents, rework and scrap, and equipment damage. This process is directed toward reducing environmental degradation and pollution.

Program development

Once strategic objectives have been generated, the process of creating broad programs begins. Program areas, broad organization-wide initiatives to carry out strategic objectives, are typically created in planning discussions among top management and ownership. Eastern began the process of identifying broad programs in a series of planning and organization meetings. Initial descriptions of program areas and appropriate projects, with anticipated global impacts, are discussed in the following sections.

Strategy 1. Secure economically priced power.

Programs	Projects and Potential Global Impacts
Program A: Secure alternative power sources through a series of long-term research projects, partnerships, and information gathering	Projects: A1: Investigate power-wheeling benefits through the use of Internet sources (explore independent power sources rather than regulated sources); for all foreign plants, investigate local regulatory and power conditions, and prepare an analysis based on the local situation A2: Develop partnership agreements with the public service commission, local and state government A3: Develop partnership with Dutch wind-generated power source company, contact Dutch wind generation company and arrange for negotiations to build wind generator to explore feasibility A4: Design and conduct an alternative power scenario with videos; prepare videos in several languages and reflect local cultural and political situations to enhance the local plant management's capacity to find alternative power sources
Program B: Develop a global database and information collection system on worldwide power source developments	Projects: B1: Identify a contractor to develop a database on platform; design contract scope to identify how databases can be tailored to local conditions in various foreign plant situations B2: Design a database and platform to reflect local needs B3: Train key analysts and users with packages in several languages
Program C: Design new aluminum production process to minimize power source requirements	Projects: C1: Outline new production concept and operating system to reduce power needs C2: Test new production concept C3: Run simulations C4: Build prototype

Strategy 2. Secure other resources at reasonable costs.

Programs	Projects and Global Impacts
Program A: Develop global supply chain program to reduce resource costs	Projects: A1: Restructure raw material procurement process to obtain lower-cost raw materials, such as petroleum coke, pitch, alumina, and hardeners, through long-term supply chain suppliers; find best suppliers globally A2: Conduct international research program to explore innovative approaches to using alternative-graded materials, such as calciner fines and lower grades of petroleum coke
Program B: Develop alternative resource sources	Projects: B1: Conduct investigation of alternative resources, e.g., optional coke products to replace petroleum coke, and get federal funding through alternative fuels program
Program C: Reduce human resource (labor) costs through efficiency measures, attrition, and retirement	Projects: C1: Design and implement a Six Sigma program at each plant in all foreign locations to reduce man-hours per ton (MPT) by 15% by the year 2020 C2: HR program to integrate job descriptions to meet job needs created by attrition and retirement; negotiate with union

Strategy 3. Cultivate customer awareness and promote customer satisfaction.

Programs	Projects and Global Impacts
Program A: Develop global market research program aimed at identifying new OEM customers	Projects: A1: Conduct profile study on current customers A2: Identify potential OEM clients
Program B: Develop new product development organization and process to meet future product needs identified in market research	Projects: B1: Create a new organizational unit in every plant in all global locations to create new product concepts from customer studies B2: Develop partnerships with foreign engineering and manufacturing associations to locate potential contributors to company's need for new products
Program C: Educate workforce globally about need to collaborate with customers to explore new product needs	Projects: C1: Enhance employees' awareness about customer and end-product satisfaction through a customer education and visit program
Program D: Achieve all necessary international certifications for customer satisfaction	Projects: D1: Achieve ISO 9002 registration and QS 9000 accreditation

Strategy 4. Create a safe working environment.

Programs	Projects and Global Impacts
Program A: Develop a safety training and research program for employees	Projects: A1: Upgrade engineering standards, safety features, and ergonomics A2: Update joint safety and health committee guidelines A3: Increase team and employee accountability for safety and health A4: Improve efficiency, safety, and job performance by reducing waste in plants
Program B: Revise incentives for safe performance	Projects: B1: Increase gainsharing payouts for accident prevention performance and leadership B2: Publish successful safety actions in employee newsletter

Strategy 5. Build a responsible and knowledgeable workforce.

Programs	Projects and Global Impacts
Program 5A: Develop a comprehensive training and development program in continuous improvement using blended learning tools, e.g., onsite and online	Projects: A1: Develop and implement online and onsite continuous improvement programs to create empowered, self-directed work team development (decision-making and responsibility) A2: New performance appraisal system for salaried employees Development planning A3: Supervisory development program

Program 5B: Develop incentives and an employee recognition program based on performance	Projects:
	B1: Assess options for incentives for improved knowledge-based performance program
	B2: Select and implement incentive program
	B3: Assess projects
	B4: Offer developmental opportunities to sustain employee education and growth through:
	    Mentoring
	    Inside training
	    Outside technical managerial training
	    Opportunities to manage

Strategy 6. Improve technology and plant equipment to produce products more efficiently and economically.

Programs	Projects and Global Impacts

Program 6A: Develop a comprehensive robotic production system	Projects:
	A1: Develop robotic productions system concept and design
	A2: Develop prototype
	A3: Select and implement a replacement for the current system
	    Consider:
	      Substation life extension
	      Continuous homogenizing furnace
	      Conversion of production line to point feed technology
	      Cast house
	      Rod shop anode cleaner
	      Bake oven rebuild
	      Production line capacity expansion
	      Rebuilt remelt furnace
	      Developed stack filter systems for metal treatment
	      Facilities expansion
	      Completed stamper upgrades for billet and slab

Program 6B: Develop a new plant performance information system	Projects: B1: Conduct a research program to ensure that Eastern adapts or incorporates improved or emerging performance measurement technologies B2: Test and implement the system
Program 6C: Develop a stronger capital project management system (CPARs) through training and other developmental assignments	Projects: C1: Develop a project management handbook and training program to improve workforce capacity to complete projects on time within budget and schedule C2: Develop a program management handbook for all multiproject managers C3: Develop and install a program management shadow system for prospective program and project managers

Strategy 7. Improve Eastern's impact on the environment.

Programs	Projects and Global Impacts
Program 7A: Identify and close gaps between environmental requirements and actual performance worldwide	Projects: A1: Develop measurement and reporting system on environmental compliance

Program 7B: Begin major monitoring and development program to exceed environmental requirements worldwide	Projects:
	B1: Prepare composite company-wide report on global compliance
	B2: Identify all worldwide environmental requirements, by plant location, and compile global monitoring system
	B3: Comply with federal, state, and local environmental regulations by:
	    Providing proactive assistance to regulators
	    Educating employees about regulatory requirements
	    Promptly reporting noncompliance and correcting any violations
	    Filing a Title V air permit application
	    Participating in voluntary activities on environment, safety, and health issues, such as EPA Greenlights, reducing greenhouse gases and PFCs, and noise nuisance reduction

Strategy 8. Reduce waste and non-value-added costs.

Programs	Projects and Global Impacts
Program 8A: Implement a Six Sigma program in each plant worldwide	Projects: A1: Design Six Sigma projects around local plant needs worldwide A2: Measure performance in cycle times and cost reductions, plant by plant A3: Develop contingency and improvement projects based on outcomes
Program 8B: Conduct a global benchmarking study to identify best practices	Projects: B1: Create a benchmarking project to identify best waste reduction and Six Sigma practices and to identify best recognition programs worldwide B2: Create a process improvement program and development of process models involving improvement teams for each type of production system
Program 8C: Target equipment damage and cost issues worldwide and address them	Projects: C1: Project to minimize equipment damage by educating employees, monitoring equipment use, and enforcing rules for properly using equipment C2: Project to improve inventory management of supplies and equipment (includes maintenance, production, and raw material in-process) C3: Project to minimize waste generation and increase recycling at all plants globally

The Alignment of Programs with Strategic Objectives

The process of creating and aligning programs with company strategies involves constant management attention as global factors and conditions change. For instance, in the 2008 global economic downturn, Eastern developed a special program under Strategy 8 to reduce waste and non-value-added costs by downsizing plant capacities to new, lower demand worldwide. This meant placing a program manager in charge of developing and implementing a series of projects that specifically targeted cost reductions made necessary by a reduced worldwide demand for aluminum products. Portfolios of projects under that program were developed through brainstorming and planning activities guided by that program manager. Projects were analyzed and prioritized using net present value, assessment of strategic alignment using weighted scoring models and risk assessment. Funding depended on available funds in each budget cycle, and program managers were expected to make their business cases for program and project funding annually.

Programs Are Organizational Commitments

While projects are temporary initiatives with short-term products and outcomes, programs are longer-term company commitments that are typically reflected in appropriately named organizational units. In other words, programs are embedded in the company structure, with business officers in charge of program units that reflect program objectives. For instance, again in Strategy 8, the company established a Six Sigma program office in the corporate office to manage the multiple projects oriented on achieving Six Sigma goals.

This chapter earlier stated five program strategies for economic recovery; here is how Eastern handled each one of them.

1. Design programs that can generate jobs.

   - All of its contingencies involving IT systems were acquired locally in target countries so that local companies could compete for the business.

   - Eastern localized its program teams to be closer to target markets for aluminum, thus generating a need for supporting businesses, e.g., facilities, communication, logistics.

2. Frame program benefit goals that align with economic and business growth.

   - Eastern developed partnerships with several companies worldwide that shared Eastern's interest in bringing the price of aluminum back up to market levels, thus creating a consortium to spur more profitability in all aluminum-producing companies worldwide.

   - Eastern promoted its new production system initiatives worldwide to align with several western countries' goals of reducing energy consumption and encouraging increases in gross national product.

3. Target programs on economic and technology goals that align with applicable foreign government objectives and recovery programs.

   - Eastern aligned its robotic production team with several foreign governments interested in developing their local R&D capacity, thus helping to stimulate local innovation and business development.

   - Eastern aligned its raw material development programs to benefit countries that had both raw material reserves and a national interest in using raw materials to improve energy efficiency and create new jobs around energy independence.

4. Target technologies that stimulate local development, manufacturing, and marketing competencies.

   - Eastern worked with selected national governments to set up marketing teams with local industries to allow an exchange of marketing strategies and information systems with locals.

   - Eastern worked with local safety experts and academic institutions in target countries to develop joint projects to improve aluminum manufacturing safety and efficiency.

5. Frame programs as recovery initiatives.

   - Eastern actually defined its power-saving program as a global recovery initiative that would benefit several national economies.

- Eastern worked with several international union organizations to educate and inform leadership in ways to improve worker efficiency and productivity while upgrading old aluminum production facilities.

Postscript

Eastern's strategic plan was designed as a guidepost for the future, a way to realize the vision of becoming more responsive to changes going on globally, more supportive to customers and employees, and more cost-effective in manufacturing processes. However, it was not a "cookbook" for success. The company recognized that management and employees would continue to have to make informed judgments each day to make the plan work, and they would have to learn better from their successes and mistakes.

Global Program Risk Management

Introduction

In Chapter 4, we addressed the strategic perspective on global program management and provided case examples of how companies align strategies with programs and projects. We found that it was possible to tie broad project benefits to strategies and to identify and integrate individual projects to implement those programs. In this sense, a *program* becomes a bridge between very broad business strategies and very narrow project deliverables, as well as an organizational framework to generate, evaluate, select, fund, and manage multiple projects from a high level in the business through global program managers.

Program Risk

In this chapter, we address risk as a program issue. We discuss some critical program risks and we also cover the PMI PMBOK standard on project risk because the standard applies directly to program risk, but at a higher level.

Risk is uncertainty and represents the potential failure of a program. For a project, risk represents what can go wrong in meeting quality, schedule, and cost goals. But for a global program, the implications of risk are far more significant because they extend over many years and are generated by worldwide factors largely out of the control of the program manager. Because the global context of business presents unique challenges to program managers—e.g., unforeseen political and/or economic disruptions—managing program risk is the key to successful program management. This chapter addresses global program risk as a *distinctly different kind of risk* than that faced in traditional, short-term projects. Uncertainty seems to magnify when played out in a foreign environment because of structural and more subtle process forces that are often at play in foreign localities.

Beyond traditional project management risk, global program management is characterized by unique risks that must be anticipated and managed in a global context. They are:

1. *Unstable political and/or jurisdictional issues.* These issues include the potential for changes in local political or nongovernmental organizational (NGO) dynamics that can affect a global program. For instance, an international cable or satellite station installation is funded partially by NGO funding, but then the NGO is displaced by the state NGO licensing body. Sponsor funding is thus reduced in the middle of the program, jeopardizing its completion.

2. *Incompatibility of technical systems, codes, standards, etc.* This risk issue is often associated with high-technology programs and projects. For instance, an avionics company building a highly technical aircraft cockpit instrumentation suite for aircraft built in another country must design the instrument to current International Standards Organization (ISO) safety requirements, but if a local jurisdiction or region chooses to establish its own, more stringent, requirements during program development, project deliverables must be changed midstream.

3. *Language-generated communication impacts on virtual team performance.* This kind of risk stems from the inevitable complications of communicating in multiple languages, with loss of meaning and intent in interpretation and attendant delays in action. The agility and responsiveness of program managers can be severely hampered by communication issues when translators are required for every personal transaction.

4. *Unstable local sponsor or client.* Local sponsors or stakeholders of global programs can create disruptions that can undermine and inhibit program management success. This is because clients and customers who are not nationals of the home country in which the project originated will respond to local pressures and issues rather than program needs when faced with difficult trade-offs.

5. *Differences in sense of urgency and work ethic.* Local program management in foreign countries is subject to different values and ethics regarding work and productivity. While schedule, quality, and scope requirements are more or less addressed consistently in one country, people respond differently to schedule expectations in different cultures. A good example is the difference between the European and American concepts of the eight-hour day. While Americans see eight hours as comprising a full, continuous workday, with a one-hour or half-hour lunch, Europeans see eight hours as a three-hour morning and a five-hour afternoon and evening separated by a two-hour lunch. While this can enhance program success, it typically represents a dilemma to a program manager when it comes to estimating and monitoring daily work.

6. *Differences in contractual and financial arrangements.* While western program managers are comfortable with written contracts, many eastern

managers do business on a handshake. When difficulties arise and potential contract provisions are violated, global program managers are likely to find it difficult to resolve contract disputes by reference to contract provisions alone. In many countries, program managers are expected to pay *special payments* to those in positions of authority, e.g., *bribes*. These costs, if necessary, must be booked and reported.

7. *Unstable supplier chain issues.* Developing a supply chain system to support global program management involves partnerships and contracts with many suppliers in foreign countries. These suppliers may commit to standards of quality and timeliness specified in a project contract, but may be more unreliable than domestic suppliers, even though they may be less costly.

8. *Logistical and transportation problems.* Simple delivery and logistics issues involving labor, supplies, or equipment may be complicated because of global factors. These factors could include lack of safety and security controls, lack of local reliability incentives, and changing logistical priorities and expectations created by local conditions such as weather, local civilian unrest, and governmental intervention.

9. *Lack of integration of projects in program structure.* Programs are made up of interdependent projects and support systems. For instance, a major telecommunications program involving many countries will include hardware and platforms, software, training, and maintenance issues. If these individual projects are not integrated, the program may well end up unsuccessful even though each of the projects may have met their quality, schedule, and performance requirements.

10. *The effects of global financial crisis.* Finally, global economic issues, such as the worldwide recession of 2008, can have a debilitating impact on program success. For instance, a critical supplier can go out of business because of lack of financing or an unanticipated downturn in local demand. Or a local government in a centrally planned economy intervenes to take over a supplier company or provides state-generated funding and *strings* that are inconsistent with program assumptions or conditions.

Risk Management

Global program management requires a high level of business analysis and risk management. Programs create risk and opportunity worldwide, thus the quality of business analysis for program risk becomes an important early consideration for program managers. In a global environment, you have the added complexity of a three dimensional risk matrix. You have the usual X- and Y-axes with risk potential, and impact, but then you have the global dimension. There is the added complexity of risk in the dimensions of distance, communication, local and global environmental conditions, world economic issues, currency fluctuations, political unrest in certain areas, and access to needed resources.

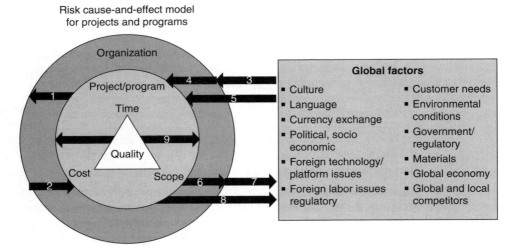

Risk cause-and-effect model
for projects and programs

Figure 5-1 Risk cause-and-effect model for projects and programs.

In Fig. 5-1, the complexity of risk is illustrated. The larger circle represents the organization where the project or program resides. The smaller circle represents the project/program that is located within the company. The triple constraints with the program or project are inside the project/program circle. The box represents external global factors that could contribute to risk management needs.

Each arrow in the diagram represents a different type of risk category that needs to be identified, analyzed, prioritized, and managed appropriately with a suitable response plan.

- Arrow 1: risk factor(s) from the project/program that affect the organization
- Arrow 2: risk factor(s) from the organization that affect the project/program
- Arrow 3: external risk factor(s) that affect the organization, which leads to arrow 4
- Arrow 4: external risk factor(s) that affect the organization and thus affect the project/program
- Arrow 5: external risk factor(s) that directly affect the project/program and not the organization
- Arrow 6: project/program causes risk to the organization, which then leads to arrow 7
- Arrow 7: the change to the organization from a project/program risk, which then causes risk to external factor(s)
- Arrow 8: project/program causes risks with external factor(s)
- Arrow 9: risk factor(s) that can occur within the project/program only

There are many examples on the types of risks that you could find with each of these arrows. For instance,

- Arrow 1: risk factor(s) from the project/program that affect the organization
    - Underestimating resource requirements needed for the program
- Arrow 2: risk factor(s) from the organization that affect the project/program
    - Micromanagement from top executives, or loss of support from top executives
- Arrow 3: external risk factor(s) that affect the organization, which leads to arrow 4
    - Cultural differences between different partner companies, which affect how negotiations or contracts are managed
- Arrow 4: external risk factor(s) that affect the organization and thus affect the project/program
    - Contract performance terms and conditions variances
- Arrow 5: external risk factor(s) that directly affect the project/program and not the organization
    - Foreign Technology Capacity
- Arrow 6: project/program causes risk to the organization, which then leads to arrow 7
    - Poor team performance or fraudulent practices, reflecting on the organization
- Arrow 7: the change to the organization from a project/program risk, which then causes risk to external factor(s)
    - Inability to negotiate successful contracts
- Arrow 8: project/program causes risks with external factor(s)
    - Inadequate product safety testing in light of local regulatory requirements
- Arrow 9: risk factor(s) that can occur within the project/program only
    - Inadequate use of earned value in monitoring the project

This risk cause-and-effect model is a very valuable tool in identifying risks that affects a global program. This information is also beneficial for a PMO that needs to support a program and provide guidance and direction in risk management planning:

Sample program risk matrix

The following is an example of a risk matrix built on the global risks discussed previously:

Program Risk	Impacts	Probability	Severity	Contingency Action
Unstable political and/or jurisdictional issues	Undermine program delivery itself Fundamentally change program assumptions	High in politically unstable regions	Can be a showstopper	Be prepared to brief new sponsors or players locally on program issues and to justify the program in view of alternative expectations
Compatibility of technical systems, codes, standards, etc.	Schedule and quality problems in program output or product performance	Low, unless program is in a new technology arena where standards are not yet developed	Medium, since changes in technical standards, etc. would typically not be applicable for several years, hopefully after program is completed	Close working relationship and involvement in all international standard-setting organizations that might affect program success
Language-generated communications affect virtual team performance and customer expectations	Delays from communication problems; affects program management and outcomes due to lack of candid and effective discussion in program design and implementation	High, given extensive communication on program issues across countries and cultures	High, given the importance of consistent understanding of reports and verbal statements of significance for the program	
Unstable or inconsistent local quality, safety, and/or regulatory requirements	Unsatisfied customers; lack of customer acceptance of program product and outcomes; product or output violates local regulations	High because of typically unstable and inconsistent quality and safety standards in many global locations	Could be a showstopper if program output violates local regulations	Place local safety and/or quality representative on program team, and integrate compliance with local requirements
Unstable local sponsor or client; potential for unanticipated change in key program clients	Can be a showstopper if change in sponsor or client changes key program scope	Low, if program manager monitors client and can anticipate changes in client structure	Could be a showstopper	Have alternative scenario and program presentations for potential new client representative
Differences in sense of urgency and work ethic	Schedule delays; changes focus from on-time delivery to quality delivery	High, since different cultures exhibit different values regarding timeliness and work ethic	Low, mostly in scheduling delays	Use critical chain concept, e.g., focus on project and task starting times, not finish dates
Differences in contractual and financial arrangements	Get templates from local jurisdictions to analyze potential problems in contracting processes and procedures	High, since contractual arrangements are different throughout the world	Low, if program manager monitors local contract processes	Develop alternative contractual templates and be ready for different interpretations of contract provisions

(*Continued*)

Unstable supplier chain issues	Schedule and cost impacts; potential quality issues in product components, configuration management systems	High, because suppliers are subject to unanticipated forces, e.g., economic, financial, social	Medium, mostly in schedule impacts from required supplier contract renegotiations	Write long-term supplier contracts and develop partnerships that can endure global impacts
Logistical and transportation problems	Schedule impacts from delays in local supply, equipment, and resource deliveries	Low, since logistical and transportation problems can be resolved in a timely way	Low, delays in project schedules	Develop alternative transportation and logistical arrangements, and write contingency contracts to be triggered by failure of key contractors
Lack of integration of projects in program structure	Project schedule, cost, and quality impacts because of lack of articulation and sequencing of projects within the context of overall program	High, since complex program and project interfaces can be obscured by narrow focus on project tasks	High, since if projects are not coordinated, the results can be showstoppers	Create a project interface council to oversee articulation of project milestones
Global financial crisis impacts	Can terminate key projects if local partners have to change their priorities because of lack of financing or customer demand	High, since current global financial and economic recessions, such as the one in 2008	High	Have alternative scenarios ready to change pace and scope of program if necessary

Programs suffer the consequences of accumulated risks in their projects. Sometimes, these project risks are more severe combined than they are when seen in the context of one project only. For instance, let's say that a program to equip a Middle Eastern region with cable for a wide variety of communication technologies has two main projects: a cable production project and a cable installation project. Problems in production and installation created by local governmental standards can disrupt the whole concept on which the program is designed. While the delays in each project can be addressed and fixed, the whole program is in jeopardy when both risk events occur.

This disruption is created when the local sponsor or client loses confidence in the business's capacity to perform. If the first project, the development of the cable product, were to have gone well and served to build client confidence in the program, some delays in installation might have been tolerated. And if the product development project were to have experienced issues but the program manager was able to accelerate the installation and create client confidence, again the client might have accepted the program. But if *both* projects go wrong, the program manager is faced with a difficult dilemma. In a domestic program, one that is being implemented in the home country, the process of resolving these kinds of issues is simpler—there are no language or culture problems, no foreign regulatory or governmental involvement. But in a global program, the client/sponsor of the program is likely to be difficult to deal with, especially if there are international issues between the target country and the home program country.

Development of contingency plans is made difficult as well, since contingency plans involving project installation may involve negotiations with local authorities that were not planned. For instance, if the cable installation requires more land acquisition or leasing than originally anticipated, the global program manager is likely to find himself or herself in the local offices of the land management department of the local governmental jurisdiction. In the home country, these discussions would be relatively routine, but in a foreign country with different legal and regulatory requirements and different land-use traditions, the acquisition or leasing of new land for cable installation could be a showstopper for the program.

Program Risk Management According to PMI

The PMI PMBOK guide on risk management applies to programs as well as to projects. The PMBOK defines risk management as the "systematic process of identifying, analyzing, and responding to project risk." The concept aims at maximizing the probability and consequences of positive events and minimizing the probability and consequences of adverse events to project objectives.

It is important to see the PMBOK as a guide, not a manual. While the following discussion is guided by the PMBOK, the reader will quickly see that each section is framed by a *reality check,* the author's personal and professional view of best practices in the real world of *faster, better, and cheaper.* The discussion presupposes a separate risk management planning process, which serves as an ideal. In practice, risk planning occurs as *an integral part of project planning.*

Issues involved with program risk

The current PMBOK has raised several issues with regard to risk.

Process.   We are learning that while a process focus is important, it misses the opportunity to integrate risk into current business and program planning and management actions. Process focus is useful for ensuring quality and discipline, but it has limitations in practical work settings.

Separate process.   The propensity to break down the project planning and control process into components misses the actual dynamic in real organizations at the program level. Risk has not proven to be useful as a separate process, but rather, is effective only if integrated into portfolio and program management processes, beginning at the highest level of the business.

Single project.   Risk is no longer looked at as a single project issue; most risk is associated with broader issues, such as global program management and the business itself, and other projects in the company portfolio.

Quantitative.   Overemphasis on quantitative tools and mathematical models suggests risk management is a science rather than an art. Most program risk management actions are full of judgment and margins of error, which make quantitative tools ineffective and intimidating.

Focus on methods, not people.   The risk process is essentially a thought process, a way or style of management that is ingrained in the way people work and solve project problems.

The overemphasis on methods and the undervaluing of the human element limits the application of the current PMBOK as it pertains to risk.

Assumes inputs.   The assumption in any process focus is that the inputs will be there; however, in many cases, the necessary inputs are not there because the whole concept of a separate risk management process is flawed when looked at in practical terms.

Unrelated to cost.   Risk and cost are inextricably intertwined; when risk events occur globally they require expensive contingency or mitigation actions that are not often included in the baseline budget (this is why all actions designed to mitigate a risk should be included in the baseline schedule as a reserve item in case they need to be implemented).

Unrelated to quality.   Risk and quality are also inextricably connected, since risk affects quality. Many risks are associated with feasibility, not schedule, and some projects may not be capable of meeting customer standards and specifications in the first place.

Ignores business risk.   Project risk is first identified and managed at the business-wide level through strategic and business planning, leading to the generation of program areas and projects.

Separate contingency planning.   The current PMBOK describes contingency planning as a separate process, but in fact, to be effective, contingency actions need to be incorporated into baseline schedules and budgets. The project manager must ensure that the schedule has buffers and contingency tasks built in to it.

Ignores risk as opportunity.   The other side of risk is opportunity—the ability to control risks creates opportunity because the competition cannot. Thus, any project aimed at capturing a market share is designed to create opportunity, and risk management is the strategic approach.

The following table provides an overview of the current PMBOK risk processes:

TABLE 5.1   PMBOK Risk Management Processes

5.1 Risk Management Planning	5.2 Risk Identification	5.3 Qualitative Risk Analysis
Inputs	Inputs	Inputs
Project charter	Risk management plan	Risk management plan
Organization's risk management policies	Project planning outputs	Identified risks
Defined roles and responsibilities	Risk categories	Project status
Stakeholder risk tolerances	Historical information	Project type
Template for the organization's risk management plan		Data precision
Work breakdown structure		Scales of probability and impacts
		Assumptions
Tools and Techniques	Tools and Techniques	Tools and Techniques
Planned meetings	Documentation reviews	Risk probability and impact
	Information-gathering techniques	Probability/impact risk rating matrix
	Checklists	Project assumptions
	Assumptions analysis	Data precision ranking
	Diagramming techniques	
Outputs	Outputs	Outputs
Risk management plan	Risks	Overall risk ranking
	Triggers	List of prioritized tasks
	Inputs to other processes	List of risks for additional analysis and management
		Trends in qualitative risk analysis results
5.4 Quantitative Risk Analysis	5.5 Risk Response Planning	5.6 Risk Monitoring and Control
Inputs	Inputs	Inputs
Risk management plan	Risk management plan	Risk management plan
Identified risks	List of prioritized tasks	Risk response plan
List of prioritized risks	Risk ranking of the project	Project communication
List of risks for additional analysis and management	Prioritized list of quantified risks	Additional risk identification and analysis
Historical information	Probabilistic analysis of the project	Scope changes
Expert judgment	Probability of achieving the cost and time objectives	

(Continued)

TABLE 5.1   PMBOK Risk Management Processes (*Continued*)

Other planning outputs	List of potential responses	
	Risk thresholds	
	Risk thresholds	
	Common risk causes	
	Trends in quantitative and qualitative analysis results	
Tools and Techniques	Tools and Techniques	Tools and Techniques
Interviewing	Avoidance	Project risk response audits
Sensitivity analysis	Transference	Periodic project risk reviews
Decision analysis	Mitigation	Earned value analysis
Simulation	Acceptance	Technical performance measurement
		Additional risk response planning
Outputs	Outputs	Outputs
Prioritized list of quantified risks	Risk response plan	Workaround plans
Probabilistic analysis of the project	Residual risks	Corrective action
Probability of achieving the cost and time objectives	Secondary risks	Project change requests
Trends in quantitative risk analysis results	Contractual risks	Updates to the risk response plan
	Contingency reserve amounts needed	Risk database
	Inputs to other processes	Updates to risk identification checklists
	Inputs to a revised program plan	

5.1 Risk Management Planning. Deciding how to approach and plan the risk management activities for a project

5.2. Risk Identification. Determining which risks might affect the project and documenting their characteristics

5.3 Qualitative Risk Analysis. Performing a qualitative analysis of risks and conditions to prioritize their effects on project objectives

5.4 Quantitative Risk Analysis. Measuring the probability and consequences of risks and estimating their implications for project objectives

5.5 Risk Response Planning. Developing procedures and techniques to enhance opportunities and reduce threats to the project's objectives

5.6 Risk Monitoring and Control. Monitoring residual risks, identifying new risks, executing risk reduction plans, and evaluating their effectiveness throughout the project life cycle

These processes interact with each other and with the processes in the other PMBOK knowledge areas. The way they interact is the key to integrating risk management with the basic project planning and control processes. The following sections discuss the salient points of integration.

Risk Management Planning

Risk management planning for a particular project is inextricably connected to how the organization prepares for dealing with risk and uncertainty in its business development and strategic planning, in its information technology investments and management of network communications, and in its organizational structure. No project manager faces risk alone—it is a company-wide issue, and it is quite likely that there is data and information on project risk in the company files.

A company prepares for risk in projects by assessing the overall risk in strategic planning. Then, in its development of a program of projects, or portfolio, the company assigns risk to a general program or product line as part of its decision to proceed. Templates for risk identification and assessment are available, and a project management office is sometimes available to support risk management by providing information templates, project review data and agendas, research findings, and historic information on various risk subjects.

To address risk effectively, there needs to be an information technology capacity in the company to assure that risk documentation and tracking can be achieved within the network and software assets available. This means a way to organize risk data, for team members to communicate risk information quickly, and for project managers to present risk data in acceptable formats.

Finally, if the company is not organized to address projects in some kind of project structure, project risk will get the same kind of attention other project issues such as cost and schedule get—very little. In a matrix structure, for instance, risk is addressed in the functional department in terms of processes and equipment through testing and monitoring technical processes. At the same time, the project manager is attuned to risk when schedules are delayed because of events or developments grounded in it.

All of this does not need to complicate the up-front risk process. *Risk is still a relatively straightforward concept of planning and controlling for things that could go wrong.* That risk management planning is tied to business planning and strategy does not need to suggest that this linkage complicates the achievement of a good risk management process. In fact, it makes project risk management easier in the sense that business planning itself provides a precursor of project risk. If I am in the avionics business, and I design and produce avionics equipment for business jets, and my overall business plan identifies major threats to profitability and success—e.g., the global availability of cheap LCD monitors, or the pending change in air traffic regulations affecting avionics—then my individual projects begin with a major challenge in those two areas. As a project manager in that scenario, I enter the project arena with a *built-in,* up-front view that I must keep my eye on both these factors and plan and schedule contingency and risk mitigation actions as part of my project plan.

The PMBOK process view emphasizes the inputs and outputs of the risk process. This is a useful perspective on risk, even though it does not really deal with the integration of risk management into the project planning and control process. The PMBOK is based on the highest level of "maturity" in an organization, a scale that is embodied in the PMI maturity model. Thus, the PMBOK process is idealized in a *mature organization,* and rarely found in all of its dazzling performance dimensions in a normal business environment.

Inputs to risk management planning

Project charter.   The project charter is an ideal project planning document that includes the business need and product description. In reality, this document is often neglected because the content for the business need is still in conceptual stages. And at this point, the deliverable is often undefined and unspecified; thus, the product description is not at the same level of detail as a configuration management document. The deliverable is defined in performance terms at the scale possible given the understanding of what is being designed and built. The purpose of the project charter can be realized in other ways, e.g., a customer requirements document and a scope of work document.

Organization's risk management policies.   If the business has a set of risk management policies and procedures, they would be used in risk planning. However, many businesses do not have such policies, nor will they; rather, they apply risk tools as an implicit part of the project planning process. Approaching this process input with a healthy skepticism, one can see the value of writing down policies and procedures, but in today's fast-moving companies, this is rarely done. The point is that a nimble mid-sized business of today that has articulated its approach to risk would expect each employee, and certainly its management, to embody that approach in the basic project management process, without a bureaucratic statement of top-down policy.

Defined roles and responsibilities.   One would hope that the basic roles and functions of the project manager and functional manager are documented, but again, this is often left undefined in order to allow a natural process of negotiating and working out roles between functional quality and project delivery interests.

Stakeholder risk tolerances.   The PMBOK is not very helpful in illustrating this input. Stakeholder risk tolerances are evidenced in modern business as "world views" of certain important people in the process—e.g., sponsors, customers, investors, top management, regulators, etc. The tolerances for stakeholders can be reflected in a spectrum of philosophies from risk seeking, to risk neutral, to even risk adverse.

A risk tolerance for an electronic instrument might be framed as a technical tolerance (mean time between failures), a performance tolerance (must perform in below-zero temperatures), or a drop-dead limit (investors will not proceed with this project if, by this time next year, there is no first article production unit) because of the anticipated rapid change in market conditions. Tolerances are often grounded in expectations; thus, it is important for a project manager to *see* such tolerances and evaluate their intensity early in the project.

Work breakdown structure.   Certainly, a work breakdown structure is necessary as a basis for identifying risk, but the WBS must be comprehensive and show all tasks before the risk identification process can really be effective. If the WBS misses some important work that is highly subject to failure, it is not a good input to risk management planning.

Issues not addressed in PMBOK

Some risk issues are not addressed explicitly in the PMBOK, but you should be aware of them to have a full understanding of risk.

Building a risk-based organizational culture.   Building an organization that protects itself from project-level risk and uncertainty through good organizational planning and management requires strong leadership. As a project manager, you have to first *feel* that you are expected by management to anticipate and deal with risk and that you will be supported in taking the time and investing the cost to build a good risk planning and control process. Risk management starts at the top leadership level with a clearly articulated vision and mission that incorporates an uncompromising commitment to quality and excellence. In doing so, the leadership commits also to a risk management process to reduce the probability of failure and to promote total quality in product design and production.

Program and portfolio management.   The management of risk in a multiproject environment, and the role of risk in selecting and maintaining a portfolio of balanced projects, is not really addressed in the PMBOK. Yet the selection of the *right* projects for the project pipeline inherently involves risk management. Projects with high risks must be identified before they enter the approved list, simply because a portfolio of high-risk projects endangers the long-term growth and profitability of the enterprise.

Interface management.   Good risk management is dependent on the availability of effective support and interface services to project managers. Risk cannot be seen simply as a project management issue; it is an accounting and cost issue, a procurement issue, and an information technology issue. Cost data must be available to project managers to assess cost impacts; contract officers must be attuned to risk and risk-sharing issues when managing contractors and supply vendors; information technology administrators must see the need for web-based, easily available, risk matrix templates and calculations software.

Risk and cost integration.   For some reason, the relationship between cost and risk is lost in the daily routine of project managers, yet it is in cost and "expected value" that future impacts of risk decisions can be made. Contingency plans add to project cost estimates, and when clear decisions must be made and crossroad trade-offs to be decided on, a support system must be available.

Tools and techniques for risk management planning

Planning meetings are important and can be fruitful, but they can also be a complete waste of time unless they are focused and deliver results. Setting agendas, facilitating meetings, and writing good follow-up notes are all useful tools.

The PMBOK treatment of risk management planning does not cover some of the most important risk management planning tools, namely:

1. *Business plan.* A grasp of the business plan helps a prospective project manager get an early start in project risk management planning. Such a plan, or a business strategic plan, will provide strategic information, e.g., SWOT (strength, weakness, opportunity, and threat) data and information.

2. *WBS.* The WBS is an early indication of the potential risks in any project, and planning for project risk requires at least an outline of the WBS to see the basic components of work involved. Risk management planning requires the project manager to anticipate how risks will be handled by looking at the WBS, or by building one.

3. *Information and network systems.* A major risk management planning tool is the company network and information-sharing system, as well as data already available from similar past projects.

Outputs from risk management planning

The risk management plan outlines the approach to how risks will be handled in the project. Frankly, many companies do not need a separate risk management plan, but they should integrate risk information into the project scope, WBS (definitions), schedule, and budget. For example, a rule of thumb for an organization could be: if the project under consideration has more than 300 tasks and is budgeted at more than $5 million, a separate risk management plan is called for.

The plan includes:

1. *Methodolog.* It is not clear from the PMBOK what the "methodology" of the plan is, but it appears that the concept says you should have a methodology. For instance, an electronic instrument production firm should use safety and reliability tools—e.g., mean time between failures—to test its prototypes to avoid the risk of performance variation and failure.

2. *Roles and responsibilities.* Here, the plan addresses who is responsible for what in a functional and project management context. This is where the role of a program management office is described.

3. *Budgeting.* This exercise estimates the cost of risk management. Frankly, it would be more important to spend time analyzing the cost of the risk event. The cost of risk management is a program management cost category, as in quality assurance and project review.

4. *Timing.* This addresses when various risk management actions will be taken in the project schedule, such as risk analyses, preparation of contingency plans, and response plans.

5. *Scoring and interpretation.* This portion of the risk management plan addresses tools such as the weighted scoring model (aligns projects for selection with business strategy, places priority weights on various strategic objectives, and scores each project against the strategic objectives), cashflow, rate of return, and net present value.

6. *Thresholds.* Thresholds address the criteria, or rules of thumb, for acting on risks or to reduce them—e.g., deploy preventative contingency and response plans for risks that could delay a project by more than 10 percent of the total project duration, unless the risk is reduced in the first quarter of the project.

7. *Reporting formats.* This provides guidance for the project manager on the formats of project reports for stakeholders—e.g., e-mail, Microsoft Project team reporting, or hard-copy spreadsheets.

8. *Tracking.* This provides guidance on what risks will be tracked and how—e.g., the acquisition of microchips for an electronic instrument will be tracked with the contractor on the basis of earned value.

Risk Identification

Risk identification should be part of the project planning process, not separate from it. Risks are identified in the development of the WBS, in estimating duration and resource needs, and in linking tasks.

Inputs to risk identification

To the extent that a risk management plan is produced, it is an important input to identifying risks. But since most companies will start the process of handling risk with the WBS and the task list, the identification of risk usually starts in earnest in the review and final production phases of the generic WBS. It is here that the project manager *reviews every task for its potential for failure.* Identifying risk involves a lot of discussions with team members and stakeholders. For instance, if a technical task in the WBS, say, achieving a given mean time between failures in a product component, "sticks out" initially because of the challenges of completing it, then it is the subject of much discussion and contingency planning early in the project.

Project planning outputs

Project charter.   If there is a charter, it should be helpful in identifying risks and confirming the judgment of the project manager on where the project's vulnerabilities are.

WBS.   The WBS is the basic source of risk identification activity, since it embodies all the work of the project—or should. The WBS should be four levels deep to give enough detail to the project profile to *see* project risks.

Product description. The product description will be embodied in a configuration management document or in some document/drawing/specification that defines the product from a performance and component perspective. This will occur in the design phase at some point when the deliverable has been fully fleshed out.

Schedule and cost estimates. The schedule and project budget (part of the schedule in Microsoft Project) will be a good source to confirm risks, but first, the project manager must prepare a risk matrix as described earlier. The risk matrix identifies the task, the task risk description, and the impact (schedule, cost, etc.).

Resource plan. While there is typically no formal resource plan, there will be a sense of the personnel, equipment, capital, space, and technology needs of the project. Ideally, this resource plan is in one place—e.g., in Microsoft Project and/or in a planning document of some sort. But at this point, if *all* the resources needed for the project are not clear, it is not critical. What *is* needed here is a clear idea of the "bottleneck" resources, those resource issues that could represent a barrier to achieving the project. These bottleneck resources might be a critical software engineer who is already spread too thinly in current projects, a piece of testing equipment that is critical to meeting quality control thresholds, or a workspace or station that is being shared with other projects.

Here, also, is the beginning of the application of the theory of constraints. Simply put, the theory states that the focus of attention in planning and control should not be on the whole project and all its tasks, but rather, on the one or two major resource constraints. The project manager protects against the risks inherent in these resources by tapping time and cost from the original estimates and withholding them for allocation when they are needed.

Assumption and constraint lists. The assumption list is a convenient way of indicating what assumptions are controlling—e.g., the assumption that a sole source contract with a foreign supplier for a key product component will last through the project life cycle. In practice, these assumptions are well known by the project team and stakeholders, but it pays to document them and revisit them in project review sessions, and to treat them as risks with contingency plans.

Risk categories

Technical. Technical risks have to do with product, process, or "technique" issues involved with designing and producing the deliverable.

Project management risks. Project management risks address the things that can go wrong with the project planning and control process, and with expected support services from the information technology source and a project management office, if the organization has one.

Organizational risks. These are the "soft" issues that a project faces. They have to do with organizational behavior and dynamics—e.g., conflicts, scarce resources, personnel performance problems, or company-wide crises, such as lack of financing or a downturn in share value. A key organizational risk is the lack of top management support, which will be evidenced by neglect of the project in the company "head shed."

External risks.   External risks are the business and global risks inherent in any business—e.g., economic downturns, trade difficulties affecting the deliverable, and the impacts of communication in multinational companies.

Historical information.   Historical information includes past project documents, lessons-learned reports, and industry information available on the competition, on demand and market issues, and on the company's performance on similar past projects.

Project files.   Project files are typically available from the company's file system, but in practice, project managers rarely look back at these, even though it would make sense to do so.

Published information.   This includes manuals, articles, and technical publications on the deliverable.

Tools and techniques for risk identification

Documentation reviews.   The key document in risk identification is the WBS. Other documents would include past project reviews and similar product performance information.

Information-gathering techniques.   These would include web-based information, electronic files with product information, and Internet research.

Brainstorming.   Brainstorming is simply having meetings with key people who know something about the project and generating ideas and options without judging them. Brainstorming generates ideas and does not filter them.

Delphi technique.   Delphi is brainstorming with key experts, who go through a systematic process of providing their views, reviewing each other's ideas, and coming up with a scenario based on the integration of their views.

Interviewing.   Interviewing key stakeholders, past project managers, and task managers helps to uncover "subtle" information that has not been documented.

Strengths, weaknesses, opportunities, and threats.   Here, we look back at the business planning processes for strategic analyses, especially including threats and opportunities.

Checklists.   Checklists are typically prepared by documentation specialists for various project and product documents. They often key into potential failure points in past projects and thus are useful in identifying risks.

Assumptions analysis.   The key source of assumptions is rarely captured in one document, but the concept of focusing on assumptions is important. A project management office is typically in charge of documenting assumptions.

Diagramming techniques.   Flow charts and diagrams, such as decision trees, are useful in identifying the various options and decisions, including expected value.

Cause and effect.   "Root causes" of risks can be identified through fishbone diagrams and other such meeting techniques.

Influence diagrams.   Diagrams that indicate cause and effect, as well as the influence of key factors, can help in identifying risks.

Outputs from risk identification

Risks.   The output of risk identification is a better sense of risks over time. In truth, the clarity of risks increases as the project life cycle progresses.

Triggers.   Triggers include those indicators or signals of risk events that become clearer in the risk identification process. For instance, in a contract negotiation in the outsourcing process, a contractor refuses to sign the contract because of a schedule requirement that stipulates a key supply or piece of equipment must be delivered before a key project milestone.

Qualitative Risk Analysis

The PMBOK separates the risk analysis process into two parts: qualitative and quantitative. In practice, companies rarely split the two. The point of risk analysis is to hone in on potentially high-risk tasks and get a more detailed picture of their impacts.

Inputs to qualitative risk analysis

Risk management plan.   The risk management plan contains an analysis of probable risks, the impact of the identified risks to the project or program, as well as response strategies to help the project or program avoid being disrupted should an identified risk event occur. Risk management plans are a living document and should be periodically updated.

Identified risks.   A completed risk matrix is required before risk analysis proceeds.

Project status.   The particular timing of a risk analysis is important because the risk impacts, particularly in terms of schedule and budget, will change, depending on when they are analyzed. The later in the project cycle, the clearer the impacts will be.

Project type.   It is important to *dimension* the risk here; a project producing an electronic instrument for a sophisticated aeronautic application, for example, will look different from a project to construct a standard building.

Data precision. The accuracy and precision of data is important; if you know that a given reliability test has a proven failure rate, then the results must be tempered accordingly.

Scales of probability. Probability is a subjective judgment, unless the product is tested many times to develop a statistical mean. In most projects, the probability of a risk occurring is the result of thinking through how many times this kind of risk has occurred in past similar projects, combined with the "gut" judgment of the project manager and key stakeholders.

Assumptions. Again, the assumptions are rarely listed, but they are apparent in the analysis process.

Tools and techniques for qualitative risk analysis

Risk probability and impact. Using the risk matrix, the project manager has already identified risks and will assign probabilities to all high-impact risks. For most risks, these probabilities are subjective and simply communicate a sense of confidence that the project manager has about the risk in question. Generally, probabilities should be stated in three forms: 25 percent, 50 percent, or 75 percent, suggesting little change of the risk occurring, substantial chance, or high chance, respectively.

Probability/impact risk rating matrix. The rating or risk matrix now is fine-tuned in the analysis process, with more data and more attention.

Project assumptions testing. Testing assumptions involves taking the time to review key assumptions and confirming that the assumption is right and that the probability assigned is in the right ballpark.

Data precision ranking. For most projects, this step is not useful. If the product is highly complex and must meet detailed performance specifications, then the data precision ranking indicates how precise the test data is compared to a common standard.

Outputs from qualitative risk analysis

Overall risk ranking for the project. Once the qualitative process is finished, two rankings are produced: how the project's overall risk is ranked compared to others (this may be completed when comparing and selecting a portfolio of projects) and how individual risks rank within the project, usually limiting the list to five or fewer.

Again, this process is related closely to the theory of constraints. A project typically faces only a few major bottlenecks or risks, and it is the job of the project manager to accurately uncover those few critical risks during qualitative analysis.

List of prioritized risks. The list of prioritized risks is incorporated in a project report to stakeholders along with supportive information, including the risk matrix and contingency plans.

List of risks for additional analysis and management. A residual list includes other risks that could turn out to be more important than they appear.

Trends in qualitative risk analysis results. Some review of the credibility of the process will uncover past analyses and how their results played out in real terms.

Quantitative Risk Analysis

Inputs to quantitative analysis

All of the inputs to qualitative analysis are relevant here, plus expert judgment. Expert judgment comes from technical experts who have knowledge at the technical level on the risk in question. For instance, here is where a project manager brings in a safety and reliability expert contractor to advise on variability thresholds in testing prototype parts.

Tools and techniques for quantitative risk analysis

Interviewing. Again, interviews with key experts on defined task risks are useful.

Sensitivity analysis. Sensitivity analysis determines how much project outcomes—e.g., schedule, budget, quality—are sensitive to particular risks. For instance, it may turn out that although a risk is only medium probability and impact, it could alter the final deliverable in a measurable way in terms of quality control.

Decision tree analysis. Decision tree analysis aims to uncover the expected value of taking one path or another when a project "crossroad" decision must be made. For instance a decision on whether to purchase land needed for a program "now or later" in anticipation of winning the contract brings with it a set of expected values for going one way or the other. If you buy the land now and don't win the contract you may lose when you try to sell the land; if you buy later, if you win the contract, you pay more for the appreciated value of the land.

Simulation. Simulations are mathematical representations of scenarios involving key project risks. This might include an equation that specifies what happens to an information technology scheme when its capacity is challenged by demand, e.g., shutdowns, performance failures, etc.

Outputs from quantitative risk analysis

Prioritized list of quantitative risks. The quantitative analysis usually places more content on the already produced list of risks from qualitative analysis. New data is presented on high risks from the analysis.

Probabilistic analysis of the project. Here, the probabilities are worked to a finer level of detail based on more analysis. A probability set at 25 percent in the qualitative phase might be fine-turned here to, say, 38 percent with more input from intensive analysis of past projects and simulations, and perhaps some experiments.

Probability of achieving the cost and time objective. A final probability is determined for meeting the project schedule and budget goals.

Risk Response Planning

Appropriately, the PMBOK places high priority on a response plan that mitigates risk. That response plan is grounded in the contingency plans already developed when preparing the risk matrix. The key point about response planning is to outline corrective action and to incorporate those actions into the baseline schedule so that they will not later be considered add-ons or changes to the project. They might later be identified with the "pessimistic" estimates of durations in the PERT analysis in Microsoft Project.

Inputs to risk response planning

Risk thresholds. Risk thresholds help identify acceptable ranges within which risks can occur without deploying contingencies—e.g., if this work is not done in the estimated time, we will give the team three more weeks before we act, since the task is not on the critical path. These thresholds come from customers, stakeholders, and technical experts.

Risk owners. Risk owners are those stakeholders who are accountable for acting on risks, or at least reporting on risk activity. If there is no designated risk owner, a risk could be unattended long after it is identified because of the tendency to avoid controversy and accountability for risks that cannot easily be controlled.

Common risk causes. Common to all industries is a set of common risk causes, e.g., government regulatory change, bad marketing information, faulty safety and reliability equipment, lack of proven competencies in particular personnel categories, etc.

Tools and techniques for risk response planning

Avoidance. One approach to a risk is to avoid it and hope that it goes away. Sometimes it does.

Transference. Transference involves turning a risk over to a risk owner, e.g., assigning a contractor the job of responding to and providing incentives for risk reduction.

Mitigation. This is a corrective action option, leading to the deployment of a contingency plan.

Acceptance. Sometimes it pays to accept a risk and deal with it directly rather than transferring it, avoiding it, or mitigating it. "Living with" the risk means plugging in schedule and budget reserves, based on the assumption that the risk cannot be controlled, and working around it.

Outputs from risk response planning

Risk response plan.   Although a formal, written risk response plan is not always feasible or wise because of the cost and effort involved, it is the *learning and thinking* process that follows from good risk response planning that gives it value. Once through the process of defining risks and planning for them and folding the results into planning documents, such as the schedule and budget, a project manager "owns" those risk responses and has incorporated them into the plan. The lack of a separate, documented risk response plan is not a good indication that risks have not been considered; it is more important that risks be incorporated into the project schedule and estimates made from "expected, optimistic, and pessimistic views," which are driven from risk analysis.

Residual risks.   Residual risks are risks that continue to exist after corrective action. Sometimes, residual risks are created by taking a corrective action that was not anticipated in the original project planning process.

Secondary risks.   These are lower-level risks that have less impact but which can grow in importance if neglected.

Contractual agreements.   The type of contract used in outsourcing work involves some explicit assumptions about risk transference. For instance, a fixed-price contract is superior to a cost-reimbursable contract in transferring risk to a contractor to control a given risk in performing contracted project work.

Contingency reserve amounts needed.   Sometimes, a project must be protected with a reserve fund or insurance program so that the company is not financially exposed due to a given risk even though it is mitigated.

Risk Monitoring and Control

The PMBOK places emphasis on monitoring risks and controlling them, but this process is again an integral part of the project review and control process.

Inputs to risk monitoring and control

In addition to the common inputs addressed previously, the new inputs to monitoring are communication and scope change.

Project communication.   Communication means exchanging information on anticipated risks so that people who have a stake in the project can assist in mitigation and adjust their expectations for the project. Communication always helps to reduce the uncertainty and "surprise" factor in dealing with customers, clients, and stakeholders.

Scope changes.   As the project is progressing and work is getting done, new information may be uncovered in the risk management process that requires a change in the scope, schedule, budget, or quality standards. Thus, scope change is a logical outcome of monitoring and seeing risks that affect the project.

Tools and techniques for risk monitoring and control

Project risk response audits.   A project risk audit looks at how effectively project management processes in general were handled and how well project risks were monitored and mitigated.

Periodic risk reviews.   Risk review occurs in the normal project review process, not as a separate process. A standard project review agenda always includes a section on risks.

Earned value analysis.   Looking at schedule and cost variance always indicates the possibility that risks are at work, since the project is not performing as planned. Corrective action when uncovering major earned value variances over 10 percent, for example, would include review of risk contingencies and impacts.

Technical performance measurement.   Technical issues could be at work in a project that is slipping or meeting insurmountable obstacles.

Outputs from risk monitoring and control

Workaround plans.   The so-called workaround plan is a manifestation of the risk mitigation process. The workaround is a contingency that should be identified in preparing the risk matrix. It sometimes takes advantage of innovative, creative options to overcome a project risk, and is often the result of "out-of-the-box" thinking that can be generated in a brainstorming session.

Corrective action.   Corrective action is the action taken when the project is not performing to plan, whether due to problems in schedule, cost, or quality.

Project change requests.   Often in monitoring, the need to fundamentally change a project scope or key deliverable occurs, thus triggering a change request.

Updates to the risk response plan.   Updates to the risk response plan result from lessons learned in the monitoring process.

Risk database.   The risk database is a documentation of risk information that will be useful in corrective action and future projects.

Updates to risk identification checklists.   Updates to risk identification checklists help keep tabs on best practices in risk mitigation as they are generated.

When risk is poorly managed there is potential failure to the project, program, and possibly the viability of the organization. Risk is incorporated in the budget, schedule, and specifications, which when successfully combined become the deliverables for a project. When you focus on the global program level, the implications of risk are far more significant because they are intertwined with the company's strategy to produce results. When risk is well managed, the desired deliverables and results can be achieved.

6

The Global Program Manager

*"Program Manager: US civilian research
and development foundation seeks a program
manager fluent in Azeri to implement and
provide support to The Azerbaijan National
Science Foundation program..."*
                    The Washington Post, November 29,
                                              2008

Introduction

In the last chapter we explored program risk management and the risks associated with doing program business abroad. Earlier chapters explored the global setting for program management, the PMI *Standard for Program Management*, the process of building a program portfolio, and the process of aligning programs with strategic plans. Now we turn our attention to the people issue: what is a program manager and how does the business grow global program managers?

The Global Program Manager

We address several models and themes for defining the *program manager* in this chapter. No one model will fit all situations so the reader is admonished to be patient with the discussion and to take from this chapter those key concepts and tools that seem to make the most sense in the reader's work setting.

   The successful global program manager is a broadly experienced and technically competent manager of people with the capacity to articulate a program vision and anticipate and respond to global program support needs and challenges. The global program manager is a leader who can hire, inspire, and motivate project managers and team members. But as the job announcement that opened this chapter suggests, sometimes, global program managers must have special qualifications and expertise—in this case, a working fluency in Azeri. The candidate chosen for this job will have to *see* globally, in the sense of *reading* program impacts and benefits across the world and mobilizing resources from anywhere in the world. This manager has a broad organizational perspective, acting as an agent or business officer, aware of the organization and target industry or sector as a whole. Furthermore, the program manager must have at least working proficiency in networks and mass collaboration systems now available on the Internet.

The successful program manager typically starts as a project manager or technical task leader, progressing to program manager in the middle or later stages of his or her career. In the evolution to program manager, new skills and competencies are learned to make the transition from leading a single project to leading a longer-range, multiproject program. While there are many gray areas in the transition from project to program, it is clear that new competencies and skills are "called up" in the move, as the focus moves to long-term program outcomes and benefits and the world becomes the playing field.

The Global Program Manager "Job Description"

Since the competency and skill requirements of a global program manager depend on the job at hand—that is, the expectations of the employing company or agency—let's explore a generic job description for a global program manager.

General responsibilities

The global program manager is responsible for the planning and delivery of program goals, objectives, and benefits. This accountability extends to the broad, intended benefits and outcomes of the program, as well as its direct outputs, products, and services. The role implies coordination with multiple projects and with the broad spectrum of agency or company support functions and business plans. The program manager is responsible for the overall integrity and coherence of the program, and reports to the program sponsor or owner, who may be within the organization or a client or customer, or both.

Specific responsibilities

Setting program goals and objectives to implement sponsor and business portfolio plans for global applications

Planning and designing the project portfolio development, participating in the selection and funding of projects and overall program support worldwide

Proactively monitoring the overall program progress, resolving issues and initiating corrective action as appropriate

Defining the program governance, management process, and administrative arrangements to assure consistency and equity across nations and markets

Establishing and implementing a quality assurance and quality control program to assure program outputs meet the expectations of customers and international industry, safety, and regulatory standards

Implementing a program risk management plan addressing global risks and benefits

Coordinating with agency or company infrastructure and functional services worldwide

Interfacing with agency or company technical and functional assets to assure consistency with industry and company standards

Creating the team of project managers and creating an innovative and disciplined team process of communication and collaboration globally

Delivering program goals and program milestones on time and within budget, with a focus on cost control and earned value—e.g., cost and schedule variances

Acquiring program resources and assets to enable fulfillment of the program objectives

Managing third-party and contractual partners in the program, including local support partners globally

Managing program and project interfaces at key program review, or "go or no-go" points and milestones

Maintaining a consistent program reporting system for program sponsors, clients, and stakeholders

Skills and attributes

Experience in the program domain and as a project manager

Leadership skills, the ability to establish a program vision and goal, and to lead program managers to work together to achieve them

Technical skills, the capacity to understand complex systems in the program domain and to see broad system performance issues and opportunities

Program management system skills, particularly, the capacity to create and enhance consistent management software platforms and approaches to project planning, task breakdown, scheduling, budgeting, monitoring, and implementation

Resources acquisition and cost control skills, the capacity to acquire resources to design budgets and cost estimates, and to control costs within the context of earned value

Communication skills, exemplary communication skills and proficiency with Internet and webinar tools for global interface

Business acumen, the understanding of business plans and strategies, and how programs affect business growth, risk, and opportunity

Diplomatic and collaborative skills to work comfortably and effectively in a multicultural environment

Global Program Management: What Does It Really Involve?

Implement business strategy: make the business case for the program and assure alignment

A global program manager is the advocate for the program, making the business case for the program and its alignment with the business or agency strategic plan. This function requires constant vigilance to assure that program planning and activity is reconciled with where the business itself is going—where the business plan points to growth. The program is seen as an investment of resources and people to further the development of the business or agency. The global program manager role implies an understanding that complex programs can get out of hand, that the dynamics of program management always present risks, and that the program achieves a *life of its own*.

Create a program workforce, team, and support system for results

The essence of any program is its people, including the direct program workforce, the program team, and its supporting cast. The job involves the selection of key project managers and key task leaders, ensuring that technology expertise is placed where and when it is needed, establishing performance expectations and goals, working with outsourced staff, and leading and inspiring *smart* work, team collaboration, and innovation. Many program people will be in other countries; thus, bringing the team together will require constant communication with the *virtual* program team. This function involves constant contact with the human resources function in the business or agency and making sure the program receives adequate attention and priority in recruitment and hiring services.

Acquire program resources and funding

Supporting and funding the program involves working with top company or agency leadership and management to give priority to the program, especially up front. This means participating in budgeting and funding decisions, making the case for the program and its resource needs, and having a good grasp of program resource requirements and cost estimates. Since most programs are not fully funded initially, the global program manager must stay attuned to the annual planning and budgeting process to ensure that the program stays visible and aligned with the business.

Interface with program stakeholders and sponsors

Global program managers are in constant contact with key program sponsors and stakeholders. This role involves the effective use of mobile communication technology and the Internet to engage—sometimes daily—top business sponsors and leaders, key customer and stakeholders. Although the global program manager can typically stay in touch with a variety of key program people through technology, a fair amount of travel is necessary because online chatter cannot substitute for "eyeballing" the situation—e.g., being there.

Ensure effective planning, scheduling, budgeting, monitoring, and review

The program manager is responsible for ensuring a consistent and effective project management methodology in all the program teams. This means ensuring that the business itself has the underlying supporting networks and soft programs so that projects are defined, chartered, documented, scheduled, and monitored using earned value and other project performance measures. If the parent business or agency does not have a "projectized" support system, the program manager must create one for the program.

Build a supporting global program network

The global program manager role involves an understanding of the necessary support system for global program success, including the global presence of key personnel, outsourcing capacity, local *intelligence* sources to monitor local socioeconomic conditions and trends (e.g., regulations, safety and security), and a system of program delivery. Program delivery may require local distribution channels, which may present challenges simply because of different cultural and language issues.

Balance program discipline with innovation and creativity

While a program is typically set firmly in assumptions about program objectives, outputs, outcomes, and benefits, an agile program team has the capacity to create new ideas and concepts that can fundamentally change the shape and texture of the program midstream. The global program manager must set limits on creativity to make sure new ideas are aligned with where the program is going. This is accomplished by disciplining task definition and scheduling, but regularly conducting program reviews at key milestones to confirm task structure and to explore ways to deliver program results faster, cheaper, and better.

Program Managers versus Single Project Managers

Program management is different from single project management. The management of single projects requires several analytic and leadership skills in addition to proficiency with project management software. Project managers must be capable of working with a customer to develop the scope of work for the project and to prepare a work breakdown structure that captures the deliverables in an organizational chart or outline. The project manager must be familiar with the technical field and changes in technology. This person must be able to put a team together, assigning team members to activity and task areas in the work breakdown structure, and lead the team through the project life cycle. The manager must have the capacity to use project management software to prepare GANTT charts, assign resources, estimate costs, produce reports, and make presentations on the project.

There is an active debate in the field on the extent to which the project manager must be familiar with the technology of the project and technical aspects of design, development, testing, and product delivery. Some say the manager need only have a cursory sense of the deliverable and the technology involved, relying on team members and subject matter experts for technical assistance. Others say the manager must know enough "not to be snowed" by the customer, team members, or suppliers and subcontractors. They indicate that the manager will have to interpret technical progress reports of team members and be able to communicate with technical counterparts in the customer organization. In any case, it is clear that the project manager must have a good grasp of the deliverable and be comfortable in the field, if not an expert. If the field is changing rapidly, it is especially important for the project manager to grasp the implications of change for the current project.

Single project managers must be wholly focused on the day-to-day dynamics of the project because the project environment is always changing and shifting unexpectedly. The horizon is short in project management. This requires that the project manager manage through the 80-hour rule—that is, the focus in each weekly review is the current status of projects, indicated in earned value analysis, and anticipating the next 80 hours of work. This way, the team is reminded of the next two weeks' work, challenges that can be anticipated and key milestones in that 80-hour period. The single project manager is typically wrapped up in the project at hand.

Project managers must have several key skills: the ability to lead a team and resolve team problems; the ability to communicate and report effectively to a wide variety of customers and stakeholders on technical and project issues; the ability to manage a number of technical assignments all at once; the capacity to deploy project management tools, such as GANTT charts and schedules; a full understanding of the project life cycle; and a proficiency in project management software. In addition, the project manager must have judgment skills to make trade-offs between cost, schedule, and quality during the progress of the project and to make difficult decisions quickly to keep a project moving.

Finally, project managers are expected to be advocates of their projects and to make effective arguments for resources and priorities based on their project needs and their project's "critical path." They are not necessarily expected to see the big picture or to make decisions on sharing resources with other project managers. They are expected to be narrowly focused on making their project goals, objectives, and deliverables, regardless of what is happening in the company. If they start compromising their focus in the context of the needs of other projects, they do a disservice to their customers and to their project team.

The Transition to Program Manager

Program management is a different kettle of fish altogether. Program management is the process of managing a "portfolio" of projects, some of which are going on at the same time and some of which are linked in a sequence of process or output enhancements over a longer period and are global in scope. The program manager's span of control is wider and broader than the project manager's is. Program managers can be responsible for a long line of products and product enhancements in one program area over time—say, over a 10-year period—transitioning from one project to another based on customer and market feedback. Program managers also can be responsible for many projects and project managers across programs or product areas, thus complicating the management process. These people often are responsible for broad corporate product lines and markets across wide technical boundaries.

Program managers typically hire and supervise other project managers. They are responsible for developing the company's project management workforce, building and advancing project managers into higher levels of responsibility. They coach project managers. Program managers serve as the interface between projects and broader company strategies and business plans; thus, they are called on to communicate broad purposes to individual project managers and to report individual project results to corporate executives in high-level program reviews.

The complex, technical, and multiproject environment that program managers face requires different skills in managing information. The program manager cannot get lost in the details of one project or one project schedule or report, but must have an "enterprise-wide" perspective. Information on many projects must be managed so that the program manager can see the big picture and make the necessary trade-offs between projects, if necessary, to resolve resource and priority problems. Program managers must be able to step away from project details, see the broader implications, and make decisions on the basis of corporate-wide considerations and impacts.

Program managers mix in the milieu of vice presidents and CEOs—both their corporations' and those of the customer. They tend to coordinate at high levels, where decisions affect broad business performance, business-to-business relationships, workforce planning and management, regulatory issues, financial performance, stockholders, partnerships, and major supplier and contract issues. These factors often bring program managers into play with issues that single-project managers may not fully understand, especially if these factors get in the way of achieving narrow project goals. This is the essence of program management: balancing individual projects with broader performance issues and implications across projects.

The focus of most program managers is likely to be change and change management. This view says that the program manager has the broad perspective to be able to manage broad company initiatives—e.g., many projects aimed at changing the company in fundamental ways—across a wide variety of company activities and divisions globally.

Global Reference of Program Management

The global reference here is the capacity of the global program manager to see opportunities and challenges throughout the world and to act on them adroitly through proven and innovative program management and procurement methods and approaches. The focus here is on global resource and personnel acquisition, outsourcing for services and goods throughout the world, tapping Internet system capacities for information and communication linkages, and the capacity to relate overall program costs and risks to long-term program outcomes and economic and social benefits.

Program Reference

As the program becomes more and more complex, the program manager must be able to promote consistent program and project management tools and techniques across all projects and supporting activities. Since programs involve business systems and supporting activities, from planning and product design to production, marketing, and distribution, the program framework is often industry- and sector-wide, and extends to the global economy itself.

Explanation of Concept

The concept here is that program managers must demonstrate a high level of leadership and managerial competency that transcends projects and tasks. Program managers are officers of the business, leaders in their industry and technology, attuned to where the industry is going globally and how the business is to grow and thrive.

Program Manager Competencies

Competencies combine the personal attributes and business functions mentioned previously. To arrive at global program management competencies, one has to look at the requirements of the role and the skills necessary to perform that role, and then integrate roles and skills into the competency.

Here are the six basic competencies that characterize the successful global program manager.

Technically savvy.   The capacity to understand current and emerging technology and systems in the context of the target program and market, and see and act on opportunities in a worldwide context.

The myth that any good manager or leader can take over any program and run it without experience in the industry is just that—a myth. Experience suggests that a broad understanding of the program sector and environment is necessary to plan and manage a complex program. This is because technical issues in most programs often escalate to program managers, who have to exercise judgment, or at least evaluate technical judgments from others. This competency includes the capacity to ask the right questions at the right time; that is the program manager's role.

Think globally, act locally.   The capacity to work in global systems of networking, technology, and program activity, but to be able to act quickly and effectively with local resources wherever appropriate. Any location that is relevant to program operations is *local* to the global program manager.

A program manager must believe in global economics and be comfortable working in the uncertainty of global socioeconomic dynamics. This is because our new "flat" world eliminates the competitive barriers between countries. Electronic networks virtually replicate local conditions wherever they are in the world. Free trade reduces the barriers to finding the most cost-effective solutions to program issues.

For instance, the global program manager in charge of producing and laying cable across the Atlantic Ocean and linking it to other telecommunications systems in other regions and countries must have broad knowledge of the global environment in which his or her program will operate. This includes the capacity to put systems together to serve and stimulate broad, global customer demand. But that same manager must be able to mobilize a strike team quickly to evaluate a problem wherever it occurs.

Focused leadership.   The capacity to articulate a broad vision for the program, to make it relevant to the program team and to stakeholders, and to inspire and motivate team members through the world to work together to achieve program goals.

This competency includes the capacity to understand the dynamic of teams and groups of people, to make the business case, to persuade people to act, to see the politics of a program and understand "whose ox is going to be gored" in a given decision context, and to anticipate market and global impacts. This competency gives the team comfort and hope when things do not go well.

Organized.   The capacity to see a program in the context of a business or agency, and as a set of individual project initiatives, and to encourage a consistent and disciplined approach to planning, scheduling, delivering, and monitoring across projects.

The program manager must know how program management system tools and techniques are applied to ensure that multiple projects within the context of the program are planned, scheduled, financed, and monitored using proven program management tools. This capacity includes understanding the limitations of soft and hard program planning tools and the ability to revolve resource and scheduling issues quickly, and to see failure points before they occur.

Business acumen.   The capacity to understand the business enterprise framework for the program, to see risks and opportunity, and to act appropriately for the business as a whole.

Any program is a major investment of business resources and thus represents both a business risk and a business opportunity. Therefore, the program manager must have a working knowledge of the parent business, understand its capacity and limitations, understand the market it serves, and know how the program evolved from the business financial and strategic objectives.

The ability to teach.   The capacity and interest in teaching, training, and empowering programs and program team people to perform. Global program managers will spend much of their time in a teaching and training mode, interpreting the world for the uninitiated, and developing project managers and team members to work in an integrated and coordinated way in a multicultural environment.

The teaching competency includes being able to articulate issues, inspire interest in subjects relevant to the program, ask the right questions, work in a classroom and online environment, provide assessment of lessons learned, and model a lifelong learning approach to business.

Training the Global Program Manager

The training of a global program manager involves direct classroom and online training, and exposure to program management and technology experiences in many global markets. This requires training in the appropriate technical domains of the program to equip the manager with a grasp of the technical and global setting of the program.

Specific training and experience includes exposure and direct training in the following areas.

Language, diversity, and codes and symbols of the global culture

Global program managers may, in any single working day, talk with people in 10 different countries, with 15 different languages, and multiple differences in basic work habits and values. This working milieu requires agility and adaptation to situations and issues around the world, often requiring more effort and time than doing business domestically. Understanding different cultures and symbols and codes is important in getting work done; thus, both exposure to multiple cultures and systems of doing business, as well as training in language and local customs is in order for a global manager.

Diversity takes on a whole new meaning in the work of a global program manager; thus, competency in working with a diverse workforce is the key to success. This often requires tolerance and patience with foreign colleagues, whose value systems and priorities may be different from those experienced domestically.

Single and multiple project management systems

Global managers must know how individual projects are planned, scheduled, and delivered in order to run a system of multiple projects that is part of program delivery. This means an appreciation for program management software and for consistent tools and techniques of work breakdown structures, scheduling, budgeting and cost control, quality control, and team meetings.

The global program manager must create a program management system and a project culture that consistently channels information where it is needed.

Global program portfolio planning, delivery systems, and issues

Training should include an introduction to the company's or agency's program portfolio system—e.g., how strategic objectives are translated into programs. This would include exposure to how the business or agency makes priority and funding decisions.

Technology developments relevant to the program

A telecommunications program manager working in a given region in the world must have a clear idea of how technology can be mobilized to achieve program goals. This means training and exposure to technology systems in targeted regions throughout the world. For instance, it makes no sense to count on mobile phone texting as a project reporting mechanism if the target market and region does not have a supporting and regulatory system for mobile communications.

Business support functions and operations

This training includes administrative and management system updates—e.g., how company's information platforms and networking systems work; how to take advantage of e-procurement capacity; and how to exercise human resource planning, real estate, and other business support functions and offices.

Managing Virtual Teams and the Technology

A virtual team is a team that does not work together physically; they are linked through telecommunications and other technologies from remote locations. Global managers should be comfortable working with such virtual teams, with key project managers and subject matter experts located in many different places in the world. Communicating and moving data to these locations requires the frequent use of web conferencing and e-mail tools, as well as specialized platforms and company networks for the exchange of data. Project General, the Microsoft Project software program for the exchange of project schedules and data, would be natural part of the team's interaction and reporting systems.

We offer one caution on virtual teams. Online exchanges and communication take on special meaning and are far more effective *after* a team has met in the same room, built a sense of the team as a social system, and has a good sense of the personalities and attitudes of each other. Be careful not to depend on virtual technology communication systems when the people involved do not know each other. This risk is accentuated in a global team, where symbols and cultural standards can be worked out early in the team's work.

When it is impossible for teams to meet face to face, the program manager needs to conduct an initial online discussion that explores each member of the team, their biographies, and team roles. The objective would be to build a sense of each other's strengths and backgrounds before trying to work together without an initial meeting.

Negotiating Skills

Roger Fisher and William Ury authored a famous book on negotiation (*Getting to Yes: Negotiating Agreement Without Giving In*, Penguin Publishing, 1991) in which they stressed four approaches to negotiation: separating the people from the problem; focusing on interests, not positions; inventing options for mutual gain; and insisting on using objective criteria. While these techniques may help in negotiations in a complex program, let's set the stage for the unique negotiations required in global program management. Let's drill deeper into the context for negotiation.

Global program managers negotiate in a number of situations in which they or their representatives are interacting with foreign nationals or governments— e.g., contract and outsourcing arrangements; regulatory, safety, product, and security standards; scheduling; cost estimating and budgeting; resource acquisition; and quality. Each requires a particular approach, as discussed in the following sections.

Contracts

Contractual negotiations in many countries are not as formal or predictable as those found in American contract negotiations. Sometimes, program contracts and local arrangements are made with a handshake or with a verbal promise instead of a written contract. Unlike the advice of Fisher and Ury to focus on issues rather than people, the global manager may have to focus on *personalities and positions* to effect contract arrangements in a personal way. Sometimes, money crosses hands in an arrangement that Americans might call bribery. Sometimes, foreign governments doing business with an international engineering or systems team will fragment responsibility in the program team so that no one country representative on the team can dominate it. The home country will play the *prime* contractor role in the program team and *coordinate* the program players from other countries to ensure local control. Therefore, negotiations within the team often involve not only the home program sponsor, but many of the team players as well, because all the players have to adjust to local conditions.

The global program manager must be aware of these variations in order to negotiate program support and facilities. For instance, a cable telecommunications supplier developing a major international cable system in the Middle East might encounter the following negotiation challenges:

1. Negotiating roles and functions for the program team as part of a larger, more international, team of telecommunications professionals and technicians, the global program manager must fit his or her team into local roles and settings. This may mean giving up controls and leverage in cable installation, such as cable access, liabilities, and maintenance costs. Entering the negotiations with this in mind, a global manager must be aware of the key priorities of the program (e.g., no competitor access) that are not negotiable and those that can be compromised in making arrangements with a local jurisdiction or prime company (e.g., maintenance costs and systems). Local easement and land use often present major issues to a telecommunications program, including local access to required rights of way for cable systems and hubs.

2. Cable installation or maintenance contracts may be written with 50 different jurisdictions, each with a different format and set of terms and conditions. Thus, standardizing and formatting contracts in a global program are often impossible. Additional time and effort may be necessary to effect contracts that take only minutes in domestic programs.

3. Relationships with partners and competitors take a different turn in international situations, requiring tailored negotiation approaches. For instance, a cable company may have to negotiate with local competitors to market cable services, sometimes even entering into compacts or partnership contracts. Global program managers must often negotiate these arrangements sensitively so that program and business assets are protected while doors are opened to program development in the local setting.

4. Governments in the Middle East may have different or no safety regulations to govern local cable installation, yet local foreigners on the program payroll can sue the program company in the United States under certain circumstances. Thus, the lack of local safety rules does not eliminate the need for safety management practice.

Using International Culture Codes

Clotaire Rapaille authored a book entitled *The Culture Code: An Ingenious Way to Understand Why People Around the World Live and Buy as They Do* (Broadway Books, 2007). In his book, he identifies the concept of culture codes, or indicators of how people behave and respond based on their cultures and national origins. Cultures create conditions for communication and agreement, but these conditions are not always apparent. Americans have several codes, but one such code is the concept of a *dream*. Rapaille explains this as a deep cultural learning about Americans fulfilling their American dreams (e.g., "you can be anything you want to be"). He describes other indicators of nationalistic or culture orientation that signal various responses and consumer behaviors in given markets.

Global program managers can take advantage of culture codes if they study and use them in management, negotiation, and follow-up program activities with foreign companies and nationals. For instance, in the Middle East, the concept of time is not a major issue in getting things done in a major infrastructure program, but the concepts of interpersonal honor and trust are extremely important as culture codes. Therefore, a program manager who micromanages a program schedule is likely to fail in making milestones, but a program manager who can earn the personal and professional respect of the sponsor or program owner will be more likely to succeed.

Global Schedule Conflicts and Negotiations: A Global Website Design Case

Different values and standards for time and delivery in various worldwide locations often require different postures in negotiation and conflict resolution. The following case is an example of a program monitoring and control situation involving one of its products: a website. The key issue is potential conflict and negotiation around a foreign contractor's performance during the design and delivery of the product. The case is set up with a brief description of the website design project. It involves the program manager's decision on whether to pay an invoice from the foreign contractor when the project manager and contractor involved differ on the work performed to date, costs incurred, and the invoice amount. The issue becomes how the remaining work in the project should be accomplished: by renegotiating and rescheduling the current work, or by terminating the contract and starting over with a new contract, or even completing the work in-house. The situation is complicated by the foreign location of the work and the interaction of American and foreign nationals and companies.

Background

This is a story about a foreign training firm that has generated a new project as part of a global information program. The program is to produce a website for virtual training programs for entry-level workers from different cultures and language bases. The project team involves a project manager, training specialist, website administrator, and documentation specialist. The project requires the services of a contract manager, working closely with the project manager to procure both website design and development, and later to renegotiate the contract.

Project schedule

The following is a picture of the GANTT chart for discussion purposes.

ID	❶	Task Name	Duration	Start	Finish	Predecessors	Resource Names
1		Website Project	473 days	Mon 11/26/07	Wed 9/16/09		
2		Procurement	15 days	Mon 11/26/07	Fri 12/14/07		
3		Procurement Solicitation	5 days	Mon 11/26/07	Fri 11/30/07		Procurement Office
4		Requirements	4 days	Mon 12/3/07	Thu 12/6/07	3	Procurement Office
5		Solicitation Documents	3 days	Fri 12/7/07	Tue 12/11/07	4	Procurement Office
6		Award	3 days	Wed 12/12/07	Fri 12/14/07	5	Procurement Office
7		Contract Work: Produce Website	458 days	Mon 12/17/07	Wed 9/16/09		
8	√	Requirements	66 days	Mon 12/17/07	Mon 3/17/08	6	Eng 1
9	√	Design	50 days	Tue 3/18/08	Mon 5/26/08	8	Eng 2
10		Develop	332 days	Tue 5/27/08	Wed 9/2/09		Contractor 1
11	√	Prototype	40 days	Tue 5/27/08	Mon 7/21/08	9	
12	√	Configure	10 days	Tue 7/22/08	Mon 8/4/08	11	Contractor 1
13	√	Document Design	30 days	Tue 8/5/08	Mon 9/15/08	12	Contractor 1
14		User Acceptance Testing	200 days	Thu 11/27/08	Wed 9/2/09	13	Contractor 1
15	🖩	Choose User Group	100 days	Thu 11/27/08	Wed 4/15/09	13	Contractor 1
16		Conduct Test Sessions	100 days	Thu 4/16/09	Wed 9/2/09	15	Contractor 1
17		Final Performance Test	10 days	Thu 9/3/09	Wed 9/16/09	16	Contractor 2
18		Deliver	0 days	Wed 9/16/09	Wed 9/16/09	17	Test Tech
19		Project Closeout	4 days	Thu 9/17/09	Tue 9/22/09	18	

Program manager's decision

The case presents a global program manager's dilemma, since the team program manager in charge of this program for website design has escalated a conflict around an invoice from the local website contractor for a progress payment. The case also raises the question of what kind of contract is most effective in enabling project monitoring: a fixed-price contract or a cost-based contract.

The program manager must investigate contractor progress using not only program reviews and interviews with the program manager, but also traditional investigative techniques (e.g., interviews, document gathering), as well as earned value tools. The definitions involved in earned value help to separate cost from schedule, since progress in one area does not automatically mean progress in the other. The global program manager understands that cost and schedule are typically *traded off with each other and with quality,* but those trade-offs are different in each local country and culture.

Key definitions used in the case:

BCWS = Budgeted Cost of Work Scheduled

BCWP = Budgeted Cost of Work Performed

ACWP = Actual Cost of Work Performed

Assumptions

The project manager has received an invoice for $200,000 from the local contractor for the contract period up through 4/15/09 (see the project schedule). At that point, the project manager already knows that the BCWS is $250,000, which is what should have been spent if the contractor was on schedule at that point based on the contract schedule. But in review of the contractor's performance, the project manager finds that the task to be completed by 4/15/09, "Choose User Group," is not finished. Furthermore, it turns out that the contractor has actually completed the work only through the "Prototype" task. The contractor cannot show deliverables for any task past "Prototype." Thus, the project manager calculates that the actual work performed has "earned" only what that amount of work was calculated to cost $175,000. Furthermore, the project manager asks the contractor what the actual costs were to date, and he indicates the actual cost (ACWP) is $200,000.

The project manager has a dilemma here, since the contractor has admitted through the invoice that he is behind schedule at 4/15/09, but has completed more work than reported and delivered to the program manager. The questions include how much to pay the contractor and what to do to ensure completion of the remaining work. What complicates this case is that the contractor is a foreign firm in a county where scheduled reporting, constant feedback, and quick communications are not consistent with the local industry practices and the local culture.

Earned value calculations

Here are your calculations:

BCWS = $250,000

BCWP = $175,000

ACWP = $200,000

While the program manager uses earned value to calculate progress in the program, thus requiring regular updates and reviews on schedule and cost variations, local sponsors and contractors are not used to this kind of oversight. Thus, the fact that the local contractor has not produced against the schedule milestones does not hinder him from submitting an invoice for progress payment on the program.

Thus, the program manager has to negotiate with the local contractor, lacking typically available information on milestones completed, to get information that will allow a determination on whether the work will be late and over budget.

The following might be a line of questions raised with the local website contractor:

1. You have invoiced $200,000; you want to be paid that amount for the work to date. Can you tell us why you are invoicing for more of the work than you have delivered to date?

2. If the program decides to negotiate an agreement with the contractor at this point in the work, what principles should govern the negotiation, given local constraints and potential conflicts?

3. What do you think should have been done at the point of review, based on the work scheduled and measured by the estimated budget for that work—e.g., if the contractor estimated the budget right (BCWS) and the work is on schedule,

they should have spent (and earned) what was budgeted for that point in the work (BCWP). In this case, the BCWS at the end of the third quarter is $250,000, based on the original schedule. But if the work is not on schedule, they have "earned" only a portion of the total amount they should have spent if the work were on schedule—e.g., the BCWP ($175,000) is less than the BCWS ($250,000) by $75,000. They are behind schedule by $75,000 worth of work. And to complicate things, if the contractor spent more (ACWP = $200,000) to be where they are than they should have based on their BCWP ($175,000), their actual costs (ACWP = $200,000) exceed their BCWP ($175,000) by $25,000; thus, they are going to overrun their budget, and their cost variance is negative.

4. The contract schedule provides milestones to give us an idea of where the website design and delivery is; can you give us some evidence that you have met the required milestones?

5. If you prefer not to use earned value, can you give us an estimate of remaining work, time, and cost?

6. To complete the work due at this point, if it is not completed to date, can you crash the project and push more work to make up time?

7. Should we extend the due date and, if so, by how much?

8. Should more funding be made available to the contractor to complete?

9. Having signed a cost-reimbursement contract for the work because local regulations prohibit fixed-price contracting, the program manager might ask, "Can you give us a listing of items you will invoice in this work up to completion?"

Negotiating from Interests

How does the concept of negotiating *from interest* rather than *position* relate to this case? In this situation, the program manager could negotiate the payment of the invoice, or part of the invoice, based on the position of the company—no payment for costs associated with uncompleted work, even though the contractor is eligible under the cost-reimbursement contract. However, this approach is likely to fail, given the strong local focus on trust and respect instead of on-time and cost. Rather, the program manager should focus on a common interest, which would be to get the program completed within budget. Time is of less interest than completion of quality work within cost. Thus, if the program manager goes into the negotiation of how to pay on the invoice and arrives at an agreement that no further charges will be invoiced for the work through the currently due milestone date, he or she can pay the whole invoice, despite the lack of evidence that work is completed.

Scheduling: What Is the Program Manager Competency?

Scheduling is more than simply aligning tasks and estimating durations. It involves committing resources and people to specific windows of opportunity to provide value to the program. Program managers must develop capacity in their programs to get commitment from the program and project teams in the scheduling process. In a global context, this scheduling process is more difficult, since in a virtual team, with communication, culture, and language issues, scheduling becomes almost impossible. This is why global managers focus more on people, tasks, and interdependencies rather than on strict, traditional scheduling based on task completions. If estimated durations are characterized in a traditional program by overestimating tasks to protect the task performer, the problem is compounded in a global program.

This means more focus on critical chain scheduling, working with starting times and not task durations; more emphasis on task dependencies and enabling more flexibility in how and when tasks are completed makes more sense in a global program. This places more responsibility on the program sponsor to determine the importance of time in delivery of program outcomes and benefits, especially when programs are not delivering according to the sponsor expectations.

Cost estimating and earned value

Cost is of major concern in global programs when local investment is involved; therefore, it makes more sense to get foreign local investment in programs in order to "buy" more engagement and local sponsorships of program outcomes. A program financed solely by American funding is likely to get less support than one in which there is local investment and commitment.

The use of earned value to calculate cost and schedule variances depends on local use of the earned value concept, which separates progress in keeping to schedules from progress in keeping to cost estimates.

Resourcing

Acquiring resources and people—e.g., *resourcing a program*—in a global program involves key resource activities at almost every level and milestone in the program. The following table shows key milestones in a new product development program, along with associated key resource acquisition decisions:

Global Program Milestones, Functions, and Tasks	Description	Resource Acquisition Issue
Establish new product program objectives	The process of identifying the broad goals, outputs, outcomes, and benefits to be produced by the program	Look at broad view of program resource needs at high level; look at labor intensity, skilled and professional worker needs, technology depth, and global resource markets
Identify program structure and domain	Structure suggests how the program will be broken down into projects and support tasks, and how resources and people will be distributed, how products will be produced, and how the program team will work together	Program structure will suggest ratio of domestic and international hires; location of resource acquisition and outsourcing potential

Global Program Milestones, Functions, and Tasks	Description	Resource Acquisition Issue
Estimate program budget	"Ballpark," high-level budget estimate identifies human resource, capital, assets, space and facilities, and travel costs	Identifies program outputs and components and resource needs for each component or stage of work
Identify program team	Team-needs assessment identifies roles, functions, and optional hires or appointments to team	Team triggers human resource process of announcing positions and filling them
Identify program support needs	Support needs (e.g., facilities, equipment, information systems) are identified; determine buy-or-make decisions	Estimate costs and sources for facilities, equipment, and information system; confirm buy or make decisions
Identify logistics needs	Logistics requires trade off between buy or make, domestic or international resource acquisition	Logistics needs estimated, including equipment transfers, freight and delivery costs, etc.
Identify new technology needs	Specialized technology (e.g., cables, computer configurations, materials) is identified	Cost out technology needs and request bids to identify potential costs
Build new product project leader pool	Special attention on finding project managers for key projects	Human resources begins search for key project managers
Define funding options	Identify sources of funding from parent company, from program sponsor and other stakeholders, and finance companies	Make decision on how program will be funded, and identify ballpark funding needs
Create supply chain and contract types	Create chain of key global suppliers to provide just-in-time material for program	Get cost estimates from supply chain partners and key vendors globally to help identify program costs
Review resources	Perform review of all resource needs and update to confirm and trigger procurement and acquisition process	Use resource reviews to confirm resource needs
Define program charter	Program charter must identify responsibility of program team to conserve resources and find cost-effective resource acquisition approaches	Charter makes clear need to control resource acquisition process to avoid big mistakes
Prepare product design and architecture	Program outputs require design and definition; produces detailed resource and component information	Fine-tune cost estimates of big-ticket items and review options and alternative sources
Identify platform issues	If program requires a key platform, search early for buy or make and for longer-term maintenance partner	Identify way to share risks for acquiring and running program platform (e.g., website, tailored network systems)

(Continued)

Global Program Milestones, Functions, and Tasks	Description	Resource Acquisition Issue
Design and produce prototype	Prototype of program outputs requires acquisition of engineering components	Find cost-effective sources for prototypes so longer-term needs can be handled by same supplier
Design to cost	Establish component resource cost targets for program outputs	Control acquisition of resources for design to ensure that unnecessary cost escalation does not get built into design process
Test prototype	Identify test process and resource needs	Determine buy or make decision on test process
Configuration management	Define key program outputs and products and how they are made to support inventory, manufacturing, and production processes	Use configuration management system to stabilize program outputs for acquisition of components and subsystems
Assure quality assurance and control	Use quality assurance to avoid costly components	Ensure components are quality-assured to avoid procurement of bad materials
Identify regulatory, safety, and public policy issues and opportunities	Political and public policy and regulatory issues, legal issues, morality and ethics questions addressed, education	Make sure program resources acquired in foreign markets are protected with safety and security regulations
Platform decisions	Permanence, aggressiveness, type of demand sought, competitive advantage, product line replacement, scope of market entry	
Packaging	Roles of packaging and options and impacts	Cost of packaging is part of resource cost estimates
Competitive analysis	Analysis of program competitors will identify differentiators, resources that might be saved in getting ahead of competition	Do competitive analysis to see how competitors resource their programs
Product and service support plan	Identify resource needs for product and service support and maintenance	Add maintenance to life cycle costing of resources
Launch management	Communications plan, alliances, strategy connections, sales, logistics, control events, product failure scenarios	Identify marketing resource needs (e.g., advertisements, commercials, promotions)

Quality

The global program manager must be versed in total quality, Six Sigma, and other quality programs to ensure that:

1. Program is customer-based—that is, signals on quality and customer requirements come from the customer or client

2. Program team is empowered and trained, especially those in leadership positions

3. Program process is reviewed, internally controlled, and improved upon; the process of program planning, budgeting, scheduling, staffing, and reporting is reviewed for waste and unnecessary steps

4. Program team is zero-defect, Six-Sigma "energized," and trained to *do it right the first time*

Managing Conflict and Risk Globally

Conflict can be managed using traditional approaches—e.g., withdrawal, confrontation, collaboration, and compromise—but handling conflict in a global team setting can be more challenging. This is because conflict in an international setting takes on more complex dimensions due to language and cultural constraints. Before you know it, a simple program-related conflict can become an international "federal case."

The easy answer to conflict resolution is to prevent conflict altogether through constant communication that raises anticipated issues and problems, and resolves them before conflict develops. This is even more important because in a global program, conflicts may break out over unanticipated and unpredictable issues and become dysfunctional before the program manager even knows about them. A good example would be a conflict created by unanticipated tension from a foreign government objecting to outsourcing program work or equipment to a company in an unfriendly or competing nation.

Another answer is to turn conflicts into debates on alternative ways to manage the program, since many conflicts begin over issues of how the program is being designed and/or run. This means being sensitive to developing opposition within the team to the ways things are being done and working to generate alternatives. Sometimes, this approach uncovers innovative and creative ways to get work done.

Program and project risks can be compounded in global operations, especially in certain markets. Companies and their program managers can pay a large price for mistakes in foreign countries. *Forbes* (June 2, 2008) reported that in Kazakhstan, ExxonMobil was fined $5 billion to compensate for project delays as part of a program consortium. This kind of penalty can result from real schedule issues or simply the result of bad relations with the parent nation—in this case, the United States. The approach to avoiding these kinds of penalties and fines is to anticipate them in contract negotiations and to monitor changes in local regulations or diplomatic relationships while a program is being implemented.

Global time management and the issue of time zones

Since global programs typically operate in countries in different time zones, program communications and information exchanges need to be managed in particular "windows" for each global market. This often requires both direct communications at odd times and the use of asynchronous discussions online, providing a platform for communication and discussing program issues when convenient in a continuing chat room setting.

The leadership function is often confused with the management function. Management has to do with planning, acquiring, assigning, and controlling resources and people. It is more focused on the use of resources efficiently and effectively. Leaders, on the other hand, can also be managers, but leaders typically focus on establishing purpose and meaning to the work, providing a vision of the future program outcomes and benefits, and making key decisions that shape the texture and process of the program. Leaders inspire people to perform in extraordinary ways, providing them with motivation and direction to *make a difference*. Tranformational leadership has to do with leading change and transforming a group or organization to move in a different direction to adjust to outside market or environmental forces.

Leadership styles include charismatic/value-based, team-oriented, participative, humane-oriented, autonomous, and self-protective. A recent study has identified these different leadership styles with different cultures. The results are instructive for global managers tailoring leadership styles to various global markets and venues (Table 6-1).

Looking at the table, a global manager doing business in Egypt and Sweden faces different challenges in leadership styles. This requires managers to employ different leadership styles based on the client base. For instance, while working in Egypt, which exhibits a highly self-protective leadership style (avoiding losing face, protecting personal reputation and image), program management must be sensitive to leadership behaviors that may threaten key local personnel—work must not be made risky or generate interpersonal conflict. However, in Sweden, leadership styles favor charismatic and participative, "open" styles that generate discussions and debate and inspire personal performance and creativity.

TABLE 6-1   How Different Cultures Look at Leadership

Country Studied	Global Market	Charismatic/ Value-Based	Team- Oriented	Participative	Humane- Oriented	Autonomous	Self- Protective
Russia	Eastern Europe	M	M	L	M	H	H
Argentina	Latin America	H	H	M	M	L	H
France	Europe	H	M	M	L	L	M
China	Asia	M	H	L	H	M	H
Sweden	Europe	H	M	H	L	M	L
USA	Anglo	H	M	H	H	M	L
India	Asia	H	H	L	H	M	H
Germany	Europe	H	L	H	M	H	L
Egypt	Middle East	L	L	L	M	M	H

Key: H = High, M = Medium, L = Low (how each country ranks seven leadership styles based on culture_
Source: Mansour Javidan, Peter Dorfman, Mary Sully de Lugue, and Robert House, "In the Eye of the Beholder: Cross-Cultural Lessons in Leadership from Project Globe," Academy of Management Perspectives, Volume 20.7 (2006) pp. 67-90.

Strategic leadership

Since alignment is so important in program management, leaders must guide programs toward strategic purposes in the longer term. They must align programs with business plans and strategic objectives, and they must be able to see opportunities to grow business activities in a strategic direction. A strategic direction is a direction that can be supported by the company's or agency's capacity to perform and is part of the business plan for long-term growth.

The way leadership connects strategy and program is by ensuring that programs and projects are linked with strategic objectives. Therefore, one key function of leadership is to see that business plans are translated into strategic objectives and that program portfolios and budgets line up with those strategic objectives. This is a process function—e.g., the generation of a process so that the business or agency annually revisits strategic objectives and programs generated to implement them.

The global significance of this process is to make sure global strategies and business planning input are included in the process if the business plans to go global. Strategies include analysis of strengths, weaknesses, opportunities, and threats on a global scale, requiring a good deal of research and information gathering about target markets and customers.

Ethics and morals in a global context

Moral conduct is typically associated with religious standards, whereas ethical conduct is typically related to honesty and other standards of integrity. There are three categories of ethical and moral dilemmas in global program management: (1) the problem of being honest and straightforward in reporting and communicating on the status of a program; (2) participating in shady practices (e.g., bribery) that might not break US or local laws but which transcend the typical ethical standards under which programs are operated in this country; and (3) the more severe problem of knowingly violating local laws or regulations in program operations.

Honest, quantitative reporting. Program managers should use standard methodologies to report progress, especially in a global program, where the risks of miscommunication are high. This means using quantitative measures of progress (e.g., earned value and cost and schedule variances) and other tools that are largely objective and consistently applied in every aspect of the program. While these measures are used, program managers must be attuned to different interpretations of progress in foreign countries and remote markets. Schedule variance may mean less in the Middle East, where relationships and trustworthiness among the program team and local control of decisions are most important and where time is of less importance, than it does in Western Europe, where timeliness and prompt schedule delivery are important.

Shady financial practice. Shady financial practices often involve using money or collateral to persuade local sponsors, partners, or foreign nationals to make decisions that favor program success or to win a contract. The way to avoid this is to develop an internal control system that monitors all financial transactions within the program and documents all local program payments religiously. This does not mean that program managers are not sometimes placed in the position of paying off local politicians or businesses, but those payoffs must be consistent with local practices. The situation is to be avoided if possible.

Breaking the law. Program managers can avoid violations of local law and regulations by training all personnel on those local laws and holding them accountable for following them. Chevron recently paid a $310 million fine in Kazakhstan for improperly storing sulfur from program operations in its Tengiz oilfield. Local regulations were apparently not clear on sulfur storage requirements, but despite the uncertainty, it was up to the Chevron program manager on location to ensure that the storage was legal before authorizing it. Global program managers pay a price if they are not attuned to the wide variations in ethical and legal standards around the world.

Administrative issues; nuts and bolts of the global program manager position

Sound administration of global programs involves organizing and administering the details and office systems required to support the program worldwide. A global program manager should have a high degree of tolerance for details and day-to-day processes and procedures to support good program delivery. These administrative functions can be broken down into nine categories:

1. *Human resource management/personnel.* This function involves a close relationship with personnel, both at home and abroad, to enable quick turnaround on personnel hiring, job and position management, relocation, and staff outplacement. Because much of the personnel management will be in foreign locations, it is important to have human resource contacts in key program areas where local labor conditions (wage scales, etc.) are clear.

2. *Time management.* The management of time involves timesheets or some kind of time accounting system that is program-coded. This means that all program and project personnel complete timesheets (preferably online and integrated with program scheduling software such as Microsoft Program) that are coded to key program and program work breakdown task codes. This allows the program manager to monitor and capture actual costs and calculate cost variance in the reporting system.

3. *Cost management.* The management and control of cost involves good estimates of program material, equipment, and personnel costs, and a system of internal control and monitoring that authorizes and oversees expenditures within preset ranges. Actual costs are reconciled against estimated costs for purposes of earned value reporting and cost accounting. In many cases, because actual cost reporting from accounting systems may be too slow to be of value in a fast-moving program, global program managers may have

to provide informal commitment registers at the project level in order to roll up timely reports on a moment's notice.

4. *Procurement management.* Acquisition forms should be provided, either online or as a hard copy, for all program material, equipment, and supply actions. These forms should be program-coded to allow allocation of actual costs to various program and project categories in the work breakdown structure. Supply chain systems and arrangements to do business with specified suppliers create standard, predicable costs over time; thus, it is important that procurement offices be attuned to program-specific supply chain partnerships.

5. *Office process management.* Office process management means that the program manager's office, the immediate staff associated with the program manager, is highly organized and orderly. Calendars for the program manager are updated and made available to team members and stakeholders.

6. *Consistent work habits.* Program managers working in a global setting must be predictably available to a wide variety of key people throughout the world on a daily basis. This means that they must be predictable in their personal availability through cell phones, etc. at all times of the day, or set clear guidelines on when they are available. Because communication throughout the team and with key clients and customers is so key in today's quick design-to-market economy, delays in important communications can cost a program manager critical time.

7. *Configuration management.* Although configuration management is usually confined to engineering products, documenting product component dimensions, suppliers, identification numbers, etc., the concept can be useful in any program. The purpose of configuration management is to preserve a product and its composition or structure throughout planning and design so the final product can be produced in volume to meet market demand. In a global setting, the production of a program output or project output is likely to be outsourced and produced in a foreign country, requiring a reliable source of information on how a particular product is acquired.

8. *Meeting and computer conference management.* In a virtual setting, meetings will be handled through the Web, requiring preparation and solid support from a web-based contractor. Because these systems often do not always perform as promised, especially when they are worldwide, global program managers should do dry runs of such meetings to make sure systems are up and running. Nothing does in a virtual, web-based meeting like the failure of basic systems to perform. Agendas are distributed before each meeting, and follow-up notes and outcomes provided to all participants electronically.

9. *Local network management.* Global program managers will maintain program and project schedules, as well as budget, quality, and status information on a company network. This network must be administered effectively so that key people have access to updates as soon as they are entered. For instance,

if a project is going to be late and resultant schedule changes will affect other projects, project managers ought to be able to depend on the network to provide almost instantaneous updates to key program people globally.

Getting reports from project managers and rolling them up for executives

Unlike project managers, who typically have hands-on contact with the work, program managers typically do business based on reports from project managers. Information is secondhand in the sense that there must be a good deal of trust and reliability in reports up the line. This is especially true when program managers compile reports into executive summaries and data for top-level decisions. In the process, if original information is bad, it becomes worse as it moves up the line.

It is said that bad news travels *down* in an organization while good news travels *up*. This means that reports from projects are likely to be enhancedwith good news, simply because there is a reluctance to report failures or problems. As the saying goes, bad news is not like fine wine, it doesn't age well. To offset this tendency, global program managers will structure reporting formats to begin the report with issues and problems, requiring project managers to anticipate problems and report them early. This means that the program manager must make it clear that failures and problems in projects are *not* going to come back to those reporting them.

Communications

Communications are important in any program, but in a global program, they are almost critical. This is because there is so much room for miscommunication across languages and cultures. Therefore, global program managers must be good communicators and provide processes and tools for communications in a worldwide context that offsets the barriers of different languages and attendant meanings. This may require more use of graphics and symbols with international meaning, such as the international symbol for litter disposal, 🗑, which might be used to indicate that certain information must be deleted from a document. Since verbal communication is sometimes the most vulnerable to error and miscommunication, verbal exchanges of program importance and program commitments are recorded and documented, and followed up with confirmations and notes, if appropriate in the local program setting.

Assembling the right team in a virtual context

The right team is the team that has the required competence to do the job and the right instincts to work in a virtual environment, where day-to-day guidance and hands-on contact is not available. These attributes are part of the initial hiring process to ensure that program staffs are attuned to global virtual work processes. The virtual management process involves a high degree of trust and delegation. Program team members are empowered to perform and to make decisions in the context of program success. This means establishing a front-end culture of integrity, reliability, enabling, and honesty among program and project team members who work comfortably and effectively in a virtual context. It also requires a mobile, web-based communication system that facilitates open and frequent communications and data exchange on program issues at a moment's notice around the world.

Reviewing/feedback to project managers/coaching

Global program management involves mentoring, developing, and coaching project managers. This is made difficult in virtual settings because of the limited time the team actually spends with each other. Thus, performance review and feedback is typically accomplished through online exchanges (e.g. e-mail), and training and development for managers is conducted through online teaching and training formats. Feedback can be provided in a number of formats, but the key is how project manager accountability is monitored and reported. Accountability can be categorized in a global program as follows:

1. *Management of the project team.* Does the manager support the team, train and develop them, provide feedback on performance, encourage diversity and innovation, and motivate the team?

2. *Timeliness and efficiency in team operations and delivery.* Does the manager act promptly and efficiently to assign and complete tasks, schedule work, and control costs?

3. *Relationships with stakeholders and understanding of local settings.* Does the manager build strong coalitions and relationships with local stakeholders, and does the manager have a good grasp of the local work setting, local language and culture, and local conditions that might affect program success?

4. *Self-development.* Does the manager pursue self-development and training activities that enhance work in a global setting?

5. *Technical development and expertise.* Does the manager keep up in the field, especially with fast-moving technological changes and breakthroughs that might affect program performance?

6. *Reporting and communication.* Does the manager report accurately and honestly on local program conditions?

Motivating people in a global program

Motivation of the program team, especially project managers in or doing business in remote settings, involves a combination of strong management skills and leadership attributes. Motivation is the process of enhancing the performance of others by giving meaning and purpose to the work at hand. Program managers motivate through incentives and recognition and by providing purpose to the program. Herzberg's motivation theories are useful here—that there are *dissatisfiers,* like money, that provide the basis for unhappiness if not adequate but that do not provide true motivation, and *satisfiers,* like recognition, training, purpose, status, and empowerment, that provide the basis for real motivation and work satisfaction. Global program managers must be attuned to both kinds of motivators, especially those dissatisfiers that are effective in a global setting, such as bad living conditions, and satisfiers such as international travel and status.

Business savvy

Understanding the business strategy is key to global program success because programs are investments to grow the business in a certain direction. The business strategy provides an avenue for guiding programs and aligning them with where the business is going, or using them to develop and deliver a new business strategy. This requires an appreciation of the global economy, the value of open trade, the essentials of the private enterprise and capitalist system, and the different views around the world on these subjects. For instance, while Americans tend to think in terms of private enterprise and limited government intervention in commercial activity, capitalism is virtually unheard of in other parts of the world. Therefore, the manager must be sensitive to different views of long-held assumptions about how business is done.

Office politics

Global program managers contend with both local and remote office politics, and must be attuned to both. Office politics involves understanding informal systems of power and territory that affect the harmony and teamwork in day-to-day office operations. This means understanding when to defer to superiors and office colleagues on issues of importance to them.

Scheduling and resource planning for outsourcing

Scheduling and resource planning for outsourcing should be accomplished collaboratively between the program manager, the purchasing or procurement office staff, and the project manager and team. It is the program manager's responsibility to develop and maintain a good working relationship with the purchasing/acquisition department, since that staff does not typically report to the program manager. Procurement is seen generally as a business function and performs against goals that might not always be consistent with the interests of a particular program. General communication, sharing of schedule and resource information, checkpoints, approvals, and feedback will be managed through a company network to the extent feasible, with meetings and hardcopy schedules arranged as necessary.

The basic scheduling procedures are as follows:

Activity 1. In consultation with the customer, the program manager defines customer requirements, product functionality, and key deliverable milestones in a project statement of work, and makes an initial make-or-buy decision on particular tasks.

Activity 2. The program manager has project managers develop top-level work breakdown structures using a standard template and flags outsourcing issues for comments from procurement.

Activity 3. The program manager works with project managers to prepare a top-level schedule showing basic task structure, durations, start and finish dates, linkages, and predecessors; links the schedule to a central resource pool; and flags contractor roles.

Activity 4. The program manager and department heads establish a project team after a review of resource constraints and consultation with appropriate staff on availability.

Activity 5. Department heads consult with the project team and prepare detailed, subtask schedules for their areas of responsibility, assign team resources to tasks, and send updated schedules with resources assigned to the program manager.

Activity 6. The program manager integrates subtasks into the detailed schedule, flags team, and asks for sign-off.

Activity 7. The program manager has project managers enter hourly, fixed, and equipment costs and produces a project budget, flagging estimated contractor and outsourcing costs.

Activity 8. The program manager gets sign-off from the business's vice president to proceed to schedule baseline and contracting activity.

Activity 9. The program manager saves the project baseline, flags team and procurement.

Activity 10. The program manager kicks off the program with project managers, hands out hardcopy schedules and individual task assignments.

Activity 11. With support from the program management office, the program manager has project managers collect and enter actual performance data from contractors on: (a) percent completion, (b) actual hours spent on tasks, (c) actual start and finish dates, (d) actual durations, (e) remaining durations, and (f) actual weekly costs.

Activity 12. The program manager prepares reports on earned value for each program and supporting contractors, estimate to complete (ETC), and estimate at completion (EAC) to departments and to the general manager on a monthly basis (or the team reviews reports themselves from the network).

Activity 13. The program manager submits schedule updates with all tasks updated and conflicts resolved to departments and the general manager on a weekly basis (or the team accesses updates from the network).

Activity 14. The program manager submits narrative, hardcopy program summaries, including issues, risks, and corrective actions to the general manager on a weekly basis.

Activity 15. The program management office analyzes resource usage from a central resource pool file and identifies conflicts, issues, and problems for current and future programs.

Scrubbing the schedule

The process of scrubbing the schedule involves four steps:

1. The program manager drafts an initial schedule, with a work breakdown structure to the second or third level, and resource assignments for all tasks. Resource assignments can be proposed by the program manager in the initial schedule, or developed by the department manager and submitted to the program manager for embedding into the schedule.

2. The program manager links the schedule into the central resource pool to identify resource issues.

3. The program manager distributes the schedule, including contractor schedules, to all department managers.

4. The program manager conducts a meeting with department managers to work out task definitions, linkages, and resource assignments, as well as the final baseline schedule, which the program manager saves as a baseline upon kick-off of the program.

Baselining the schedule

Establishing the baseline schedule is a significant action in the program management process, signifying the official kick-off of the work and indicating a strong commitment to the schedule and resource plan, including all outsourced tasks. The baseline is the point of departure for monitoring and tracking process. When a program is baselined—that is, saved as a baseline schedule in the project software program and posted on the network—the project schedule is complete. Here are some rules of thumb for baselining:

1. The purpose is to get to a baseline schedule that captures all the work to be done. The baseline schedule does not change unless the basic scope changes. The baseline schedule represents the end point of the planning process. Once agreement is reached, the program manager confirms the baseline by saving it and making it available on the network *as the baseline*. There is no uncertainty regarding what the baseline is and how to access it.

2. The baseline schedule is the agreed-upon, scrubbed schedule for the program, linked to the resource pool. The baseline shows all interdependencies, linkages, and resource requirements; includes all tasks necessary to get the work done; and shows impacts on parallel programs and resources. All procurements and test equipment are covered in the schedule.

3. The baseline schedule is resource-leveled—the schedule can be implemented with current, available resources. Assigned staff are aware of the commitments and have "signed on" to complete their tasks to meet the schedule milestones.

4. Getting to the schedule baseline involves collaboration between the program manager and all departments and staff involved in planning and

implementing the schedule. A baseline meeting is held to arrive at a final agreement on schedule and resources committed before the baseline is saved to the network. The program manager facilitates the meeting, and all department managers attend and come prepared to commit their resources to the final, agreed-upon baseline schedule.

5. The final review of the schedule at the baseline meeting involves reviewing all stages and tasks, linkages, and resources assigned, line by line.

6. The baseline schedule is monitored weekly, with actual percent complete data and changes in start and finish dates entered weekly and reported on at program review.

Managing schedules on the network

The basic objectives of the network management of program schedules are to (1) enable the program management department to control schedule updates and schedule versions, including contractor activity, and (2) provide department managers, contractors, and staff with a way to provide input to the scheduling process and have easy access to up-to-date schedule information.

Schedule access. Contractors will have limited access to schedule files. The program management office will have "write" access to the schedules. Department managers, systems engineers, contractors, and other engineering and manufacturing staff will have "read" access to program schedules.

Schedule updates. The program management office is responsible for maintaining and updating program schedules on the network. Once the program manager and department managers agree on a proposed schedule and/or update, the schedule will be linked to a total workforce resource file, and resource conflicts will be identified and resolved. The program manager will then save the schedule as a baseline schedule. The baseline schedule will be the only version of that schedule housed on the network and will serve as the source of "planned versus actual" tracking information.

In preparation for the weekly program review and weekly report, the program manager will update the schedule in consultation with department managers, contractors, and the program team. To update the schedule, the program manager will update percent complete for all tasks and update all start and finish dates as appropriate. All updates will be made to the version of the schedule on the network.

Department and functional managers will be responsible for maintaining their own department schedules and resource pool files, and assisting program managers in updating program schedules. Program managers and department managers will share schedules through e-mail or hard copy until updates are agreed upon.

Resource planning for outsourced activity

Good program and project management requires that there be a central resource pool to identify impacts of changes in contractor and project schedules and to assess the overall efficient utilization of the in-house and outsourced workforce. The resource pool information on the network is shared with management staff and all team members to allow each team member to evaluate the work assigned to him or her and to provide guidance on task definition, durations, start and finish dates, and interdependencies.

Resource conflicts with contractors are identified through the resource pool, typically indicated in project software as hourly totals in red. Project managers are responsible for identifying and anticipating contractor conflicts and root causes, and involving the team in resolving issues. Corrective actions are taken by the program manager, including resource leveling—the process of enabling the software to reschedule project tasks to resolve resource conflicts identified in the resource pool.

The process of collaboration between program managers and functional managers is encouraged through a matrix-type organizational structure in which the functional managers are responsible for quality and technical capacity of the department, while program managers are responsible for mobilizing teams from the functional departments to produce deliverables within budget and schedule constraints that meet customer expectations and specifications. Procurement and purchasing staff are considered part of each project team.

Preparing staffing policy and plans

The objective of this policy is to initiate a planning process to prepare and implement staffing plans based on forecasted product development workload requirements. Staffing plans are produced through a process of forecasting staffing needs, developing staffing standards, projecting workload requirements, and building staffing levels and mixes to meet program and workload requirements.

It is the responsibility of department managers, in consultation with the director of product development and program managers, to review staffing and resource requirements for their departments quarterly and to develop a staffing plan to meet department program requirements. Based on these plans, department managers are jointly responsible for acquiring the necessary staff resources and building their capacity to meet projected needs.

The planning process involves six steps:

1. Determine department staffing levels and assignments

2. Develop staffing/workload standards

3. Forecast future requirements

4. Develop department staffing requirements

5. Develop department staffing pattern

6. Prepare staffing plan

Determine department staffing levels and assignments.   This step describes and documents current department staffing levels and staffing mixes, and relates current staffing levels to current assignments. The purpose of this step is to describe the current department workforce, functions, and assignments.

Develop department staffing/workload standards.   Staffing standards help compare staffing levels with performance goals. This step involves developing staffing standards that relate staffing levels and mixes of technical expertise to specific workloads or performance goals. Standards compare staffing with particular product development workloads, schedules, activities, documents, and other outputs. This allows the department manager to forecast what kind of staffing increases are necessary to meet future needs for timely product development. Functional areas, workload factors, and standards should be tailored to the department's functions, but should relate performance—e.g., meeting a functional requirement within a given schedule requirement—to staffing levels.

The following format is recommended:

Staffing Level	Functional Area	Workload Factor	Task Duration	Standard
Three software certification engineers	Software certification documentation	Concurrently prepare all software certification documentation requirements for three program phases	45 days	Meet certification documentation requirements in 45 days for three programs concurrently, with three engineers

Forecast future department staff requirements.   This step identifies future program and project requirements for the department—e.g., what workload the department is likely to be required to do in the future—and forecasts the appropriate staffing level and mix to meet those requirements. Future requirement information will be obtained from program planning forecasts and other information on future program demands, as well as analysis of current assignments and durations.

Develop department staffing pattern.   This step involves developing a department staffing pattern that defines the level and mix of staff required to meet current and future needs, and places them into a department organization chart. The staffing pattern identifies core positions—e.g., positions that are essential to carrying out the department's role and function—as well as support and junior/training positions as required.

Develop staffing acquisition plan.   This step produces a plan to acquire the levels and mix of staff necessary to meet future department staffing needs. This step will determine various sources of the expertise needed, various employment arrangements (e.g., employee, contractor, etc.), and outline a hiring and/or internal career development approach.

Prepare staffing plan. A department staffing plan will be developed that will include the results of the previous five steps. It will include current staffing levels, mixes, and assignments; staffing/workload standards; a forecast of future program requirements; future department staffing requirements; a department staffing pattern; and a staffing implementation plan.

Resource planning: rules of thumb

Resource planning implies that program managers have adequate workforce data and make accurate assignments of resources to program tasks. The following rules of thumb apply:

1. Workforce and manpower planning require that managers plan out job assignments and needs over a 60- to 90-day horizon, assuring that *their* current and projected workforce is fully occupied with direct, scheduled work and/or support assignments such as training and development. This requires that managers in charge of supervising staff take this responsibility seriously.

2. Manpower planning is not the same as scheduling work. Labor planning requires that staff have job definitions and assignments, which go beyond scheduled work—dealing with other department assignments, training, etc.

3. The scheduling process can be a good workforce-planning tool if all work is scheduled and all assignments are updated.

4. Program managers cannot be expected to manage the whole workforce assignment process. We are expecting the program managers to fully plan out the assignments to staff; but their job is to get product out the door, not to fully encumber staff. The departments are responsible for managing their workforce and project out their assignments, keeping them busy and occupied with a variety of work and development.

5. Performance metrics determine whether a corporation has the capacity to meet its product development goals. In other words, the reason we are working to improve our scheduling and manpower planning is to increase our capacity to produce quality products on time and within budget. How will we make the business case for better scheduling and manpower planning?

6. The matrix environment requires a balanced team approach to scheduling and resource planning. Responsibilities need to be made clear between functional departments and program managers. This clarifying process starts with the commitment to do it.

Preparing program budgets

Program budgets will be built from individual work breakdown levels and integrated with baseline schedules. Once resource requirements are determined, standard labor rates, component costs, and fixed costs associated with the program are entered into the scheduling software. Once the schedule is baselined, actual labor costs are entered as "actuals" to determine cost and schedule variance at program review.

Roles in the Program Management Process

The program manager is ultimately responsible for designing and delivering the program benefits, outcomes, and products. The program manager controls the "what and when" of the program, leading the program management process and organizing and facilitating each project team. He or she produces time-phased schedules for each program, anticipates future programs and other major engineering activities, tracks progress and anticipates future impacts, assures linkages with related programs and projects, and produces resource-leveled schedules to avoid resource conflicts.

The program manager controls and schedules work and resources, working with department managers and project teams to accomplish product development and program management goals. All products be managed through the program management process, carried on cooperatively by program managers and department managers. Program managers are authorized to establish and manage project teams to accomplish these goals. Department managers are responsible for supporting program managers in the staffing of project teams and in the development of technical capacity in their departments. Program managers are accountable for achieving cost, schedule, and performance requirements.

The program manager is responsible for customer interface, planning, scheduling, tracking, and coordination of a program, and is responsible for maintaining program documentation and updates as design and plans change. To sum up, the program manager:

- Establishes a program plan that meets all of the objectives identified for the program
- Creates a detailed program schedule, in conjunction with team members and department managers, that meets all program objectives
- Provides direction to all team members and any/all departments for the purpose of meeting program objectives
- Tracks progress of a project or series of projects
- Reports progress against the plan and schedule to management
- Identifies and resolves problems associated with the program/project
- Identifies and mitigates developmental risks

In collaboration with the department and functional managers, the program manager (PM) is responsible for the overall management and coordination of the program/project during each of the stages of development. In order to accomplish this, the program manager is required to conduct periodic program review meetings with the program team (including design, certification, manufacturing, and procurement personnel) to ensure a unified and informed effort. The PM is responsible for ensuring that corporate management approves any changes in requirements. As program requirements change, the program manager ensures that the program plan and schedule and other applicable management documents are kept current and distributed.

From initiation of program planning through completion, and transition to production, distribution, and marketing, the program manager ensures that all program requirements are identified, implemented, and verified. All changes are tracked and incorporated; and all management documents are produced and maintained throughout all stages of the product development and production. Along with the departments, the manager is responsible for determining the impacts of any design changes to products that are in either the product development or production stage. Along with the director of product development and corporate management, the program manager decides what changes are to be made.

In the event of issues and resource conflicts that arise in the program management process and cannot be resolved by the program manager and department managers, they are responsible for raising them to the level of the director of product development for problem resolution and decision.

The program manager has the primary responsibility of creating a program plan and scheduling tasks and milestones associated with successful achievement of all program objectives for a given program. The program plan will be created with support from the program team members and department managers. The PM is required to keep the program plan current, track progress, and incorporate changes as required. The program plan must include, at a minimum:

1. Program goals, benefits, and outcomes

2. Overview of customer requirements and program scope

3. Identification of major schedule tasks and milestones

4. Basis of program (new development, modify existing product, etc.) and strategy to meet objectives (e.g., qualification by similarity—the product is like one already tested—or perform qualification testing)

5. Identification of test equipment and components needed for the program

6. Identification of any special tests required for the program

7. Procurement requirements for development efforts

8. Manufacturing requirements for development efforts

9. Outsourced test facilities needed

10. Outside integration

11. Estimate of required materials for test assets, test equipment, outside facilities, travel, and any other pertinent costs

12. Risk assessment and risk mitigation plans

The PM has the primary responsibility of creating and maintaining a detailed program schedule that meets all program objectives. The schedule must contain, at a minimum:

1. Tasks and milestones that correspond to all major program objectives contained in the program plan

2. All activities and tasks required to execute a given program, including systems design, detailed design, certification, test equipment, reliability, safety, design reviews, manufacturing, procurement, test assets, etc.

3. Tasks detailed to the lowest practical level; activities and tasks should generally be identifiable to a single resource

4. Resources assigned to activities and tasks, and leveled to reflect a realistic workload

5. Identification of labor requirements for the program

The program management department is responsible for reviewing the resource impacts of all program schedules and anticipating resource conflicts and problems. This is accomplished by assessing each proposed program schedule in terms of its impact on current resource commitments and negotiating resolution where necessary.

Departmental and Functional Manager Roles

Department managers for administrative support, systems engineering, mechanical design, electrical design, software certification, and other areas hold the resource and process keys to success of the program management process. They are essentially responsible for the division's capacity to perform its basic functions and the quality of technical processes. Here are some key functions of these managers:

Organize department. The department's organizational framework must be made clear: how staff relate to each other and what the key functions, jobs, and reporting relationships are. Assignments are made directly and performance is monitored.

Develop technological capacity to perform. The department's technical capacity to perform is a function of its people and their ability to keep up in their fields. Technical training is an integral part of the department's job.

Develop and execute budget. A department budget is used to guide and schedule implementation of the business plan. This includes labor and equipment acquisition.

Motivate and evaluate performance and develop career tracks. Department managers are responsible for motivating staff and evaluating their performance and taking corrective action.

Forecast future workforce need and prepare staffing plan. Matching future demand on the department with current workforce capacity, department managers maintain a rolling workforce plan so that they can make the business case for new resources in the budget process. Staffing plans are developed for the department as a framework for hiring.

Work with program managers in scheduling work and manpower. In a matrix organization, department managers work with program managers in bringing program schedules to baseline, determining resource requirements, and supporting delivery of program products.

Risk Management Plan

The risk management plan will include a risk analysis, a risk mitigation plan, and an outline of all inherent program assumptions. Risks include technical and administrative factors, and are defined in a risk list, along with likelihood of occurrence, relative risk, weights against various project objectives, and current status.

Risks to be addressed include regulatory risks (e.g., will not certify for procedural reasons), technical risks, timing and scheduling risks, cost risks, resource risks, and computer and system-related issues.

Program review and monitoring

Program managers should hold weekly program review meetings to discuss broad program issues; detailed technical, resource, and schedule problems; project team performance; and risks. Project managers are responsible for preparing for these reviews and anticipating key agenda items. Department managers attend program review meetings as appropriate.

The program manager is required to track the progress of the program on an on-going basis and updating the program schedule on a weekly basis. The program plan is to be updated as changes in plans warrant. Program managers are required to hold periodic reviews with the program team in preparation for reporting and to support task assignments and feedback, either as a single meeting with all functions represented or as a series of meetings with major functional areas represented at each meeting. The program managers are required to report progress to management at a minimum of biweekly intervals.

Problems that significantly affect cost, schedule, or product performance must be identified, investigated, and reported to management so that decisions can be made on a fully informed and timely basis. As part of the program tracking and reporting, it is the program manager's responsibility to identify those problem areas and to assume responsibility for resolving them.

Program team roles

Each team member is responsible for understanding the individual tasks to which they have been assigned and for providing general support to overall team performance.

Program administrator and the program management office. Either as part of a formal project management office or as a staff function, the program administrator provides support to program managers and departments with general support, including scheduling, reporting, and workforce activities. The program administrator analyzes all program schedules and resource utilization to identify and resolve conflicts, and reports on new program resource needs and impacts. In addition, the administrator:

1. Supports the general manager, program managers, and project managers in planning, structuring, scheduling, resourcing, and monitoring programs and projects to meet cost, schedule, performance, and quality objectives

2. Develops a resource planning process to anticipate and meet future project resource needs

3. Builds commitment in the organization to use standard project management tools and techniques to schedule and implement projects and tasks

4. Provides training and information support to project teams in the use of project management systems and software

5. Supports the development of the matrix organization, promoting coordination between functional and project/task managers

6. Assists in developing performance metrics for assessing program and project performance

Specific project administrator tasks

1. Define a consistent approach to project plans, scopes of work, work breakdown structures, schedules, budgets, and other elements of the product development process.

2. Build capacity to anticipate and flag project issues and help resolve them, and to report effectively on corrective actions and progress

3. Produce project schedules, budgets, and special analyses as requested by the general manager and program managers

4. Build project team capacity to estimate, schedule, plan, and monitor through information and training support

5. Develop a longer-term project resource planning process and resource database to identify and acquire necessary resources to implement future projects

6. Develop tools to support each team member in documenting and accomplishing assigned tasks, to ensure that everyone knows what has to be done when, who has to do what, and what each job priority and sequence is

7. Assist the finance department in developing performance metrics

8. Assist in documenting and reporting on project progress

Program review and monitoring and the senior management role

Program review is not an ad hoc system of probing into a particular project whenever problems occur; the review process requires a basic and high-level program management approach, training and empowering project managers to make the right decisions and consistently reviewing milestones and anticipating issues. There is a clear differentiation between the role of the project manager and the role of the program or senior business manager.

Such a program management system uses a consistent approach and set of tools and techniques for planning and implementing task-oriented system development projects or professional services efforts, with definable start and end points and distinct deliverables. Deliverables can include products as well as services. When implemented across the company, a project management system provides a corporate-wide way of thinking and a set of professional tools and techniques for identifying and meeting customer requirements in a predictable way. Formats for planning, information exchange, project review, and decisions are governed by distinct project phases; reporting requirements; and schedule, cost, and quality controls. Project managers are trained and developed in a consistent way, both in technical project techniques and team leadership, and manage their projects using standard tools. Project teams are established as formal groups, with charters in writing and performance guidelines and criteria for evaluating team members. A project management manual is used to communicate policies, procedures, and support systems, and the system is continuously improved using business process reengineering tools. A common language governs communication about projects, both across the company and vertically through the hierarchy. Senior management is an integral part of the project management system, serving to create the conditions for success and using project information for business development.

Organizations face six key issues in their upgrade of the project review processes:

- *Find a balance in project information needs.* How can senior management arrive at an effective balance between increasingly excessive project task structures for complex projects and broad, corporate information for senior managers?

- *Conduct effective project reviews.* How can senior management obtain actionable information from project managers during project reviews without inhibiting the project management process?

- *Develop consistent application of project management tools and techniques.* How can senior management develop a disciplined project management system that project managers follow without creating expensive, time-consuming, and detailed reporting systems?

- *Build a project manager pool.* How can senior management identify core competencies and build a reliable project manager cadre?

- *Coordinate projects across functions.* How can senior management ensure that projects are cross-functional and avoid insulated, company "silos?"

- *Use project management tools and techniques to develop professional services.* How can senior management use project management practices to expand its professional services and outsourcing business?

Senior Management Roles: Creating the Conditions for Project Success

In a firm committed to improving program success, senior managers are responsible for establishing and supporting a project management and professional services system that accomplishes corporate goals. However, projects are also a vehicle for building customer relationships and partnerships. The development of a professional services business relies on building relationships through successful projects. Project managers can identify new business development opportunities during the process if they are attuned to where the company is going and senior managers take interest in "actionable information." Actionable information is information that is produced during a project, product, or service development process that helps senior management accomplish *their* key strategic and business development goals.

According to authors Englund and Graham (Robert J. Graham and Randall L. Englund, Creating An Environment for Successful Projects: The Quest to Manage Program Project Management, Jossey-Bass, 1997), senior management is responsible for creating an environment for successful projects through the following 10 roles:

1. Leading the transition to a fully project-based organization
2. Aligning projects with the company's strategic business plan
3. Understanding the impact of senior managers on the success of projects
4. Developing a core project team process
5. Ensuring an effective project management organizational structure with clear lines of authority and responsibility
6. Developing a project management information and project review system that produces "actionable information"
7. Developing a plan for identifying project manager core competencies and a selection and development process
8. Developing a learning organization, stemming from project management documentation and processes

9. Developing initiatives to improve the project management system

10. Developing senior management's ability to manage project managers

Later , these 10 roles will be addressed in further detail.

The challenge lies in the area of potential "project overkill," avoiding the laying on of inordinate reporting and detailed planning requirements and software that drive project managers into a far more detailed picture of their project tasks than even they need to manage. Furthermore, the risk is that detailed reporting requirements will detract from a project manager's ability to "rise above" detailed tasks and subtasks to produce broad, actionable, business development information for senior management's attention.

The timing for the development of a consistent project management and review system is best suited when the following conditions exist:

- The company is already embarked on improving its project planning and status reporting process and project documentation as part of a broad quality program.

- The company is heavily committed to education and training as a way to equip its people for global competition.

- The company is moving in the direction of more standardization of products and processes, thus the consistency with a disciplined project management system.

- The company is going through change and transition, and recognizes that employees at all levels need to know how to link their project work with the company's changing strategic direction.

- Change itself will be facilitated by internal, change-oriented projects, managed like product and service projects.

- The transition to a professional services format for customer relations will require a focus on front-end definition of project deliverables, outcomes, and results.

- The company's customers themselves are going through major structural change; thus, the company needs to help customers organize new marketing, product development, and project management processes.

What Are the Typical Project Management Windows for Program Review?

Whether product- or service-oriented, program management processes should involve distinct, predictable windows for review. A standard approach to phasing allows all project and senior managers to understand project progress in terms of progress through common phases, providing predicable "gateways" for communicating actionable information.

While there are always variations on the theme, the four basic project management phases include:

1. Concept

2. Project definition

3. Design and implementation

4. Project close-out and follow-up

The following table shows the major activities in each phase of a typical project for review:

Phase 1 Concept	Phase 2 Project Definition	Phase 3 Design and Implementation	Phase 4 Close-out and Follow-up
Identify and clarify customer requirements	Project definition	Design concept	Financial performance
Planning and project Selection	Establish project team	Engineering studies	Project documentation
Generate new project concepts: product and professional services	Detailed scope of work	Prototype and test	Lessons learned
Screen and rank-order of project concepts	Professional services plan	Installation plan	Obtain customer feedback
Broad scoping	Project plan	Delivery of services	Close out books
Establish and maintain potential project list	Define work breakdown structure	Earned value	Follow-up on potentials
	Develop task list with interdependencies		
	Develop critical path		
	Develop responsibility matrix		
	Enter baseline data		
	Develop risk analysis		

The senior manager role is different from the program manager role.

Introduction to phase 1

While senior managers must play a role in all project phases, the senior management role in the concept and strategic planning phase is critical. This is the point at which senior management can encourage project managers to think through the kinds of project/product/professional services mix appropriate to each customer and target market. Here, senior management sees that early project and professional services concepts are grounded in legitimate customer needs. The role here emphasizes that projects will be selected on the basis of needs, pay-offs, and the company's strategy to develop a broad professional services practice.

Senior management keeps the project pipeline full through a process of encouraging new services and product concepts and designs to be developed to meet customer needs. A customer-oriented project firm will stress the development of a long-term planning process to ensure that the projects with the highest pay-off are undertaken.

Identify and clarify requirements.   This step requires senior management to focus on aligning projects so that they are consistent with corporate strategy and are grounded in customer requirements. This can be accomplished by requiring that proposed projects define customer requirements clearly up front in the planning process. Concepts should include both product development and professional services.

Generate new program concepts.   A company that encourages continuous improvement, out-of-the-box thinking, and a strategic focus will generate new project and service concepts to generate the project selection process. Senior managers need to encourage a certain amount of "free spirit" thinking to create the conditions for innovative, forward-thinking project formulation. They can accomplish this by establishing incentives and recognition programs for new project concepts that align with company strategy. Senior management is responsible for "inspiring" and generating breakthrough concepts by developing incentives, removing barriers, and encouraging project developers to challenge the current way of doing business toward a mix of product and professional services.

Approvals.   Approval of projects is a key senior management function, again ensuring that projects are aligned with strategy. The process of rank ordering and selection should include formal presentations, preparation of proposals and justification material, and consistent criteria. These criteria will include pay-offs, revenue and business development potential, and contribution to the field.

Senior management must be careful to apply consistent guidelines to ranking and selecting project concepts to avoid the appearance of "pet" projects and to ensure that there are processes and incentives for proposing and justifying new project and product ideas. Ranking should be accomplished using corporate criteria, which are made clear company-wide.

Broad scoping.   Program managers should be encouraged to develop the capacity to broadly scope product and services projects in one- or two-page narrative formats that allow senior managers to make preliminary decisions to commit company resources. This is called "broad scoping" the project.

Establish and maintain potential project list.   Potential projects are kept alive through a listing of potential projects, which is updated frequently. Senior management is responsible for choosing projects for implementation.

Introduction to phase 2

This phase further defines the program and project, once approved, and "fine-tunes" the concept, schedule, budget, and team. Here is where the company integrates the project into its project "pipeline" and schedules the work and resources required in detail. Typically, the budget process serves to facilitate project selection, but senior managers need to develop a way to transition from budget decisions to an "approved" listing of projects to be undertaken that is communicated widely.

Senior management's role here is to ensure that product development and professional services projects are defined and costed out clearly in terms of work breakdown, deliverable specifications, schedule, budget, and quality requirements. A consistent approach to project definition, scheduling, and baseline data entry will ensure that projects can be evaluated and reviewed effectively and that senior managers can concentrate on gleaning actionable information from projects rather than second-guessing project managers in their day-to-day project activities. Senior managers must be careful not to lay on reporting requirements that drive project managers into too much detail, at the same time expecting them to see the big picture as well.

Program definition and project scope statements. The project scope statement is a brief description of the product deliverable or cost-reduction improvement. The statement includes the following information:

- Background customer information
- Description of work: What is to be done?
- Unique requirements
- Product and professional services information
- Special considerations and issues

Establish a formal program team. Teams should be formally established, authorized, and chartered in writing by senior management. The project manager is responsible for proposing the composition of the project team. The team will consist of the appropriate technical staff and representatives, as approved by senior management. Establishment of a team indicates company commitment to the project.

Conduct kick-off meeting. The program manager runs the kick-off meeting to communicate the requirements of the project and to discuss assignments, commitments, resource questions, and other issues relevant to completing the work.

Work breakdown and task list. For complex projects, a summary work outline or breakdown structure and a detailed task list unique to the project are developed. The work breakdown structure is a top-down outline of the steps in the project process. The task list, built from the work breakdown structure, indicates task name, duration, start and end dates, and interdependencies, or predecessors. No project should be detailed in the WBS to more than five levels down.

Responsibility matrix.   The program manager and the team create a responsibility matrix, or organization chart for the project, once all tasks are identified. This matrix is a formal assignment structure, which indicates workflows and unit responsibility for each task.

Develop a broad program schedule.   A broad program schedule is developed in consultation with appropriate stakeholders in the work, using the task list and project management software if available. The schedule is graphically displayed as a GANTT chart or a PERT arrow diagram. The critical path of the project is shown to indicate the combination of tasks, which, if delayed, would delay the overall project. A project baseline is established, as is a project schedule and budget, which is the point of departure for proceeding into work performance and monitoring.

Develop bottom-up project budgets.   A project budget is prepared indicating the total cost of doing the work and ensuring the assigned profit margin. The budget is entered into the database system. Project budgets are built from the bottom-up by costing out tasks and rolling them up, "loading" direct costs with appropriate indirect overhead, general, and administrative costs.

Develop risk analysis.   During this phase, a risk analysis is prepared to anticipate the risk involved in the project and to prepare contingency plans. Risk analysis focuses on feasibility, technology, costs, quality problems, instability in customer requirements, and forecasting information that might affect the project.

Senior management must be attuned closely to risks in a project because project teams tend to become advocates of work initiated and sometimes lack the perspective to see increasing risks that threaten program and project success.

Introduction to phase 3

Phase 3 is the design and implementation of the program outcomes and deliverable(s), including products and professional services. It includes execution of the project scope and schedule, and the utilization and consumption of project resources. Phase 3 involves monitoring actual performance against planned estimates for schedule and cost.

Design.   Design involves a detailed fleshing out of the outcomes desired and product and services design. It includes a full description of the performance specifications for all products and systems, as well as a description of the kinds of professional services that will meet customer requirements.

In prototyping and testing, designs are "dummied up" and tested with customers, and final testing is accomplished.

The installation plan is the detailed description of how a system is going to be installed in the customer's operating environment.

The senior management role in this phase is to be sure that the project manager is providing accurate performance data and information during design and testing of the product, and that the customer is satisfied that the design is appropriate.

Implementation.   Projects are implemented according to the schedule, sequencing tasks and monitoring progress at convenient gateways in the schedule. Senior managers should focus on critical issues during this process, including:

- Accomplishments (what has been accomplished, especially unexpected gains)
- Earned value—e.g., variations from schedule and cost estimates (see tools and techniques in the next section)
- Customer feedback (what the actual client of the product or service is saying)
- Performance feedback (team's data from testing and implementation in terms of performance to customer requirements)
- Quality issues (product defects, service inadequacies, process problems)
- Other actionable information (see the section on project reviews for actionable information)

Introduction to phase 4

Phase 4 should be seen as an opportunity to deliver the product and service and continue the relationship with the customer. It should not be seen simply as "project close-out." Thus, the key issue in phase 4 is finding follow-up business development potential from the project deliverable, documenting lessons learned, and moving to the next level of customer partnership.

Financial performance.   The project's financial performance is reviewed at this point, ensuring that revenue and profit margin goals are going to be realized. Corrective action includes reestimating cost to complete and finding cost-cutting measures if appropriate.

Project document project.   Software is documented and a project history is prepared according to a template, including:

- Project planning documents, scope, etc.
- Full files from each project phase
- Financial data
- Scheduling data
- Quality data
- Team performance information

Lessons learned.   The project manager is responsible for documenting lessons learned in a special database of information updated by the project management team. Documented information includes:

- Unique lessons from this project
- Avoidable mistakes and failures
- Communication problems
- Organizational issues
- Technology information

Obtain customer feedback.   A customer survey form is presented to the user or customer of the project. Senior management ensures that customer feedback is obtained and routed to the appropriate senior and project managers for appropriate quality control and management action.

Explore follow-up potential.   Here is where the program manager and team are expected to explore follow-up opportunities with the customer. Senior managers should lead this process, focusing project managers on marketing targets and helping to build partnerships with customers to further professional services opportunities.

From the senior manager's perspective, the key project planning management tools include the work breakdown structure, task list, GANTT chart, calendar plan, task usage chart, and earned value analysis. Senior managers should ensure that project managers are trained in and can use these tools to manage their projects and report on them.

Because Microsoft Project 98 is a useful and popular project management software program, illustrations of these tools in this manual are taken from that software. What follows is a brief discussion of each tool and the senior management perspective on them. Care should be taken, however, with any project management software to avoid too much task detail and structure simply because the software allows it.

A sample of each of these tools is shown following this discussion.

1. *Work breakdown structure (WBS).* The WBS involves top-down planning of the project, beginning with the product or services requirement. It decomposes the project into greater and greater levels of detail. Roll-up of the WBS completes the project, so its purpose is to plan down to four or five levels of detail in an "organization chart" of the project. Remember to keep your project managers focusing on the integration of tasks horizontally and using the WBS to identify the "big chunks" of work at the top of the WBS early in the project that need to be completed and integrated before the project is successful.

2. *Task list.* The task list shows the tasks required for the project, with durations, start and finish dates, predecessors, and resource names. The task list is the basis for preparing the GANTT chart and is the basis for scheduling and resource allocation.

3. *GANTT chart.* The GANTT chart is a bar chart showing each task, its dependencies, and its durations (both planning and actual) against an actual calendar

to permit scheduling and resource planning. This chart is a convenient way of showing senior managers, and the customer, the whole project and its key milestones, gateways, and interdependencies.

4. *Calendar plan.* An actual weekly calendar showing program and project task activity

5. *Earned value analysis.* Earned value analysis is used to identify when the overall project is using resources at a greater or less than planned rate and when the project is ahead of or behind schedule. Because it is dangerous to look separately at the schedule or budget for an indication of how things are going, earned value combines the two. Earned value is important to senior managers because use of the concept provides a simple indicator of *both* schedule and cost variance using one measure: dollar value. From earned value indicators, senior managers can ask questions and assure themselves that the project manager knows how the project is going.

Some definitions:

- *Budgeted cost for work scheduled (BCWS).* This is the estimated cost of the project and each task.

- *Budgeted cost for work performed (BCWP).* The cost contained in the project budget for completed work, plus a proportional share of the budgeted cost of work partially completed. This is the actual earned value—it answers the question, "Of the work you have completed, how did its cost compare with your plans?" In effect, it is asking whether the project "earned" the budget devoted to the work completed or whether the work is actually behind schedule.

- *Budgeted cost of work performed (BCWP).* This is the actual expenditure to date.

The senior manager is interested in:

Cost variance (CV = BCWP – ACWP)

Schedule variance (SV = BCWP – BCWS)

Estimate at completion [EAC = (ACWP/BCWP) × total budget]

As an example, if a project is showing BCWP = $200K, BCWS = $400K, and ACWP = $400K, the project is badly behind because the value of work performed (BCWP) is far below what it should be (BCWS) at this point in the project, but the project has spent project funds (ACWP) as if the work were on schedule.

Narrative report and presentation.   Each project manager should be expected to report on accomplishments, issues, actionable information, and corrective plans in a consistent format across the company. Project reviews, however, do not have to consume large blocks of time, and most review information can flow without the need for meetings.

Sample tools

The purpose of this section is to place special emphasis on focusing on the right information—termed actionable information—for project reviews. Actionable information is information that creates opportunities for senior management to develop technology and marketing concepts and accomplish corporate business development goals and objectives. Project reviews and reporting systems should focus on actionable information as follows:

- Status of project, issues, and plans
- Overall progress compared to plan
- Accomplishments
- Issues
- Plans for corrective action
- Customer feedback
- Marketing information
- Developing customer requirements
- Business process outsourcing potentials
- New product and market information
- Successful presentation approaches
- Key contacts at high levels in the customer organization
- Lessons learned in project management
- Organizational support issues
- Standard templates
- Tracking software successes
- Project management process improvements
- Financial and costing controls
- Successes in identifying customer requirements
- Successful communication and reporting strategies
- Change order management
- Breakthrough manufacturing or product development technology
- Software innovations
- New equipment performance information
- Industrialization opportunities
- Transferability
- Information on competitive challenges
- Intelligence on competitive developments

- Lessons learned by customers from competitors or stakeholders
- Financial and productivity data
- Project performance
- Successful approaches to leading the project team
- Team performance feedback from customer and from team members
- Team core competency information, e.g., feedback from team composition and skill needs

Information on Competitive Challenges

- Intelligence on competitive developments in financial and insurance professional services
- Lessons learned by customers from competitors or stakeholders
- Financial and productivity data

Project team performance

- Successful approaches to generating excellent team performance
- Feedback on team interaction with the customer
- Insights on core competencies needed for project leadership and team membership
- Successful organizational approaches

Opportunities for Customer Partnering

- Shared interests with customer for alliances
- Business process outsourcing opportunities
- Opportunities for developing professional service relationships
- Joint project team opportunities

Earned Value

- Tracking data showing major schedule problems and cost variances
- Cost to complete
- Plans for corrective action

Senior management is responsible for obtaining actionable information from project reviews and avoiding costly project review sessions, which sometimes second-guess project decisions without adequate information.

Summing Up: 10 Key Roles of the Program Manager

Recounting the 10 key roles of program management addressed previously, here are more details on specific senior program management actions to be taken:

1. Leading the transition to a fully project-based organization
    - Establishing a sense of urgency on project management success connected to company performance and individual compensation
    - Creating networks of project sponsors and support staff
    - Communicating the changed vision, including the emphasis on professional services projects
    - Empowering project managers to perform
    - Connecting project management to strategic initiatives, e.g., professional services business development

2. Aligning program and portfolio projects with the company's strategic business plan
    - Getting involved early in project generation, planning, and approval, and reflect the movement toward professional services
    - Making sure project managers understand the business plan
    - Raising issues early on misaligned projects

3. Understanding the impact of senior managers on the success of projects
    - Getting feedback from project managers on the effectiveness of project reviews and other senior manager involvement
    - Communicating with other senior managers on impacts
    - Listening to project managers

4. Developing a core project team process
    - Encouraging teams and teamwork
    - Officially chartering teams in writing
    - Providing team incentives

5. Assuring an effective project management structure
    - Delegating authority in writing
    - Clarifying reporting relationships
    - Giving authority with responsibility
    - Building capacity to manage
    - Focusing on big picture
    - Avoiding conflicting messages on project detail and reporting needs

6. Developing project management information system
    - Communicating project review information needs

- Benchmarking good project review approaches
- Providing software and training
- Investing in project management information systems
- Controlling level of detail in project planning

7. Developing core competencies
   - Identifying what makes a good project manager
   - Developing project manager profiles
   - Publishing core competencies
   - Integrating into hiring

8. Developing a learning organization
   - Deciding on documentation needs
   - Creating a sense of urgency on lessons learned
   - Identifying costs of repeated mistakes
   - Assigning responsibility

9. Developing project management improvement initiatives
   - Looking for opportunities to improve the process
   - Asking project managers and customers for ideas
   - Giving priority to internal improvements

10. Developing capacity to manage project managers
    - Communicating with project managers on their issues
    - Focusing on leadership capacities and core competencies
    - Preparing individual development plans

What Does the Program Manager Expect from Project Managers?

The project manager is responsible for knowing the project management system and appropriate roles and functions. This person is responsible for the establishment and performance of the project team and provides support and leadership.

The project manager is accountable for meeting schedule, quality, and cost goals of the project and satisfying customer requirements. This person establishes teams; works with other engineers, managers, and support personnel to schedule and budget projects; serves as a contact point for both the client and internal team members; and monitors project performance.

The project manager is expected to resolve problems and communicate with senior management on resource or performance issues and to be aware of and respond to the actionable information needs of senior management. He or she is expected to plan and run project meetings efficiently and to assure good project documentation.

What about project team members?

The team member is responsible for completing tasks assigned within the project work breakdown structure, accepting responsibility for assigned tasks and completing them on time and within budget, attending meetings of the team, and working cooperatively on project tasks with other team members. Team members are liaisons between the project team and their functional departments. They must work collaboratively with each other for the project to be successful.

Program Manager Review Agenda

Program managers review program and project status through program review meetings and reporting processes.

Here is an example of a typical agenda for such a meeting.

1. Overview of program, schedule, and major accomplishments and activities
2. Current performance to plan
   - Changes
   - Variances
   - Percent complete
   - Estimate to complete
   - Plan updates
   - Recommendations for corrective action
3. Issues and concerns
   - Scope
   - Quality
   - Resources
   - Risks
   - Technology
   - Staff and team
   - Costs
   - Organization
4. New program opportunities
   - Plans for the coming week
5. Program review survey

It might be useful for businesses contemplating improvements in program review to benchmark the process with other leaders in the field. Here are some of the questions that might be asked in such survey.

Sample Program Review Questionnaire

1. In your experience, executive or top (program) management reviews of project status:

   - Are very useful to you as a PM and should continue
   - Are not very useful to you as a PM
   - Are disruptive to you as a PM
   - Don't happen

   Comments?

   ----

2. Rank in order of importance the following agenda items for a project review by top management: 1=high; 2=middle; 3=low:

   - Schedule status
   - Cost status
   - Quality status
   - Problems and issues
   - Accomplishments
   - Technology lessons
   - Marketing opportunities
   - Team performance issues

3. As a project manager, list the key project review agenda items you think top management should be using:

   ----

4. If you are an executive or top manager overseeing project managers, list the review agenda items you think are appropriate:

   ----

5. When should top management project reviews be held?

   - Every week
   - Every month
   - Every quarter

- At the end of key tasks
- Other

6. Who should participate in top management project reviews?

- Customer
- Project team
- Project manager
- Top management
- Other

7. Are top management review agenda items distributed before the meeting?

- Yes
- No

8. Other comments. Do you have more comments on executive or top management roles in project review?

Program Management Information System

A program management information system provides project information to all team members and stakeholders on a network, which ensures easy access and communication. Program managers and department managers will have access to all schedule data on the network. Microsoft Project is preferred for developing schedules, saving baselines, and providing monitoring information because it is compatible with other Microsoft programs, including Word, Excel, and PowerPoint.

Procedures for Schedule Tracking and Reporting

To set the stage for effective tracking, goals to support program review must be set. Here is a suggested list of such goals:

1. Time-phased schedules for all ongoing programs (available on the network). Program managers working to scrub schedules where possible.

2. Resource-leveled schedules for ongoing and new programs to avoid overburdening resources (we are not resource leveling as yet because our resource pool is not yet accurate; working with departments to reach a point where we can act with confidence on an "overallocation").

3. Predictions of changes and how they will affect future milestones (the program managers are doing this daily, but not systematically; linkages across projects not there as yet).

4. Staff are given clear assignments from current schedules so they know what they are expected to do and when (when the resource pool is cleaned up, each staff member will get a weekly task list from their department manager).

5. Departments use resource information as s manpower planning tool.

6. All schedules are linked together to resolve conflicts and to predict effects of changes in one program on another program in terms of costs, manpower, scheduling, etc. (the program managers pick up these conflicts and work them now, but not through a system; they are not looking at costs and manpower impacts as yet).

7. Common schedule and task format; common work breakdown structure, with dictionary.

8. Budget and track actual labor costs (you have indicated that tracking hours would be sufficient). I am trying to work out a way to capture both, but with emphasis on hours.

9. Integrate test equipment into the scheduling process (no progress as yet).

10. Earned value, BCWP, ACWP, etc. (can't do until we start entering costs into schedules and save as baselines).

11. Internal program plan documents for each program, including overview, scope, schedule, budget, etc.

12. Reports:

   ■ Weekly updated schedules with conflicts resolved (you have updates weekly, but real conflicts will not be apparent until resource pool is cleaned up)

   ■ Weekly program summaries and risks

   ■ Monthly manpower reports, by group, with rolling 12-month forecasts (again, when we get a cleaned-up version of the resource pool report, we can report manpower needs)

   ■ Monthly earned value reports (need to decide how to get there)

Program tracking information

Program managers manage the tracking system, including:

1. Tracking starts when a schedule is saved as a baseline and linked to a central resource pool.

2. Criteria for tracking system:
   a. Simple
   b. Flexible
   c. Regular
   d. Tracks the right things
   e. Focus on exceptions and variances from plan

3. Tracking updates project and/or selected tasks as follows:

   a. Actual finish (when tasks start and finish on schedule, Microsoft Project will automatically update schedule)
   b. Percent complete

    c. Actual duration

    d. Remaining duration

    e. Costs (if applicable)

    f. Updates to be entered weekly

4. Tracking determines whether a new baseline is required, if developments in the project indicate that a major summary task should be added or overall structure changed. For example, a new baseline may also be required if a change in scope is directed by top management.

5. Procedures for tracking:

    a. Tracking and reporting will be integrated into weekly report system

    b. Department contacts designated to gather tracking information for their department, or individuals doing the work provide tracking information

    c. Measures tracked:

        i. Percent complete

        ii. Changes in start or finish dates

        iii. Resource assignments

        iv. Minor new tasks

        v. Major summary task changes

    d. Options for gathering data:

        i. Program management office requests progress information weekly by e-mail from department contacts to program management office. The people actually responsible for doing the work should provide progress. This information will be obtained as part of the weekly team meetings.

        ii. Departments have read-write access to folders on the network and update schedule weekly, entering their own updates. (This would probably not work well; relying on department managers for program management information is liable to create problems since they are not as close to the work as the program manager is.)

6. Access versus reports

    a. Program plans and schedules in planning stage are on program management office admin directory, updated as the program managers and departments build to the baseline.

    b. Once baseline is saved, file is saved.

    c. Weekly updates are made to the schedule, with tracking GANTT showing variances.

    d. Current schedules are on program management office admin directory, with departments.   ·

    e. Program managers will keep all historic versions of schedules in a separate folder labeled "obsolete."

7. Program management office concept: Program planner and program managers seen as one office, working as a team: division of labor between program managers and program planner in tracking and reporting:

a. Program manager

   i. Get update information in weekly team meetings
   ii. Change schedules (or ask program planner to change schedules)
   iii. Assessment and interpretation of updates
   iv. Identify conflicts; facilitate resolution
   v. Corrective action and report to general manager as necessary
   vi. Restructuring schedule, if necessary

b. Program planner

   i. Flag current and new tasks for the week (should not replace program manager's critical responsibility to track tasks)
   ii. Hand out resource assignments
   iii. Identify conflicts; facilitate resolution (Jeff feels this is critical function of program planner)
   iv. Request update information
   v. Print out following reports for general manager, program management office, and department managers:
      1. Weekly hardcopy updates of all schedules
      2. Summaries of variance between planned and actual by schedule (text and graphics)
      3. Problem areas with a plan for correcting the problems
      4. Risk management report
      5. Twelve-month lead time resource assignment reports, summary plus department functions (software engineers, electrical engineers, test technicians)
      6. Labor utilization and earned value (where possible–longer-term goal)
      7. ETC (estimated time and budget to complete–longer-term goal)
      8. EAC (estimate budget at completion–longer-term goal)

8. Meetings

a. Weekly team meetings
b. Weekly program management office meetings (Ray, Jeff, Steve, Bruce)

9. Coordination with purchasing, manufacturing, and APC

a. Separate reports to flag purchasing, manufacturing, and APC on requirements and expectations (parts, LCDs)

10. Finance reporting

Finance is going to send us monthly reports on hours charged against job numbers, but is looking for guidance on whether and when we go to subaccounts. They can also rack up all costs monthly against job numbers if we want to see it.

Program Review Meetings and Reports

Program team meetings and reviews occur weekly with agendas set by the program manager.

Reports, meetings, and documentation

Periodic reports to the general manager:

- Weekly schedule updates with all tasks updated and conflicts resolved
- Weekly summaries (text and graphics) of progress in each program area, with issues, risks, and corrective actions
- Monthly earned value (schedule and cost variance) reports for each program phase, including estimate to completion (ETC) and estimate at completion (EAC) indicators
- Monthly workforce reports on manpower resource utilization, including a 12-month forecast broken down into individual functions, e.g., software engineers, electrical engineers, test technicians, etc.
- Weekly corporate report
- Weekly program planning meetings

Issues in program management improvement

Continuous improvement in program management practices requires that organizations:

1. Be consistent across programs
2. Scrub schedules the same way
3. Treat baseline, current, and actual the same way
4. Monitor and report on progress the same way
5. Treat the whole program management improvement effort as a team project

Design and follow consistent planning procedures

1. Create draft schedule
2. Assign resources
3. Link to resource pool
4. Scrub draft schedule
5. Level resources
6. Enter costs
7. Save baseline
8. Put in network folder

Monitor baseline #1

1. Save baseline #1 to PC
2. Gather data weekly on *actuals,* percent complete, change in start and finish, etc.

3. Report (against baseline #1) on earned value, slipped tasks, costs, etc.

4. Report on resource impacts

5. Revise schedule as necessary

6. If minor changes, save new current version

7. Revise schedule if major changes

8. If there are major changes in schedule, save new baseline (interim)

Monitor against baseline or current

1. Save interim new baseline or current to PC

2. Gather data against new baseline or current

3. Report data on new baseline or current

Report consistently

1. Integrate into weekly report

2. Common format

3. Reliable reports

4. Add costs

Network Information Exchange

Find an acceptable way to provide the network team with baseline and current versions, actual performance information, and resource tasking in a predictable way.

Program managers and workforce planning

Long-term workforce planning begins with some understanding of what your current program and project workforce can do—e.g., its capacity to put out the work, especially in a global setting. Standards are developed that relate capacity to performance. The process involves relating workforce levels and/or staffing mixes to standards for work and output. For instance, you might define what a given level of software certification engineers can produce based on past history—e.g., five engineers have supported three concurrent product development projects with turn-around on software certification documents of three weeks. Then these standards are used to estimate what various alternative staffing levels might produce in the way of more capacity to produce certification documents sooner or more certification documents in a given period. The issue in this case would be to see what new staffing levels would be required to put out twice as many software certification documents in the same three weeks.

The more immediate workforce decisions are made on the basis of schedule conflicts created by current scheduling impacts on the current workforce. Whenever you show a current conflict created by current and planned schedules and have to put off some program step or activity because of scarce resources, you know that you could avoid schedule slippage by hiring new staff. But you don't know how much staff you need until you have done the previous exercise.

Good workforce planning involves the staff in the planning process. Providing team members with information on assignments helps to give them ownership on key project issues such as task interdependence and resolving resource conflicts. It also allows them to assist in estimating levels of hiring necessary to raise the capacity to improve performance to various levels to meet future demand. To provide access, program managers or the program management office provides the central resource pool, reflecting assignments for all scheduled programs to team members. If staff confirm that these resource plans accurately reflect the work actually going on, that would tell the manager that he or she is on the right path in the scheduling process. That is, there would be reasonably good alignment between what is being scheduled and what people are doing. It would *not* tell the manager whether the scheduled tasks were the *right* ones, or that the team had everything that needs to be done captured, just that people are doing scheduled work that is aligned with expectations. Positive staff reaction to this information would tell the manager that staff need more information on expectations and assignments than they are getting.

Feedback

Here is typical feedback from such a system:

1. All staff appreciated what we were trying to do in tightening the scheduling process, relying more on the schedule baseline as the "official" starting point, working out conflicts, and informing them of scheduled work assignments.
2. Most found the data pretty accurate, with major exceptions.
3. Most objected to the way tasks were described, and many could not immediately put the cryptic descriptions of a task in context of a program.
4. All had widely different views of how to improve the process, but all appreciated the information.
5. They confirm that we are *not* capturing a lot of work in the scheduling process, work that is actually going on, including troubleshooting and integration.
6. Not as much as I expected on conflicts.
7. As an aside, several staff appear underassigned, but extremely busy doing unscheduled things.

A sampling of typical comments that might be expected from such a survey includes:

1. I didn't know I was supposed to do these tasks.

2. My current work in software integration and troubleshooting is not captured anyplace.

3. Managers should be reviewing the assigned tasks from the master schedules; it appears as though slippages are not being accounted for.

4. This information is accurate, but I know that slippage is not reflected.

5. It was nice to see only my time and schedule; the format of one person's hours and tasks is good because we rarely see personalized tasks.

6. Information not accurate and doesn't capture general support in failure reporting and corrective action system work.

7. My work is difficult to capture because it goes across programs; these assignments reflect only program work, not department work; we do a lot of across the board work not captured in schedules, e.g., support to manufacturing.

8. This was very useful; I didn't know I was doing anything other than my current responsibilities.

9. Work not mine; assigned to someone else; inaccurate.

10. This information contained a task I had never heard of; checked with program manager and I guess I will be doing it, but later than shown on the schedule.

11. Some of the work hours are worst-case scenario; I cannot accurately assess how long it will take to complete a document; they vary from program to program. I tend to keep my eye on the project deliverable due date as a general reference point for my work. You might show us endpoints with this information.

12. This was useful, but it may be beneficial to indicate a priority order based on long lead time tasks, critical path items, etc. to show what we should work on first, second, third, etc. to optimize the schedule.

13. This was useful, but I have been reassigned as of 9/1 and so this schedule is out of date.

14. I was surprised to see all of the work projected for me, but this is the first time I have seen it all. Need to be introduced to it.

Training and professional development

Training in program management approaches, tools, and techniques is provided to all program managers, as well as team facilitator training and project management software training. A typical training and development program consists of workshops on:

1. The program management organization

2. The global setting for program management

3. Job functions and competencies

4. Program planning

5. Scheduling and resource management

6. Negotiating skills and conflict resolution

7. Staffing skills

8. Team facilitation

9. Change management

We addressed several models and themes for defining the program manager in this chapter. No one model will fit all situations. However, a strong understanding of the models allows a person to select the right tools they need for their organization and the challenges of their position to successfully deliver the desired outcomes of a program. Program managers set the vision for the program and articulate it to the team. Program managers focus within a continuous improvement environment where they grow and develop their team and build and improve upon program business processes. The successful global program manager has depth and breadth of knowledge and experience in project, program, portfolio, and business management.

7

Partnerships, Contracts, and Procurement

Introduction

Chapter 6 covered the functions and roles of the program manager. One of the key roles of the program manager is to negotiate and manage contractors and work with partners in directing global programs. This chapter explores some of the key issues involved in global program management partnering and contracting.

The chapter addresses the issue of how program and executive managers approach the partnering, outsourcing, and contracting of programs and projects globally, and how their challenge is different from traditional project management. Rather than dwell on the mechanics of contracts and procurement that are more the concern of project managers, this chapter is aimed more at the issues and decisions involved in partnering with organizations and companies to leverage intended program outcomes and benefits. Programs are seen here as *targeted business operations with goals that are designed to change systems, customer or client behaviors, or change the business, or to change social and economic conditions in a foreign country.* In this chapter we are describing functions and activities associated with good global business and program management, not necessarily "what program managers do." We recognize that some businesses engage in program management practice without calling it "program management," and others call what they do program management but do not really carry out global program management practices. The reader is cautioned not to assume that position and organizational titles actually portray what businesses and people actually do; it is the activity that defines the program management function, not the words.

Innovation and Outsourcing

We do not start our conversation on program management and partnering, contracts, and procurement with a cookbook approach. These are matters that involve the mechanics of contracting and procurement. It is not that program managers don't care about the micro-aspects of procurement projects—it is rather that program managers enable their project managers to deal with these processes. Program managers must rise to the 30,000-feet perspective that enables them to see global strategic principles and issues. These principles and issues include:

1. Program managers should see global e-procurement as a series of systems for partnering, collaborating, negotiating contracts, and completing procurement processes to access talent, innovative ideas, and lastly materials. Materials and equipment are not the keys in this process; it is creative and innovative ideas on customer requirements, business opportunity, new technologies, and economic trends.

2. Know what is out there. Many global companies already provide platforms to help program managers access local and actionable information on procurement and outsourcing. There are many systems out there for accessing information—e.g., Skype and Google. Programs survive and thrive first on ideas that come from talented professionals; if you cannot access these professionals, you cannot take advantage of their ideas.

3. It's getting personal. The personalization of information has created operating vehicles for collaboration—e.g., Facebook, the social networking platform; Wikia, a content-sharing website that creates wiki-based communities; Basecamp, project management collaboration software; WebEx, Cisco-driven on-demand web collaboration software for web and video conferencing; AdMob, mobile advertising marketplace; and LiMo, Linux for Mobile platform to build mobile applications.

4. You may have to reorganize and restructure how companies access and use partners and suppliers, moving from traditional vertical integration to initiating programs to access specialized global suppliers. As we indicated in the introduction, program management has become a core business activity that shapes current functions of the business through targeted projects. The company may have to develop the administrative and process capacity to deliver on those outcomes through disciplined project management and strong support.

5. These new platforms for electronic collaboration and virtual teams may require reshaping program goals, benefits, and objectives to achieve more ambitious outcomes. New program goals will reflect the capacity to reach any part of the world with products and services.

Developing partnerships and supplier relationships

As we indicated in Chap. 6, contractual negotiations in many countries are not as formal or predictable as American contract negotiations. Sometimes, program contracts and local arrangements are made with a handshake or with a verbal promise instead of a written contract. Unlike the advice of Fisher and Ury to focus on issues rather than people, the global manager may have to focus on *personalities and positions* to affect contract arrangements in a personal way. Sometimes, money crosses hands in an arrangement that Americans might call bribery. Sometimes, foreign governments doing business with an international engineering or systems team will fragment responsibility in the program team so that no one country representative on the team can dominate it. The home country will play the prime contractor role in the program team and coordinate the program players from other countries to ensure local control. Therefore, negotiations within the team often involve not only the home program sponsor, but many of the team players as well, because all the players need to adjust to local conditions.

The global program manager must be aware of these variations in order to negotiate for program support and facilities. For instance, a cable telecommunications supplier developing a major international cable system in the Middle East might encounter a number of negotiating challenges:

1. Negotiating roles and functions for the program team as part of a larger, more international team of telecommunications professionals and technicians, the global program manager must fit his or her team into local roles and settings. This may mean giving up controls and leverage in cable installation, such as cable access, liabilities, and maintenance costs. Entering the negotiations with this in mind, a global manager must be aware of the key priorities of the program that are not negotiable (e.g., no competitor access) and those that can be compromised in making arrangements with a local jurisdiction or prime company (e.g., maintenance costs and systems). Local easement and land use often present major issues to a telecommunications program, including local access to required rights of way for cable systems and hubs.

2. Cable installation or maintenance contracts may be written by 50 different jurisdictions, each with a different format and set of terms and conditions. Thus, the standardization and formatting of contracts in a global program are often impossible. Additional time and effort may be necessary to affect contracts that take only minutes in domestic programs.

3. Relationships with partners and competitors take a different turn in international situations, requiring tailored negotiation approaches. For instance, a cable company may have to negotiate with local competitors to market cable services, sometimes even entering into compacts or partnership contracts. Global program managers must often negotiate these arrangements sensitively so that program and business assets are protected while doors are opened to program development in the local setting.

4. Government regulatory constraints. Governments in the Middle East, for example, may have different or no safety regulations to govern local cable installation, yet local foreigners on the program payroll can sue the program company in the United States under certain circumstances. Thus, the lack of local safety rules does not eliminate the need for safety management practices.

Licensing and other program issues overseas

While domestic programs often face major licensing and permitting issues, programs doing business in other countries experience more complicated problems. This is because it is difficult to keep up with changes in foreign governmental regulatory processes. Authorities in foreign countries can also change standards and regulations quickly to respond to program issues such as delays and quality variances. The following case is illustrative.

The global avionics program management case

A French aircraft manufacturer, Aeronomics, needs to procure a suite of avionics instruments for its business aircraft and seeks out bids from potential instrument developers because it does not have the capacity to produce instrumentation. (A *suite of instruments* is designed for a given aircraft cockpit configuration.) An American company, Basic Systems, Inc., responds to the bid offer and wins the bid based on its program management experience, cost control, and quality performance. The contract with Basic is for a fixed price at $50 million for a series of instruments, despite their attempt to negotiate a cost-reimbursement arrangement. Basic contends that the standards for avionics instruments internationally, unlike those in the United States, change so often and are open to such a wide variety of interpretations that they cannot predict the cost of compliance in design and production of a suite of instruments. Aeronomics insists on a fixed-price contract with a clause that allows for renegotiation if standards change. There is a large incentive fee, $2 million, if the program is completed on time. There is a $10 million limit to Basic's liability for the program.

Because of a number of issues, design changes, scope creep in component production, late material from suppliers, and staff performance problems at Basic, they fall behind in the development phase. Several projects aimed at particular instruments (e.g., altimeters) do not meet current standards set by the civil aviation agency in several countries in Asia where the aircraft will operate. Several instrument projects in the program are finally completed six months late, delaying Aeronomic's aircraft production schedules by a year. Cost overruns exceed $10 million.

Basic delivers the instruments and gets technical product approval, but files a contract claim for an additional $5 million for excusable delays, blaming the delays primarily on their suppliers and instability in local avionics standards. Aeronomics hires a law firm to defend its decision to reject the claims. The law firm indicates that they will need to review the technical and legal requirements of the contract to determine whether Basic satisfied the requirements. The contract states that disputes must be submitted to an international arbitration panel.

Aeronomics refuses to pay Basic on the last $3 million invoice submitted for final work completed. Basic contends that it has completed all milestones in the program, but Aeronomics argues that the last three milestone deliverables have not been delivered. Basic also contends that currency rates had changed since negotiation of the contract, at their expense.

Things get more complicated when Basic seeks help for its claim from the U.S. government, contacting the National Institute for Standards and Technology (formerly the National Bureau of Standards) as the U.S. representative involved with the International Organization of Legal Metrology (OIML), which is charged with developing the standards code under the U.S Trade Act of 1979. Basic also argues its case with the American Society for Testing and Materials (ASTM) and the American National Standards Institute (ANSI), which is charged by the trade legislation to represent U.S. interests in nongovernmental international standards organizations. Aeronomics counters by bringing in its own French avionics regulatory authority to defend changes in avionics standards based on recent aircraft accident data.

Here are some of the issues in this case from perspective of the program manager's job:

1. Had the program manager for this work been chosen by senior company management when the company chose to bid this work?

2. Did the program manager work to develop a close relationship with company executive management responsible for this program area?

3. Did the program manager participate in the decision to bid for this international work?

4. Did the program manager participate in the negotiation of this fixed-price contract, and was there any discussion of promoting a cost-reimbursement contract?

5. Did the program manager have a close relationship with the company's acquisition/procurement officer so that the program manager was brought into company discussions on the bid proposal?

6. Was the program manager aware of the technical challenges in designing and producing this suite of instruments?

7. Was the program manager aware of the national and international standards that applied to the products of this work and how they were changing?

8. Did the program manager attempt to find out about Aernomics to determine if there was a potential cultural and technical fit with this French company?

9. How did the program manager handle delays in several projects when they occurred; was earned value monitoring used and regular program reviews conducted, and, if so, why were program problems and earned value red flags not addressed?

10. Did the program manager approve the final invoice submitted to Aeronomics, and was the program manager aware that Aeronomics would not pay it based on lack of deliverables?

11. Did the program manager develop a good working relationship with the counterpart French Aeronomics program manager, and were they in communication through an effective reporting system throughout the program on schedule, quality, and cost issues?

12. Was the program manager aware of the company's attempt to bring in American government regulators and standards institutions to the case and how governmental intervention might affect the program, and even the business itself?

Postscript

While doing business internationally offers exciting business opportunity, it also opens up new risks and issues for a business, especially in designing and delivering new programs. After all, programs are creative activities focused on change and improvement, and on new products and services. Change is the core characteristic of program management. And while it often appears that the devil is in the contract details in an outsourced program dispute, experience suggests that conflicts and disputes result first from ineffective partnering relationships and from the lack of continuing communication that results.

This issue—development of working relationships—is the domain of the program manager, as both the official representative of the company and as the leader of the many project managers and supporting managers involved in a global venture. This is made more difficult when business is being done in foreign countries and where communications are often through web-based products and services. And while a premium is placed on the virtual project team in these cases, the critical success factor in a program is rooted in the lack of a working relationship between the leadership of both the owner/sponsor and the contracting agent.

Doing good with program management: collaborative programs targeting international social development

Worldwide social and economic programs.   C. K. Prahalad's *The Fortune at the Bottom of the Pyramid: Eradicating Poverty Through Profits* describes private-sector initiatives that use targeted programs to improve social and economic conditions (e.g., reduce poverty, improve public health) throughout the world. These programs leverage their business services and products in association with partners and collaborators in a local market where the company hopes both to bring about fundamental social or economic change and to profit from the effort. They make money doing good. A case in point is an Indian company, Hindustan Lever Limited (HLL), that decided to leverage its soap brand to eradicate family health problems in India and countries in Africa and Asia. It is in these cases where the concept of program management is best understood as a broad, benefits-oriented initiative involving cross-sector collaboration over an extended period and with significant costs.

What is a program in these cases?

In cases of social program development, a *program* is a collaborative partnership between a private enterprise, government, nongovernmental organizations (NGOs), and academia to achieve a broad social outcome or benefit. While some projects within such a program focus on building systems and infrastructure to support the outcome (e.g., communications, product development, team development), the program itself is aimed at fundamental change. Needless to say, programs often go off course and are difficult to manage, and require a healthy capacity to work with ambiguity and paradigmatic change.

Program strategy

The HLL program strategy was to focus on price performance, innovative and hybrid solutions, scalable solutions, conserving resources, new product development, process innovation, deskilling work (e.g., taking into account low skill levels in targeted markets), customer education, anticipation of hostile environments, interface management, and concentration on the broad architecture of the system or platform. In addition, HLL brought a capacity for accountability through "careful evaluation of investments in projects to ensure success"—in other words, good program management.

Benefits management

Benefits intended in this program involved fundamental change in the personal behavior of citizens in the use of soap in their daily lives, since many public health issues in developing countries result from poor personal hygiene. Thus, the benefits of the program were not in the volume and profitability of soap sales as much as in the visible and measurable improvement of public health in a local target market.

Program governance

Program governance in this case was not strictly through a classic "program management" office and methodology. The company saw this program as a mainline activity to be carried out through its corporate structure. The aim of program governance was to target business resources and evaluate results in terms of the broad health goals of the program.

Strategic partnerships and stakeholder management

HLL's program stakeholders included the London School of Hygiene, U.S. Centers for Disease Control, the World Bank, trade associations, country-specific state and municipal jurisdictions, and nongovernmental organizations. A partnership with all of these institutions was necessary to accomplish the program's broad goals. What made this especially difficult was that this was a private-sector, not government-sector, initiative, thus requiring government entities to accept leadership from a profit-making company.

Program design

Program design was accomplished in the context of a three-legged stool for changing consumer behavior in targeted countries and localities: (1) initiation and information, (2) large-scale propagation, and (3) reinforcement and preparation for sustainability.

Programs were to be carried out through small teams of specialists who would scale the work and collaborate with local communities to set the groundwork, research the conditions, promote and sell the products, and evaluate results, all with strong corporate support from HLL.

Lessons learned

Program management in the context of broad social programs is different from narrower program management involving the design of new systems and platforms. HLL and other companies engaged in broad programs have learned to:

1. Foster a communication system that is two-way

2. Leverage the existing infrastructure

3. Avoid program stovepipes

4. Accept that software is not a system (e.g., an integrated system requires a broad approach)

5. Accept that technology alone will fail without change management and capacity building

Classic program manager role in procurement

Programs like HLL's require both strong management focus and a supporting cast for partnering and contracting. The program manager sets the standards for contractor negotiations and performance for all projects within the program, and can contract directly for support to the program from a prime or overseeing contractor. However, the program manager cannot micromanage project contracts. Establishing the performance requirements involves ensuring that all contractors are aware of the quality and administrative requirements of the program and can produce and report within a consistent format. Supply chain management can be employed to provide for a planned sequence of contractor support and to ensure consistent contractor performance.

Outsourcing to a prime contractor to support the program manager involves development of a request for proposal that identifies companies specializing in support to top management, assisting in the day-to-day administration and management of projects and in monitoring performance.

In addition, the global program manager provides leadership in:

1. *Make/buy decisions in a global context.* The program manager approves all project decisions to outsource work overseas to ensure that risks and implications of doing business with foreign companies are properly assessed.

2. *Procurement process variances.* The program manager approves any variance from the standard procurement process (e.g., exceptions to quality and cost analysis, etc.) to ensure a clean outsourcing process.

3. *Standards and requirements across countries.* The program manager approves any variance from written standards for project components, products, and services.

4. *Contractor-based disputes with global impacts.* All disputes are escalated to the program manager if they involve global impacts and/or parties.

Procurement process and program manager roles

The program manager sets the ground rules for how contracts and procurement in general are handled in all projects in the program. The program manager does not micromanage such transactions, but handles issues and escalated problems that arise out of the process. One of the major roles of the program manager is to be sure that the company or agency procurement and acquisition systems work for the program. The process looks like this:

- Define project plan, including goals, customer requirements, business case, team composition, scope of work, schedule, budget, and risk management plan.

- Define deliverables, including products and outcomes, in a clearly stated manner.

- Define the bill of materials which is typically provided through some kind of configuration management system, defining the profiles and sources of all product materials and components.

- Define goods and services to be produced in-house, including the make/buy decision.

- Select bidders and complete requests for proposal. The program manager may participate in identifying potential bidders, especially if the company is participating in a supply chain system in which certain partnering contractors have already been chosen; make sure the project managers appreciate the added benefit of input from a broad spectrum of functional experts to ensure that the solution chosen will suit the company's requirements.

- Prepare bid proposals, including how bid proposals are formatted.

- Evaluate bids and award the contract, with the program manager participating in this process.

- Contract management, including the kind of contract used (e.g., fixed price, cost reimbursement), time and materials, how contractor performance monitoring and payments will be carried out, and how earned value will be reported consistent with in-house work.

- Close out the contract; procedures are provided that ensure consistent processes for closing out contracts after the books are closed and the contractor has performed satisfactorily.

Organizational interrelationships and interfaces

These can be keys to success:

- The program management structure includes how the program management office (PMO) will be organized and how the program manager will report to upper management and be situated with project managers and functional managers.

- Top or department management includes special relationships with program sponsors in business leadership and ownership and key stakeholders.

- If there is an engineering or design function and a matrix structure, establish strong lateral relationships with functional managers who are typically at the same level as program management and sometimes integrated with program management.

- *Licensing and permitting.* This includes making sure that interfaces with permitting and licensing are keyed into global aspects of the program and the local variances are identified.

- *Marketing.* Since the program manager may be responsible for not only delivering program products, systems, and outcomes but also distributing and marketing those products, he or she must have a strong working relationship with marketing and sales, which are typically separate in the organization.

- *Manufacturing.* Program products and outcomes usually involve production and inventory, so it's important to build a strong relationship with production management and engineering.

- *Construction.* If construction is involved with the program, ensure that it is integrated into the program in terms of implementation and quality control, where problems sometimes arise.

- *Testing.* The design of new program products and outcomes will include testing; ensure that all projects use the same international testing standards and reflect local conditions where the product will be used or consumed and that outcomes are monitored.

- *Operations.* A strong interface with operations and services of the company is needed, since many operational aspects are involved with program development and delivery, and some operations may change because of the program.

- *Quality.* Because project managers tend to develop their own concepts of quality, sometimes in close association with the customer but sometimes not, the program manager has to establish a quality assurance and control program for all projects.

- *Accounting and finance.* Again, because accounting must produce cost data in a timely way to the project managers for earned value calculations and for associating invoices with completed work, and because accounting systems often are not structured to respond to cost capture and reporting on a program and project basis, the program manager is responsible for establishing a close working relationship with accounting and finance.

- *Scheduling.* Since scheduling can be accomplished in many ways, ensure that all project managers use the same scheduling formats and software so that stage-gate decisions and earned value determinations are consistent when looked at by the program manager.

- *Legal.* Since projects can get into legal difficulties, ensure that legal representatives are involved early in project planning and especially in contract negotiations. Ensure that contracts include all necessary terms and conditions, including surety bonds.

- *Risk management.* Ensure that every project manager prepares and uses a risk matrix that identifies all project risks, defines them, identifies impacts and probabilities, and prepares risk contingencies.

- *Procurement.* Since purchasing, acquisition, and procurement activities are not typically within the control of the program manager, make sure there are good working relationships between purchasing officers and project managers, that both are served by the other, and that procurement actions do not get delayed because procurement officers are delaying one-time buys because of company volume discounts that do not affect the program.

- *Business case.* Focusing the project managers on using a project management memo to accompany the project plan can help the program manager make the business case for funding projects and ensure that each project is aligned with company business plans and strategies.

- *Ethical standards.* Provide all project managers and team members with a clear statement of the company's ethical standards and criteria for outside contacts, gifts, etc.

Interface with procurement

The interface of program management and procurement is the key to the success of programs. Program work breakdown structures are to include specific product components, parts, and test equipment requirements, and program managers are expected to trigger the necessary procurement and purchasing actions as soon as schedules are baselined.

Procurement is a separable activity and must be included as a scheduled task, with linkages to other scheduled tasks. Each schedule will incorporate the tasks necessary to procure test equipment, fixtures, cables, etc. as needed to produce the deliverable. These activities will be integrated into project schedules, including key hardware configurations and design master charts for target and test equipment, as well as software configurations and design activities.

Other program manager procurement issues
in a global context

While global procurement is not fundamentally different from domestic procurement, several issues in the global process present special risks:

- *Pricing approach.* Pricing in a global world is difficult, since prices need to reflect local conditions and economics. This requires a program manager to have the capability to access local economic situations through the Internet to help fix prices for foreign contracts to prevailing rates in the area.

- *Litigation.* If litigation is a possibility in a foreign location, the program manager should have access to corporate legal assistance that can work consistently in the appropriate location.

- *Negotiations.* Contract negotiations in a foreign environment often require sensitivity to local cultures and values in order to avoid violating local norms.

- *Contractor capacity.* To ensure that local contractors have the core competence to do the work, a local due diligence is often necessary to check out companies, backgrounds, etc. This "vetting" process is sensitive and should be conducted by a local, trusted third party, such as an auditor.

- *Supply chain management.* Global contract management is conducive to supply chain management, but long-term agreements with preferred contractors may violate local laws or regulations.

- *E-procurement.* All procurements globally can be conducted through e-procurement using various software tools.

- *Partnering.* Project partnering, "transforming contractual relationships into a cohesive, cooperative project team with a single set of goals and procedures for conflict resolution"—e.g., moving beyond adversarial relationships.

A case: airport designer dumped

In August 2005, Kirsten Tagam of the *Atlanta Journal-Constitution* reported the following:

> The general manager of Hartsfield-Jackson International Airport fired the design team working on the future international terminal Monday, saying the firms had failed to complete their design on time and within budget.
>
> The design team, in turn, blamed airport officials for micromanagement and inflated demands.
>
> Ben DeCosta, the Atlanta airport's manager, said the "setback" will be delayed until 2010. Also, the airport has spent $34 million on designs it may never use.
>
> DeCosta said the design team, led by the well-known architectural firm Leo A. Daly, never warned airport managers that the price tag for their plans exceeded the budget by at least $140 million. He called the plans "overdesigned" and "luxurious" and said the team wanted to use more steel than necessary.
>
> "They designed it to be far more muscular than it needed to be—at our expense," DeCosta said.
>
> But the chief operating officer of Leo A Daly, John Whisler, said the designers were under "intense oversight" from the airport's many consultants and employees in charge of the expansion project.
>
> Whisler said the airport officials kept requesting expansions and additions—such as a new maintenance terminal for the underground train system—but told Daly to keep the costs "below the line," or hidden.
>
> "The folks who report to the airport director wanted more than they could afford," Whisler said.

"The price rose also because of huge increases in the market cost of steel and concrete due to an unprecedented building boom in China," he said.

During the past two years of design work, the airport increased the terminal's proposed area from 930,000 square feet to 1.2 million square feet and expanded the number of gates from 10 to 14.

Even so, DeCosta said airport officials were repeatedly reassured by the design team that the building would come in under budget.

But Daly said DeCosta was "fully aware" of the additions and rising costs.

Whisler said he believes the design team is a scapegoat for the airport's growing concerns about Delta Air Lines' ability to help pay for the international terminal, which is part of a massive expansion project that includes a fifth runway opening next year. The budgeted cost of the international terminal is $688 million.

The airport expects to repay the terminal bonds with a combination of passenger ticket charges, federal grants and commitments from Delta. But the Atlanta-based airline is fighting to stay out of Chapter 11 bankruptcy proceedings, in which a judge could cancel that obligation.

"I had words from Mr. DeCosta to the effect that a slow-down in the decision-making on this terminal could be beneficial in light of the situation at Delta," Whisler said.

Hartsfield-Jackson has paid the team $34 million for their designs, and is scheduled to get the completed plans next week.

The drawings call for a huge glass wall framing the Atlanta skyline, roadways and an entrance that would create a new "front door" to the airport. The new terminal on Hartsfield-Jackson's east side, adjacent to Concourse E, would eliminate the need for people arriving on international flights to recheck their luggage and exit at the other end of the airport.

DeCosta said he would rebid the project and hoped the new design team could use the existing plans. But "if it turns out they can't salvage them, they will be scrapped and the new team will start from scratch," he said.

DeCosta declared the design team's contract in default in late June. Negotiations to salvage the contract were unsuccessful.

Leo A Daly has done other airport projects, including the striking Cesar Pelli-designed terminal at Reagan National Airport in Washington.

Daly was joined on the project by three minority-owned companies based in Atlanta: KHAFRA Engineering Consultants, Anthony C. Baker Architects and Planners, and Browder and LaGuizamon & Associates.

Case discussion. Public-works programs like this one are often characterized by these kinds of disputes since the work is planned and bid in an open process with great transparency. Since public-works projects are rarely fully funded and are subject to scope creep and inaccurate cost estimates, each time the contractor proposes changes in scope or costs, the process is laid out for public scrutiny. If this case involved a global contractor at an airport outside the United States, the process could be even more complicated. This situation would require a local, hands-on program manager.

A good example of the lack of effective global program management in this case is the contractor's inability to anticipate huge increases in the global market cost of steel and concrete due to an unprecedented building boom in China. Here is a good example of the need to anticipate global economic and pricing activities in the proposal process and for both the owner and supplier to examine price assumptions in the original contract negotiations. However, how does a program manager look ahead three or four years in the planning of a program of this magnitude and estimate material costs from a foreign country like China? The answer is in contingency planning. This is a risk management issue—estimating the price of steel in China in four years. A contingency would have handled this in the following way. The program cost estimate would have been made contingent on a certain price, with options for possible best- and worst-case scenarios. A contingency could have been provided in the original program proposal to allow the contractor to acquire the steel from another source.

Make-or-buy decisions

How does the program manager set the tone for make-or-buy decisions to be made, and how can the program manager take advantage of economies of scale for outsourcing similar tasks across projects? The answer is that the program manager sets the guidelines for make-or-buy decisions and intervenes if there are potential economies of scale in procurement, contracting, and/or partnering that are possible only at the program level and that no one project manager will see. Potential criteria for make-or-buy decisions include:

1. *Access to talent*. Outsourcing provides human resource and/or management competencies that the company does not have.
2. *Cost savings*. Outsourcing proposals demonstrate that significant costs can be reduced because of lower wage and/or equipment costs overseas.
3. *Logistics*. Transportation and logistics costs make it impossible to build components in the home country.
4. *Cultural differences*. Recognizing that the local culture and social conditions require that the work be done locally.
5. *Capacity and competencies of the company*. The company simply does not have and cannot easily acquire the capacity to perform this particular work in-house.
6. *Political and partnering reasons*. Sometimes it simply makes sense to spend program resources locally to support the local economy and to engender goodwill locally.

Global contract reviews.   Program reviews should be conducted at least weekly with project managers, either virtually or in one location, including the following agenda:

1. Changes in local factors, events, regulations, etc., that could affect projects
2. Status of local program and program stakeholders in host country

Country-of-origin product instructions. All product components, containers, and field-replaceable units produced or procured in any projects within a program must be documented in a configuration management program that marks the component with the country of origin and the source for the component. For all components, the markings must be legible and permanent.

Source information that is regularly updated to ensure that the source company is accessible for producing quality inventory.

Developing suppliers with a global view. Partners in foreign nations must demonstrate an understanding of the countries and markets that are targeted by the program, and these suppliers must have a "worldwide view."

Encouraging overseas manufacturing. Program managers must be sensitive to their own foreign policy issues and ensure that their programs are consistent with foreign policies and procedures regarding program products. This would include technical information interchange and contributions to social and economic activities in overseas communities.

Types of contracts used in local markets

Program managers must set clear policy on the types of contracts used in all projects within the program. Contracts, in effect, share risks with suppliers in different ways.

A cost-reimbursement contract places most risk on the owner or contracting organization, since there is no limit on costs as long as they are eligible. Such a contract implies that there is not enough information about costs and/or that the product is not definable, as in a research and development or new product development effort. Time and materials contracts are similar, covering eligible material and labor costs.

A fixed-price or lump-sum contract, on the other hand, implies that the outcome or product is definable and all components and associated costs can be estimated. In this kind of contract, the supplier takes most of the risk since they must assume responsibility for any unanticipated costs.

Doing business internationally

Program managers face unique problems and opportunities when doing business internationally, especially in new product development programs. The following sections discuss some of the issues involved.

Global analysis: public policy analysis. This is the *global* view of the product, looking at the economic and social factors that will affect product success globally. If the product concept is subject to governmental or regulatory requirements in any country in which it will be marketed, an analysis is made here of the potential safety, environmental, economic, or local impacts. The aim is to identify anticipated regulatory and safety constraints, by agency and by country, if applicable.

This environmental scan is likely to capture other outside forces as well, including potential current events, which could influence the success of the product in a given market setting. For instance, a political change in a given country may influence the potential success of a given product or service. This process should not be seen as defensive simply to protect the company; often, these outside standards provide new ideas on how to improve product performance in a given customer setting.

Intellectual property analysis. If the new product has potential intellectual property value, an analysis is performed on the risks and processes associated with intellectual property control and potential loss. This analysis requires a full understanding of current and projected intellectual property (IP) law and regulatory structure, and thus, should be performed by IP specialists.

This will include a look at opportunities to take advantage of patent, license, and copyright opportunities, and other ways to preserve and protect intellectual property. A risk assessment is completed on risks associated with failure to control intellectual property.

Market demand and other impacts.   This is a rough assessment of market demand. The concept is looked at in terms of the dimension of potential market demand and in terms of financial performance for business growth given that demand. What is the demand likely to be for the product, and will it generate adequate revenues in the marketplace to cover development costs and produce profitability and market share goals, and over what time period?

If the new product is an internal system or process improvement, the activity becomes an assessment of whether the improvement is really needed and whether it promises to produce continuous improvement in business operations and effectiveness.

Other impacts of the product or system on corporate or agency operations are also reviewed. For instance, a product may replace another or affect the performance or market share of another. These impacts need to be injected into the decision-making process.

This is the beginning of the process of developing a case and rationale for market launch. Most data addressed here are preliminary and do not require in-depth market analysis, but there needs to be enough of a "case" for the product to suggest moving on. On the other hand, if the product or system is a breakthrough improvement, the issue is how much the product will induce new demand—that is, how will the product change the current demand picture simply because it will create new value for customer, user, or client?

Here is where the "voice of the customer" is sought out; here is where potential users, clients, and/or customers, as well as stakeholders, are identified and dimensioned using focus groups, simulations, and other tools.

Product functional specifications and global impacts.   How will the product work, and what are its functional specifications? The concept phase fleshes out technical specifications for the product in high-level terms. In this phase, the product is defined in terms of function and performance, not necessarily in terms of design. In other words, the focus is on how the product will serve customers' needs in the user setting. Specific design issues are left open for the development phase, although models can be developed to understand the look and feel of the product.

The product's functional specification is outlined in high-level terms. Functional specification describes how the product will perform and what design specifications enable it to perform at its highest level of efficiency. Not all the data is available at this point, but enough information is gathered to allow management to gain a picture of the product.

Table 7-1 shows an example of a high-level product functional table.

The definition of functional specifications for a new product is a tricky process in this initial concept development phase because the tendency is to go into too much detail on how the product will work and how it should be designed before analyzing its potential value to the customer and its business value to the company at a high level.

The management challenge in product functional specifications is to find a balance between detailed and graphic representations of the product, which are often time-consuming and expensive to produce, and an inexpensive and workable model or prototype of the product that can be tested with customers and users. Engineers will sometimes go too far in detailing a product before its feasibility and customer value is established, spending valuable time defining tolerances, risk analyses, design options, and networks of a product that may not survive the test of customer and business value.

Global commercialization analysis.   Here is where the product is looked at in terms of how it will be commercialized. Commercialization is a concept that embodies the successful volume production, distribution, marketing, sales, and support system that is required to be a winner in the commercial sense. A strategy for commercialization will require a marketing plan and a partnering agreement with other companies in the supply chain, defining direct and indirect costs and barriers to market success, and identifying the capacity of the company to bring the product to customers and to support it over time.

Here is where logistics costs and opportunities are identified, including distribution channels, trends in economic and social developments that might affect logistics (e.g., availability of a labor base) and competitive forces that might inhibit free flow of the product *pull* to the customer.

TABLE 7-1   **Product Functional Specifications**

Desired Performance Attribute	General Functional Specification	Customer's Priority	Comparison to Industry or Competitive Standard	Comments
Helps cell phone users use phone while doing other things	Must create place for cell phone in various mobile environments	High	Current cell phone holders	Current holders do perform well over time

Global competitive analysis. This is the search for competitive differentiation on a global scale (e.g., value unique to this company). Differentiation is the value of the product and the company's capacity to deliver and support it, as compared to the competition. It is relational, based on the opposition. It requires some speculation on how the competition is planning to meet the customer need and where the competition is in the new product development process.

This analysis relates directly to the company's strategic plan and *risks, threats, and opportunities* assessment. The analysis also looks at what is liable to happen once a product is made public—that is, how the market will react to the product in terms of competition, barriers, and risks. Technological developments are reviewed here since new product development often spurs technology development and the generation of new companies and even new industries. Sometimes, this means looking for partnering opportunities early to offset potential competitive forces; sometimes, it may make sense to sit down with the competition to work out a partnership in a growing or new market.

Table 7-2 shows what a competitive analysis might look like.

Finding drivers of competition worldwide

What are the forces of competition, and what are the key sensitivities in the market that open up opportunities for competition? Drivers of future competition include all the critical success factors and forces that combine to create competition for the target product market. The value in identifying such drivers is that they can be monitored and, in some cases, offset by effective product development and marketing approaches. Drivers include:

Technology development. New products create technology challenges that many companies cannot handle; thus, a major force in competition is the relative technological strength of other companies in the industry.

Visibility of future markets. Future markets may not be visible to competitors, depending on the effectiveness of their marketing activity.

Timing. Some competitors will not be in a position to take advantage of new markets, even if visible because of inconsistencies and timing of their own competitiveness.

Financing. New product development requires funding, often from venture capitalists; thus, the availability of financing (e.g., interest rates) determines entry for many companies.

Scale. The scale of the markets—and the costs of entry—may be too large for many competitors to enter.

TABLE 7-2   Competitive Analysis

Competition	Pros and Cons	Potential Response	Potential Contingency	Comments
XWZ Company	Has a product in development like ours, but its performance is suspect over time	Competitor is liable to get out of this market if challenged	Partner or merge	

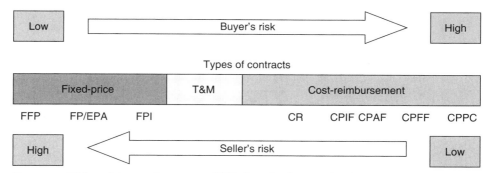

Figure 7-1 Risk and types of contracts. FFP: firm fixed price; EP/EPA: firm fixed price with economic price adjustment, e.g., linked with inflation; FPI: fixed price with incentive; T&M: time and materials; CR: cost reimbursement; CPIF: cost plus incentive fee; CPAF: cost plus award fee; CPFF: cost plus fixed fee; CPPC: cost plus percent of cost.

Program contract management and risk

Figure 7-1 shows how three basic types of contracts, firm fixed price, time and materials, and cost reimbursement, each spread risk between the buyer and seller in a contract relationship differently. Note that fixed price contracts usually reduce the risk to buyer (or project sponsor) while cost reimbursement contracts reduce the risk to the seller (or project contractor). Time and material contracts fall in between the two.

But in a global context, because of the exigencies of global markets and changing foreign political and economic conditions, and the different and often unpredictable values placed on contractual relationships, program managers should depend on local team members to negotiate contracts and agreements in the context of the targeted region or country. It is quite common for foreign governments, jurisdictions, and sponsors to change contract requirements and processes frequently and to place more emphasis on personal rather than legal tools to conduct program work.

Risk and Types of Contracts

Given project risks and the need to contract project work that is at risk, develop a strategy for using appropriate contract types to reduce and/or share risks with project contractors.

Some of your biggest risks may come with managing contractors who are doing some of your project work. Thus, it's important to choose the right kind of contract type and process to make sure the contractor shares in the risks of completing the work. The more you can get the contractor to share risks, the more likely the contractor will do a good job and flag risks early. We will explore fixed-price, unit-price, and cost-reimbursable contract types and their impacts on risk management.

When you think about it, the decision to contract out work changes the risk "equation," since work done in-house does not have the same "leverage" that contracting out work provides. First, you can rarely "incentivize" (provides incentives such as added compensation) in-house personnel to manage risks successfully on a particular task, while a contract allows you to do so. Second, contracts allow you to document shared risks—that is, to share with the contractor the costs of a particular risk occurring and affecting the project.

Various kinds of contracts have different impacts on contractor performance because the nature of the contract sets the conditions for the work. Contract type can affect financial objectives, contractor involvement, costs, schedule, and quality, as well as customer satisfaction. Here are the four basic contract types and their implications for risk:

Lump-sum contracts. Lump-sum contracts encourage the contractor to cut corners, since there is no process for getting more funding. The risk is that if the deliverable is not well defined, the contractor may not produce to the project design. Lump sum puts all the risk on the contractor; therefore, the contractor acts accordingly.

Unit-price contracts. Unit price is based on paying for each unit. This kind of contract shares risk with the contractor, encouraging the contractor to reduce the cost of volume production.

Target-cost contracts. Target cost is the ultimate shared risk contract, since both parties estimate a cost and work together to achieve it, leaving the door open for more funding if the target is not achieved.

Cost-reimbursable contracts. Cost-reimbursement contracts do not share risks very well. The contractor can repeatedly claim more costs based on continued work to refine the deliverable and cover past mistakes. The project manager is left with the basic risk in a cost-reimbursable contract.

The process of negotiating a contract brings out the risks issues because it is in the interest of both parties to avoid surprises and anticipate and mitigate risk. Financial and costing considerations become the vehicle for sharing risks. In the process, the contractor attempts to minimize risks by pressing for more clarity in requirements and definition of the deliverable, and by negotiating a contract type and price that protects the contractor from unforeseen risks. The contracting company or agency attempts to minimize risks by transferring them to the contractor to the extent feasible. The contract relationship helps to clarify the implications of risk and how risks can be avoided.

Global e-procurement

The playing field for global procurement and contracting has changed radically with the Internet and the high levels of connectivity between continents and nations. Program managers can reduce the time and cost of procurement through these tools. Internet-based bid packages now facilitate and manage collaborative bid systems and procedures between program managers and contractors. Software products now provide program managers with structured and uniform responses from potential bidders so they can make more informed decisions and compare bids. Packages also help contractors throughout the world receive clear descriptions of program needs and requirements. Tools also include information on various pricing mechanisms in various parts of the world. These systems now provide information and documentation on a global company's services and goods in a secure environment. This means that a program manager can now easily check out companies anywhere in the world as they plan and execute outsourcing decisions.

These systems also facilitate change requests and change management for programs in a global context. This means that as program and project scopes change throughout the program life cycle, both the program managers and contractors have a consistent way to request and implement changes. A by-product of this process is that e-procurement systems can serve as a platform for new ideas and innovations from contractors. E-procurement can also provide an effective platform for developing longer-term supply chain relationships so that when program outcomes and benefits are integrated into the business, there is good documentation on how and what suppliers helped them get there.

This would mean, for instance, that if a major IT program and system change is implemented over a five-year period globally, the resultant system can be documented and maintained using configuration management information on system components directly from Internet-based procurement sources.

Downside risks of global program e-procurement

Program managers must also see the risks associated with e-procurement—e.g., security and safety, intellectual property, and fraud. The Internet opens up potential problems with data and information security because RFPs and other contract documents sometimes contain important information that must be secured. Intellectual property issues can result from opening sensitive company information to the bidding process as well. The best way to handle these risks is to incorporate them into the program risk management process, develop contingency actions (e.g., purchase tailored security software to protect e-procurement actions), and implement them in procurement actions with those risks.

Open-ended global collaboration on the Internet

Companies such as IBM, Linux Second Life, and Amazon have developed global strategies that focus on what is called peer production. Business webs open up opportunities to work with peer companies, even competitors, to open up a discussion of new ideas and concepts.

These companies use their web systems to develop communities of participants that increase as new information is developed. Research and development programs are especially suited to using this kind of platform for the generation and evaluation of new ideas in a given industry. Collaborative networks of companies now design and develop programs, projects, and products globally.

Dan Tapscott, in *Wikinomics: How Mass Collaboration Changes Everything* (pp. 237–238), reports that BMW, Boeing, and the Chinese motorcycle industry have learned lessons from this open-ended approach to generating new programs and projects—namely:

1. Focus on the critical value drivers, looking for new competencies evolving out of the global economy that can create program opportunities.

2. Add value through orchestration, targeting an extended partner base that takes advantage of special competencies of partners in creating programs.

3. Instill rapid, iterative design processes, speeding up new product development by collapsing product cycles through concurrent engineering and collaboration.

4. Harness modular architectures, meaning the process of inviting global discussions of best practices in a given industry, targeting platforms and systems that work and enhancing current programs.

5. Create a transparent ecosystem—e.g., developing partnerships in a supply chain that grow and thrive instead of creating disputes and conflict.

6. Share the costs and risks, meaning use the right kinds of partnership agreements, contracts, and collaborative processes to make sure no one partner is carrying inordinate risk.

7. Keep a keen futures watch—e.g., make sure program managers are constantly speculating and envisioning how developing business ideas (e.g., modularization) can create opportunities for new program and project development.

Summary

This chapter covered the basic strategic issues involved in negotiating global business contracts and managing global outsourcing and procurement. Various conditions and local values are addressed to alert the global manager that *international business is not business as usual.* The key message here is that the program manager cannot count on conventional procurement and contract tools and techniques to establish collaborative arrangements and buy materials. Personal relationships and a full understanding of local values and contract requirements are essential.

8

Federal Program Management

Introduction

The previous chapter addressed partnering, contracting, and procurement as a part of the program management process. Now we address a specialized topic, federal program management, because so many federal agencies employ program management tools and techniques but often fail to control their programs.

We devote this chapter to federal program management because of the rising concern with ensuring cost-effective administration of rising public-sector programs and resources. We start with an analysis of what is wrong with federal program management and recommendations for needed change.

What's Wrong with Federal Program Management?

Since programs in the public sector are typically long-term, costly initiatives, often with broad political and social goals and associated earmarks or legislated entitlements, they are subject to different management and administrative criteria and are open to public exposure and criticism. Many of these programs have broad, ambiguous, and conflicting objectives embedded in their parent or authorizing legislation, reflecting the lack of consensus in the Congress on intent.

But they need not be managed any differently from private-sector investments if structured in a corporate framework addressing classic program management tools, including controls on cost, schedule, and quality. To operate successfully, federal program managers must be disciplined with strong and clear management guidelines. Also, program managers need to be protected from dysfunctional political intervention, especially by Congress and narrowly defined interest groups.

Federal program and project managers should be trained career professionals, not political appointees or "accidental" managers promoted up and into their level of incompetence because they excelled at something other than program management.

What do these federal programs look like? They fall roughly into four groupings:

1. Formula driven
2. Direct program and product delivery
3. Regulatory and oversight
4. Service

Formula driven.   These programs are typically apportioned to the states or local government, as in the federal highway program. The federal program is actually a collaboration process with states in setting guidelines for state-operated, public-works programs.

Direct program.   Direct programs include National Aeronautics and Space Administration (NASA), Federal Aviation Administration (FAA), federally managed Department of Interior parks, and National Institutes of Health, in which programs are planned and implemented by the federal government.

Regulatory.   These federal programs include Securities and Exchange Commission (SEC), Federal Deposit Insurance Corporation (FDIC), Federal Reserve, and Federal Communications Commission (FCC), and involve the control and coordination of private-sector activity in a given media or domain.

Services.   These are recurring services, such as the U.S. Postal Service, Health and Human Services (HHS) welfare services through states, Housing and Urban Development (HUD) rent subsidies, and the Smithsonian Institution.

What is wrong with these federal programs is that they are not really run as *programs,* with distinct improvement goals, outcomes, and benefits. *Rather, they are typically administered as continuing, open-ended services that are funded annually in a resource-oriented environment in which federal program managers are rewarded for spending all their funding each year and acquiring more resources, instead of for meeting specific quality and project goals on time and within budget.*

What is wrong?

Federal program managers, especially those involved with international and global missions such as defense, research and development, housing and urban development, space exploration, and transportation, face difficult challenges because of open-ended mandates, partially funded budgets, unclear objectives, ineffective procurement systems, and a lack of definitive end-points. Rather than being managed by professional business officers in a typical corporate setting with bottom-line considerations, they often report to uninformed and unequipped political appointees who typically do not have the technical expertise in the program area, or managerial experience and competence to direct complex agency programs. And while there are good examples of successful federal program management—e.g., NASA and the Space Shuttle and international space station program, Department of Transportation (DOT) interstate highway program, and Department of Defense (DOD) logistics programs—there are few examples of successful systems-type programs to improve federal services or to upgrade federal administrative and information systems. These programs are often contracted out and are typically late, over budget, and usually do not meet requirements when completed.

It is well documented that federal contracting has increased as a tool to design and implement federal programs, partly because of the lack of in-house agency staff and partly to encourage the privatization of federal functions.

Here are some of the key weaknesses of the process:

- Mission-critical functions are being outsourced, compromising successful delivery of intended benefits and outcomes.

- Federal contracting procedures are subject to federal procurement regulations that allow most program and project managers to award contracts without competition (so-called no-bid contracts).

- Federal contractors often manage other federal contractors because agencies cannot.

- Federal contracts are partially funded, not fully funded; thus, cost overruns are common.

- Some federal contracts integrate project plans and schedules into contract terms.

- Many federal contracts are cost-plus arrangements in which most of the risk and cost is carried by the federal contracting agency and program managers.

- Contract oversight is often undisciplined, with no standard practice across agencies.

- Inspector General staff audit projects and often come up with critical findings simply to keep their jobs.

- Fraud, waste, and abuse are relatively common in many contracts.

- Contracts are short-term and often do not allow the development of longer-term, trusting relationships and partnering.

- Federal program managers often leave government and go with the contractors they have awarded, thus creating potential conflicts of interest.

As indicated, one of the major weaknesses of the process of contracting for program functions is that many program managers compromise mission-critical roles by outsourcing, either because they do not have the staff to manage the contracted functions and tasks in-house, or because they do not participate in the agency decisions to "make or buy" and cannot make a business case for what is mission-critical and what is not. To address this function, program managers should determine what program functions are mission-critical as part of program planning. *If there are not enough funds to man mission-critical functions, the program should not proceed.* Here are some examples of agency mission-critical functions:

- NASA should not outsource operations support and communications functions with its Shuttle program.
- DOT should not outsource air traffic control functions.
- Interior should not outsource park ranger or park security functions.
- HUD should not outsource subsidized housing administration.
- Treasury should not outsource financial regulatory tasks.
- Office of Management and Budget (OMB) should not outsource budget analyst functions.
- Education should not outsource education reform programs.

What kind of change?

The significant change needed here is a new concept of federal program management. The change would be from open-ended, annual appropriations-driven programs to disciplined, targeted programs as investments in improvements with short-term products and outputs and long-term benefits and outcomes. Programs would be run by professional (career) program managers and would be evaluated regularly in stage-gate reviews in which the business case must be made after each phase. This kind of transition would mean, in practice, that all federal program activity would be framed in shorter-term, goal-oriented projects and that federal program managers would be held accountable for achieving limited, realistic goals.

Congressional staffs intervene in programs as a matter of course in Washington, D.C.; a key to the success of current program managers is how well they relate to Congressional staffers and key chairmen in their program areas. This kind of regular intervention would need to change, and the key to change would be to restrict communication with congressional staff on programs except at key points of progress reporting and evaluation. Further, the White House would set up a single point of congressional contact for all federal programs in OMB in a professional management czar, as described in a following section.

What Is Real Program Transparency?

Real transparency would be achieved by opening up information on program and project progress and allowing users and clients of the program to participate in evaluation. Transparency is not simply giving public access to how money is spent; it is opening up programs to evaluation at key milestones and providing for user or client feedback at key, stage-gate reviews.

Coordinating Federal Program Management

The federal government manages billions of programs, but has no standard approach to program management and no "vice president for program management." No corporation would run its programs like the federal government does, without top management oversight and control. A *management czar* is needed as a central focus for program management *reform and performance budgeting*. The position should be created in the Executive Office of the President and preside over an *Office of Federal Innovative Management,* a small staff that would take over the managerial functions of the current Office of Management and Budget and guide federal agency and program management reform, delivery, and continuous improvement. The case for a management czar can be made simply on the basis of the need to coordinate the disparate and often conflicting objectives and programs of the federal agencies, and to serve as a lightning rod for best practices and business applications to public management.

The management czar should be a recognized, international authority on public and/or private management and customer service, with experience in both sectors. This position would serve to demonstrate to the country and to the federal agencies the president's commitment to change in the management structure and performance of federal programs. The management czar would bring to government a new emphasis on best practices, benchmarking, and managerial training and development for federal public managers. The position could also evolve into a focal point for improving intergovernmental management of programs and crises, and for a closer partnership between public and private managers engaged in similar or common programs.

A new Office of Federal Innovative Management in the Executive Office of the President

Creating "change" in federal government performance will involve both organizational change and management change. But most of all, it will require top-down leadership and support at the program level, where most federal decisions are made. Policy pronouncements cannot compete with the *tyranny of small decisions.*

First, it is the purpose of this chapter to make the case that this kind of fundamental change can come only from the president and the Executive Office of the President through a new Office of Federal Innovative Management. Second, it will require real and effective support of program managers throughout the federal government; to do so effectively, the office must *earn* the respect of agency officials and program managers, much like the Office of Management and Budget has earned its role in federal budget development.

Global Reference

The global significance for an Office of Federal Innovative Management (OFIM) is in the potential of this small staff in the Executive Office of the President of the United States to assist program managers in aligning their programs with presidential policy and coordinating the impacts and benefits of programs worldwide. Many federal programs have impacts far beyond their immediate program products and outcomes. For instance, if the office can ensure that federal transportation programs are supporting the president's energy-independence policy goals, and these impacts help lower the dependence of the United States on foreign oil, then the case can be made for such an office in global terms.

Program Reference

Program management does not occur in a vacuum, whether in the public or private sector. Programs have major impacts; the new significance of performance budgeting brings into play the longer-term benefits and outcomes of programs.

Programs feed on the energy of top leadership and on expressed vision in organizations and companies that create the conditions for innovation and program success. The model of top management providing direction and support can be realized in the federal government through a small, effective Office of Federal Innovative Management.

Explanation of concept

The concept is to establish a small coordinating office that serves the president by helping to coordinate programs across agencies to ensure that they are aligned with presidential policy. In addition, the office will encourage innovative approaches to program development and management, and be staffed by senior, career program managers.

There are many models for the notion of stimulating innovation and change in the private sector, but not many that have been effective in the federal government. James Cash, Michelle Ead, and Robert Morison in the *Harvard Business Review* (November 2008) wrote an article entitled "Teaming Up to Crack Innovation." They propose Enterprise Integration Groups that manage the corporate portfolio of integration activities (p. 97). They serve as a "corporate creator of expertise in process management and improvement, large project management, and program and portfolio management." Their capabilities in program management include "ensuring that complex initiatives and changes dovetail with one another through coordination of objectives, resources, and interdependencies."

Furthermore, the authors articulate a new, innovative role for the IT department. This department would have a major role in integration involving "experience with program management—the planning, coordination, and measurement of the many projects and activities involved in a multidimensional change initiative."

In *Fast Strategy* (Wharton School Publishing, 2008), Yves Doz and Mikko Kosonen talk about strategic agility, the capacity to create direction while remaining flexible and responsive to change. They say programs are a way of broadening initiatives so that they cannot be "owned" by one unit of the organization. They see programs as another vehicle for dissociating resource "ownership" from business results across company programs (p. 104). They advocate a program planning team to prepare the business case for the programs, including business benefits, schedule, and resources.

The business case is an important aspect of federal programs as well, since programs create not only product and service outputs, but also impacts over the longer term in the form of *streams of economic costs and benefits.* Since agencies typically operate within narrow legislative and regulatory guidelines and constraints, they cannot always be counted on to see the broader impacts of federal programs. Facilitating continuous review of program impacts and costs and benefits would be the role of the new Office of Federal Innovative Management.

Structural Change

As we stated earlier, there are at least four models of federal activity, and each deserves a different model for evaluating performance. They are formula driven, direct program and product delivery, regulatory and oversight, and service.

Placing agencies with similar delivery systems in working teams or committees using these categories makes sense based on my previous experience. Federal agencies do not coordinate well with each other without incentives. And as you know, intervention from OMB will not be welcomed unless earned. Your success will be measured on how many agencies come to you for assistance.

Performance should not be confined to performance budgeting, as has been the case in the former administration; performance should be focused on cost effectiveness, products and outcomes, customer satisfaction, and quality. Application of traditional project management tools should be encouraged across the board with compatible software (e.g., Microsoft Project). Agencies lack cost control and earned value tools used widely in private enterprise; that needs to change. We need more standardization—each federal agency should be structured and managed from a uniform model of program management and budgeting so that when management inevitably changes, the wheel does not need to be reinvented.

Some tailoring is in order to reflect unique agency programs, but all programs should be structured with program plans, risk management plans, scopes of work, schedules, cost estimates, and earned value monitoring. Process improvements can be made in federal programs, but the process approach has its limitations, since its focus is on identifying waste and unnecessary work and not on outputs and outcomes. We would focus on career senior executive managers to hold them accountable. Change does not occur easily in federal programs simply because agencies operate within legislation that is often ambiguous. If change in program management is a priority, then the vision for change in programs must come from the top. There is no organizational element in the Executive Office of the President to foster such change and innovation—thus, the proposal to establish one. The concept of this kind of top-down control and support of programs and program management could be useful in the public sector, specifically in the federal government, where impacts are often global in nature.

If the president is to deliver on the promise of change, he must create a way to control and coordinate federal agencies from the top while providing them with wide discretion for innovation and efficiency within their own boundaries. For instance, if federal agencies were to perform together to support national energy policy, there would need to be a fundamental change in the way federal agencies work on a daily basis. A change in control of the federal government structure and agency program and regulatory performance will require a new "presence" for agency management in the Executive Office of the President. The change and control problem is not simple. Many authors have addressed the dilemma of public management and, specifically, federal government management, but they do not address the policy and management structure from the perspective of the president. But the key to change and control lies in how agencies are handled from the Executive Office of the President and in Cabinet-level leadership decisions. The resolution of the federal change and control problem lies in how the agencies are grouped and controlled at the top.

Past approaches to management reform

First, the federal government has no central management arm; the OMB is a policy and budget agent, but not a management agent of the president. Past attempts to make OMB a management agency have failed because of its makeup and culture, and because of opposition from agency heads and interest groups. OMB's focus is policy and budget, not consistent delivery across agencies. Second, the iron triangle of agency/congressional committee staff/interest group continues to control agency performance in the direction of their constituents and the Congress, not always in the direction of presidential policy and management objectives. Third, the appointment of unqualified political executives and managers in the agencies at all levels complicate the job of management delivery, typically getting in the way of professional career management.

Fourth, the career assistant secretaries for management/administration are typically career figureheads who do not get into line decisions and who are mostly administrative in character. They are housekeepers at best. There is usually no single line manager of any agency below the secretary, so agencies are divided and conquered easily by special interests.

Fifth, agencies act without anyone knowing the impacts of their actions on key presidential policy areas. For instance, the Department of Transportation is likely to initiate programs and legislation that would undermine the president's priority for national energy independence by encouraging more single person cars. The only program that has successfully provided for impact analysis before action has been the environment impact statements (EIS) that have served to adjust many federal decisions that were not supportive of presidential policy. The EIS system actually worked. Sixth, many presidents have attempted to control spending by impounding funds (impounding involves not spending appropriated funding), but they have not been successful because of long and drawn-out litigation. Any president interested in controlling agency spending would have to exert impoundment in order to stop unnecessary and wasteful outlays, and would have to test the courts again on this issue. Chances are the courts might treat this issue differently in an age of trillion-dollar deficits. Thus, a key transition issue is government change and control. The president would have to take six actions to ensure immediate control of federal actions:

1. Establish an Office of Federal Innovative Management in the White House (Executive Office of the President) that would act as an agency and program management office (PMO) for the president. Place the vice president in charge, and populate the agency with career senior executives on annual performance contracts, who know the agencies they are assigned to and review agency management and performance regularly. Focus on performance budgeting—the planning, delivery, and evaluation of programs to compare costs and outcomes.

2. Group the federal agencies in clusters and manage them as a unit through a "super-policy" executive similar in function to the chief operating officer of a major corporation. The CEOs of each cluster would report to the vice president.

3. Select Cabinet members based on their proven management and business capacity and their support of the president's policies (e.g., energy independence). Do not appoint anyone to an agency position unless through a merit system that matches the job with qualifications and experience (staff experience with congressional members would not be considered management experience, per se). Make sure Cabinet secretaries understand the role of the new Office of Management.

4. Quickly (in four months) develop and distribute federal policies in key areas, foreign and domestic. For instance, issue a federal energy independence policy that would provide an overall framework for federal agency action, and require agencies to prepare a strategic plan to implement that policy. Establish an energy independence impact (EII) statement program/process to be run by the new Office of Management, and require all agencies to complete EII statements that must be approved by the "super-policy" executive for that agency's group. These statements would be prepared for all agency and program decisions, with potential impacts on energy independence. The Office of Federal Management would issue guidelines for implementation of the EII program within three months of the president taking office.

5. Prepare similar policies for education and health, and require implementation plans.

6. Rewrite ethics guidelines for federal agency officials that require them to report all contacts with interest and constituent groups, or congressional staff.

Resurrecting a Reagan administration proposal

Here is what a presidential panel of the National Academy for Public Administration said in 1983, during the Reagan reform movement in the federal government, on the subject of an office of federal management:

The concept of creating a coordinating management arm in the Executive Office of the President was recommended in the Reagan administration as part of several initiatives, including the Grace Commission, to bring business practices to federal administration and program management.

Since World War II, the pivotal role in federal management matters has been assigned to the Office of Management and Budget and its predecessor organization, the Bureau of the Budget. BOB/OMB has attempted to maintain a general management capability along with its primary role as the president's budget staff. During several periods of its existence, the management organization has performed exceptionally well. It has successfully carried out management projects of importance to every president and has attempted to maintain a general stewardship over management activities throughout the federal establishment.

BOB and the early OMB tried to serve as a key source of government-wide initiatives for keeping federal management modern and up-to-date. In recent years, however, there has been a growing concern that even while OMB continues to be capable of occasional excellent performance, it has irretrievably lost its overall effectiveness as government-wide leader in management matters. Students of BOB and OMB have always recognized that its primary concern on behalf of the president has been the federal budget process. This had been regarded as both a strength and a weakness in terms of OMB's management role. It is a strength because the budget gives the OMB director a powerful voice in all government activities, with direct and continuing access to the president and powerful leverage to deal with agency heads from a position of strength. It has always been hoped and expected that these strengths would be brought equally to bear on the managerial agenda from time to time, depending on the president's own interests and the director's sensitivity to management problems. [It does not] appear likely that, in the foreseeable future, the budget preoccupation will decline; therefore, proposals and arguments for building up and revitalizing OMB's "management side" seem less hopeful and less realistic.

Meanwhile, the need for strong sustained management reform leadership becomes more pressing. The Reagan Administration's ambitious Reform '88 Project and the recently emerging findings of the President's Private Sector Survey on Cost Control serve only to emphasize the breadth of the management reforms, which are needed, and the paucity of the resources available to take implementing action.

The Office of Federal Management would inherit from OMB such important functions as the drafting of non-budget-related executive orders, management circulars, and other means for conveying presidential direction on management subjects. It would include the present Office of Federal Procurement Policy and would coordinate responses to [Government Accountability Office] GAO audits of government-wide applicability. It would also administer the requirements of the Paperwork Reduction Act. It would represent the president in Congress in cooperation with the White House congressional office on government organization and management matters, and would coordinate preparation of administration management-related legislation.

The role described for the Office of Federal Management extends from strategy to implementation. It is intended to be the one organization to which the president can turn that can develop an overall, long- and short-term strategy for government management.

However, it must also be able to convert that strategy into action, largely in the form of specific management improvement programs and projects. Reform '88, the increasing volume of GAO and [Inspector General] IG reports, the reports of the

President's Private Sector Survey on Cost Control, and this National Academy of Public Administration NAPA study once again demonstrate that there is a backlog of unresolved management problems and a wealth of management improvement ideas and opportunities, all begging for the kind of broad-ranging, experienced, highly competent leadership that an Office of Federal Management could provide. Its institutional role must be to loosen up the system, dislodge entrenched interests, motivate the federal workforce to greater effectiveness, and recreate the capacity to innovate. Its role should be to lead, to promote, and to assist—far less to regulate, control, or enforce. To fulfill these functions, its role must be clearly defined as a broad range of responsibilities of sufficient importance that the president will give its director direct and continuing access. The NAPA Panel believes that the proposed Office of Federal Management should assume management leadership responsibilities of OMB and continue to serve as the single agency that can represent the president on all federal management matters. There is no reason to believe that the new Office of Federal Management would impair these other central agencies; each would continue to be fully responsible for its policy areas as defined by the president. Said another way, the president would look to the Office of Federal Management for management policy objectives for the whole government.

The whole pattern of conclusions and recommendations of this report rests on the premise that, if the federal government really wants to achieve general and broad-based reform of its management and create an environment in which managerial excellence is demanded and expected, then it must place top priority on its managers rather than its systems. A large number of recommendations are made that would greatly unburden systems and make them more useful and flexible for managers to use. But the panel recognizes that as systems are redesigned to place greater reliance on managers, there must be a recognition that, in many cases, managers themselves are ill-equipped and ill-prepared to fulfill these more demanding roles. There must, therefore, be a parallel and equally serious effort to upgrade managerial skills and create a new climate of motivation and reinforcement at the highest levels of government.

The panel also recognizes that many efforts for management reform founder because of the failure to implement action. It is not unreasonable to point out to these same federal managers that they can become the most powerful means through which such reforms are accomplished. Much of what is proposed can be achieved if managers are willing to create their own initiatives and become a stronger force within government for solving the problems about which they have so long complained.

Although the notion of a federal Office of Management was grounded in control and coordination of management change and reform efforts back in the 1980s, its purposes would now be different, given current policy and program conditions. The difference now is the need for change and innovation, not change and control. The federal management office is needed even more now, but now its role is fundamentally different. Its purpose is to encourage and promote innovation and alignment with presidential policy.

Functions of OFIM

The basic objective of OFIM is to promote interagency coordination, innovation, and exchange of information on basic program management delivery and evaluation, and to help agencies ensure that they are supporting presidential policies and objectives globally. The office would achieve its purpose through the following units, each staffed with career policy and program managers:

1. *Establish and monitor a standard program management process and discipline.* This standard would be published and made available online, and would serve as the basis for a federal program management certification. No federal programs would be run by managers who were not certified, and agency heads would have to identify a program manager, who is responsible for program management performance in their agencies.

2. *Benchmark program information and innovation.* Seek out best practices and help agencies find innovative and proven techniques and tools to design and deliver federal programs. Benchmarking would involve sharing information on lessons learned and best practices in program management to avoid "reinventing the wheel."

3. *Program and performance budgeting review.* The office would conduct regular program reviews to help agencies evaluate program impacts in a global context, with particular attention to interagency and global impacts, as well as alignment with presidential policy. The focus is on performance budgeting and longer-term impacts and global outcomes.

4. *Program information exchange.* A central website would be established and facilitated for the exchange of program information and discussion among federal managers in both synchronous and asynchronous modes. The website would facilitate synchronous "threaded discussions" between program managers on common issues and challenges, and live, synchronous webinars and chats between program managers on urgent program issues.

Program planning and evaluation

Program evaluation is a lost art at the federal level because of the difficulties of evaluating multiple program objectives and the lack of interest in assessment of outcomes. OFIM would stimulate assessment by providing technical and managerial support for program planning and evaluation of impacts. Operating somewhat like the Congressional Accountability Office but generated in the executive branch, OFIM would publish results and facilitate exchange of best practices and lessons learned from federal programs. Particular attention would be given to alignment with presidential policy.

Common program and project management methodology

Federal programs are managed with no clear and coordinated methodologies. Program plans, schedules, and budgets are not coordinated and often cannot be reviewed and evaluated using common measures. The OFIM would promote consistent planning, scheduling, and budgeting tools and techniques, utilizing such programs as Microsoft Project to allow exchange of program and project data across agencies.

Energy independence: a case example and an introduction to the concept

The coordination of federal programs at the top of the federal sector will require a broad process and program focus. This means that the OFIM will have to achieve a working balance between coordination and support on the one hand and control on the other. Much like a corporate office of program evaluation, the OFIM will need to earn the respect and support of federal program managers, who have been operating autonomously over the years.

Federal program managers do not seek out the guidance and "intervention" from their own Cabinet offices, let alone guidance from the Executive Office of the President, unless *they see direct value in the process.* Federal program managers typically seek visibility and resources for their programs, but they also seek out useful information and best practice assistance in running their programs. In other words, top-level guidance for federal program managers must be framed in legitimate processes that *earn the right* to participate in program development and delivery.

Global reference

Many federal programs—e.g., the federal highway program, the federal alternative fuels research and development program, federal housing programs—are designed and delivered with narrow goals and objectives. Legislation setting up these programs does not always consider impacts on other federal programs, on the American economy in general, and global impacts.

What Happens in the Real World of Federal Program Managers?

Typically, career federal program managers have difficulty managing programs because of vague and ambiguous goals, conflicting objectives, and political interference from congressional staffs and interest groups. They need support and insulation from outside influences that dilute their ability to achieve program objectives in the context of executive and presidential policy. Furthermore, agencies often do not practice portfolio program management, the process of generating and selecting programs, goals, and projects, and managing them consistently across the agency. Agencies often start programs and complex projects, which turn sour, but they cannot stop them—they lack the staged, "gateway" process, which ensures that bad projects are recognized and terminated, when necessary, before they consume major resources.

James Brown, in the *Handbook of Program Management: How to Facilitate Project Success with Optimal Program Management* (McGraw-Hill, 2008, p. 205), tells a story:

I once spoke with an organizational leader at a government agency who was moved to another part of the agency. He was surprised to learn that there was a lack of solid business goals and targets in this new part of the agency, and this lack of goals created program managers who were not guided properly. The result was that some of the program managers had done nothing new for years, while others were all over the map, spending resources and implementing projects that were way out of the scope of the organizational mission.

Program reference

A *program* suggests complexity and interdependence, a series of coordinated projects designed, developed, and managed in concert to achieve a broad program goal, such as energy independence or a broad education reform objective. This kind of aligned program management framework often involves cooperation across companies, economic sectors, and even across private enterprise and government boundaries. Program suggests partnership and consortium, with broad purpose and private and public benefits. For purposes of illustration, this discussion covers the development of a federal energy independence policy and program framework that enables the development of a clear fuels market. This kind of federal guidance across agencies and sectors has not worked well in the past because the federal government has not been adept in formulating programs that enable private enterprise to work efficiently *to support the public interest.*

Thomas Friedman, in *Hot, Flat, and Overcrowded,* has suggested the need for a coordinated federal/national program to stimulate a private-sector market for clean fuels, all in the interest of energy independence and growth of an American clean fuels industry. This case is about how such a program would be developed and how it might work as a public/private program.

Good programs, whether private, public, or nonprofit, succeed because they are framed in a solid strategic and policy framework. The development of an energy independence program first will require the development of a federal policy framework on energy independence that addresses the federal role and relates that role to the generation of a market for clean fuels. The policy also must address the intergovernmental issues—e.g., collaboration among federal, state, and local governments. And perhaps most importantly, those industries who want to *play and succeed* in the energy business must see how they can *both* serve the national interest articulated in the policy framework *and* grow and generate profitability for their stockholders and stakeholders. That will require the federal government to use its leverage, program funding, regulations, policies, program management delivery decisions, and research and development in a coordinated and coherent way.

As we have seen over the years, the coordination of federal programs to support a national energy independence policy and program cannot be left to the agencies. Because of the *tyranny of small decisions,* agency management loses control over delivery issues unless there is a comprehensive system to check the impacts of key decisions on the vision for energy independence.

**Energy independence/clean fuels program strategy
and program objectives**

An energy independence and clean fuels policy could have five key strategic objectives:

Strategic Objective 1: Through the effective management of the development and supply of alternative fuels and changes in national energy demand through price signals for private industry, the United States will reduce dependence on Middle Eastern oil by 50 percent by January 2014, and by 100 percent by January 2018.

Strategic Objective 2: The federal government will strive for a balanced energy program by 2018, including major participation of wind, natural gas, electric, oil, nuclear and coal-generated energy sources. Agencies affecting these energy sources would develop coordinated programs and plans to support the achievement of national energy goals.

Strategic Objective 3: The federal government will do its part not only in proactively supporting the development of alternative fuels, but also through an energy impact statement program that will ensure that key federal program decisions do not undermine the achievement of energy independence.

Strategic Objective 4: American industry is called upon to identify ways to support this national policy while maintaining and growing business profitability; federal involvement would be through broad national strategies and goals and technical and informational support systems.

Strategic Objective 5: The academic community is requested to support this energy policy with directed research and curriculum development toward energy independence, with federal support.

OFIM Program Role in Energy Independence

These strategic objectives would evolve into five key program management objectives in the new Office of Federal Innovative Management:

Program Objective 1: Work with federal agencies to review current programs for energy impacts and lessons learned using documented program evaluation and impact information.

Program Objective 2: Work with appropriate federal agencies to develop an energy independence strategy and program management approach to leverage those programs in the direction of energy independence.

Program Objective 3: Develop a cadre of innovative federal career professionals in each agency who will develop and coordinate energy independence policy and program management for their department administrator or secretary.

Program Objective 4: Develop a program tracking system that follows federal program management results and outcomes, and evaluates progress toward achievement of the federal energy independence policy.

Program Objective 5: Collaborate with Congress to develop energy independence legislation that facilitates effective program management to achieve national energy independence.

OFIM Involvement in Heath Care Reform

OFIM will improve the capacity of the federal government to evaluate programs in terms of costs and benefits. An example would be health care. The Office of Management and Budget will undoubtedly promote the application of performance budgeting to health insurance programs and tax policy.

Performance budgeting relates a program's budget to its effectiveness in achieving its objectives. The Congressional Budget Office director said in 2007 that there had been very little analysis of whether health care spending is generating corresponding gains in the health of enrollees.

The director goes on to say:

> Both federal health insurance programs and tax expenditures could be subject to performance budgeting. The executive branch and the Congress generally engage in various forms of effectiveness reviews through their annual budgeting process, but the performance budgeting seeks to make that consideration more systematic and more consistent among programs... Even with a systematic and well-designed assessment of performance, determining what specific policy steps to take as a result may be difficult. Unlike discretionary appropriations, the costs of federal health programs and tax expenditures are not set directly by policymakers, but, instead, depend greatly on the actions of individuals and firms. The effects of any reforms aimed at improving performance, thus depend on the responses to policy changes. OMB is unequipped to ensure that agencies plan and deliver programs from the standpoint of performance budgeting; it can promote and enhance agency capacity, but it does not have the staff or interest to serve as a management support system for agencies.

Federal Agency Portfolio Management

OFIM would work with federal agencies to develop a common approach to portfolio development—e.g., the process of generating, selecting, proposing, funding, authorizing, and managing programs. A common approach to portfolio program management would ensure a central depository and documentation of all programs, categorized in terms of intended outcomes, both domestically and globally.

Federal Agency Program Management Systems, Tools, and Techniques

Federal programs would be encouraged to use common and consistent tools and techniques to allow an open exchange of program and project information across agencies, with state and local governments, industry groups, and other stakeholders. Common program planning, budgeting, and scheduling templates would be developed in concert with agency program managers, and websites established for program information exchange and dialogue.

Promoting innovation in federal programs

The essential ingredient in successful new programs, products, and services is not disciplined process and analytic tools. It is, rather, a vibrant and energetic agency organization that enables its creative people and its partners to think innovatively about what they are doing, know where the agency is going and what it can do, and feel free to generate and integrate new ideas and new concepts and processes into the system. In the best companies, new program development is a mainstream activity.

In these companies, agency associates are not risk-averse. They know *and feel* that their professional careers are enhanced by generating new ideas in their domain, but they also know that their advancement is not solely determined by the success of a new program or service. This frees them from the fear of communicating bad news—if appropriate—or terminating a project because it turns out that it makes no sense in the market. The best ideas for new products come from both those who do the regular work of the company and from market research and new product teams.

This targeted organizational environment is a *culture of ideas,* the organizational *tone* that encourages free flow of feedback *into and out of the agency* on current programs, products, services, customers, and operations from those closest to them. This does not happen naturally or easily because new ideas are not always welcomed in companies working hard to develop and market current products and services, and because new ideas tend to challenge the current way of doing business—they change processes and strongly held paradigms.

New organizational structure for new programs

We live in a global economy in which the turnaround of structural and technological change is rapid and sometimes unpredictable. Federal agency processes are changing even as we speak, and traditional jobs and work settings are becoming virtual, e-driven, and mobile. This atmosphere of change creates the need to structure the agency organization so that (1) agency people can create new programs and make executive and legislative proposals that make sense and can be funded, and (2) the agency can design, develop, and produce new services (or redefine old ones) better, faster, and cheaper and get them to citizens. Development life cycles have gone from years to months to days, and federal agencies need to catch up to market forces.

Traditional ways of organizing agencies around narrow functions (e.g., highways, parks) does not seem to encourage new thinking about agency services in a broader context, nor does it easily surface global impacts. How does the agency organize in order to avoid being captured by narrow views of the mission determined by traditional organizational structure? The business answer is that organizing around processes appears to be helping successful twenty-first-century businesses to generate creative thinking because it is in process framework that companies can best see their customers in a system. Processes can describe various systems and interactions that lead to product success, thus encouraging people to think in terms of *process domain,* which gives them a better chance of coming up with new processes and products to serve customers.

Curbing new unaligned programs

Not all innovation is helpful in the public domain. In the public sector, creativity and innovation in the pursuit of unauthorized programs, or programs that serve narrow interests, can be a major diversion for federal agencies. Therefore, innovation must be channeled and disciplined by policies, boundaries, and processes to transition ideas and concepts into viable programs that support presidential and Congressional policy.

New programs and outsourcing

Sometimes it pays to outsource creativity and front-end new program development to capture domains that the agency cannot create itself. That is, to the extent that outsourcing increases the probability of partners *seeing* new processes and products differently, outsourcing can increase the chances of successful programs. This conclusion must be conditioned, however, on the presumption that the relationship with outsourced partners is more than a simply contract—that, indeed, a partner is sought out that represents a committed, long-term associate in a given sector and therefore can be trusted to act in the interest of the partnership and the public.

Organizational learning

We have seen that successful new public program and project rollouts occur in agencies that learn in the process, evaluate programs in terms of outcomes and benefits, and document their insights. The federal Head Start program is a good example. In this case, organizational learning means that there is a measurable process within an agency that captures insights, experiences, lessons learned, and an understanding of their domains, and uses this learning to get a leg up on the competition. New product development and program management effectiveness seems to improve when the agency brings past program experience and insights to creative people so they can generate informed changes and improvements. This results in a powerful combination of individual and team creativity joined with agency insights into particular development avenues.

Six key strategies to create innovation in federal program management

Professional program managers from the corporate world respect the importance of innovation and creativity in program and project management. With clear objectives and the incentive to complete programs on time and within budget, federal program managers would generate exciting new ways to manage programs.

Six key strategies can ensure that the agency creates a culture of ideas. Agency leadership must:

1. Remove barriers to the generation of new ideas in the agency, particularly new ideas on how to plan, deliver, and evaluate currently authorized programs

2. Provide a system of information and feedback on current programs, products, and services

3. Create a positive, perhaps virtual, place for new ideas to incubate

4. Generate a process of filtering, evaluating, and transitioning new ideas into a portfolio and new program development process

5. Manage and control new program development

6. Demonstrate that new ideas can produce programs that align with agency and presidential goals

Remove barriers to the generation of new ideas

Removing barriers to the generation of new ideas in the federal bureaucracy involves removing the fear of failure and providing incentives and organizational platforms for good ideas on new programs and services. Agency employees, associates, and stakeholders see opportunity in their day-to-day work settings simply because they see what works, what does not work, and what would work in given situations to produce new product or service opportunity. They see opportunity and risk on one screen. These opportunities include process improvements that could substantially improve company efficiency and effectiveness, product improvements that could enhance marketing and sales, and service improvements that could help sustain product life and longer-term customer relationships. However, these opportunities will not surface to management and agency leadership unless employees are encouraged to generate them as an *integral part of their jobs* and without fear of rejection.

Promote return on innovation and creativity

Agency people need to know that there is a *return on creative solutions* that achieve both public- and private-sector goals. They should know that assessing the return on creativity is the objective of the program management process and that employees will be rewarded for coming up with new concepts that create value and intended outcomes. On the other hand, they should also know that if the return is judged not to be worth the cost, the decision to terminate a program midstream will not reflect on those who came up with the idea.

Paradigms as barriers to new programs and services

Thomas Kuhn referred to paradigms in his classic *The Structure of Scientific Revolutions*. He said that paradigms are concepts or structured views of the world that restrict scientists from seeing new data that do not agree with the prevailing mindset. Paradigms operate to filter out information so that strongly held concepts are not challenged, but rather are continuously confirmed despite data and information to the contrary. New ideas challenge locked-in views of a service in an agency, and thus often face difficult challenges because management cannot see the potential success of an idea because it is inconsistent with their paradigm for the agency. To avoid paradigmatic barriers, management must allow new concepts to receive the light of day outside the normal chain of command, and sometimes outside the operating systems of the agency. Often, the best ideas come from operating levels of the agency and are never reviewed by the leadership.

Provide a system of information and feedback on current programs and services

How do employees generate new ideas on agency programs and services? The system must provide a visible and working channel for employee ideas through e-mail, suggestion forms, project reporting, lessons-learned meetings, and hard-copy recommendations from every level of the organization. There must be transparency in the agency. If the system is not visible and working—and taken seriously by management—it will quickly die under its own weight.

The process begins with empowering agency people to be themselves and to communicate openly with each other on improvements and opportunities they see. Furthermore, employees are invested and engaged in the agency's success.

Create a positive place for new ideas to incubate

New program administration and service ideas must often incubate in the organization before they can be conceptualized and defined. This incubation process requires a place, a website or physical office or staff, that represents the company's incubation "laboratory." This is where ideas are placed for initial review and development by a staff of associates whose primary responsibility is to incubate new ideas from the company workforce.

Generate a process of filtering, evaluating, and transitioning new ideas into a portfolio and program development process

Ideas are filtered, evaluated, and transitioned by a dedicated innovations staff into potential program features and project definitions. This would include risk assessment and risk response, using a risk matrix to categorize new concepts and identify their risks, impacts, probabilities, severity, and contingencies.

Create real agency program management success stories

There is nothing like success, visible to all. In other words, good agencies publicize new program successes as stories and reports, making sure that all agency stakeholders see that the process of generating, developing, and marketing new programs can work to achieve intended public policy outcomes. Working hard to create those stories is part of the energy or "cognitive" aspect of the agency workplace. The cognitive aspect suggests a high level of energy and engagement, a feeling of motivation and purpose.

Organizational agility is the metaphor for agency responsiveness and energy. Leadership creates this attribute by injecting excitement and purpose into the agency. This process begins with ownership, leadership, and management, which must be trusted to articulate a new program mission and integrate it into the fabric of the agency.

Management "walks the talk" by reinforcing the value of innovation and creativity, and, most of all, by focusing the program on cost-effective results and performance outcomes.

The Structure of a Federal Energy Independence Program

Developing an innovative and effective energy independence program will require a high degree of thinking "out of the box" about federal agency role and function. And the concept of *program* implies a high level of complexity, both in process and in organizational relationships; thus, creative thought must be empowered with good analysis. At the federal level, programs involve a variety of complicated intergovernmental and public/private relationships, particularly when it comes to energy. Because of this, coordination and cooperation are key elements of effective federal program management. This is especially true if the program involves guiding the market system toward a national effort to meet a broad goal, such as fuel independence.

A Customer-Driven Public/Private Partnership

The basic premise in the achievement of energy independence is that the citizen/customer makes most of the key decisions that will drive future energy demand, and American industry responds to the customer in our private enterprise system. That system must be preserved in any federal energy program; thus, the achievement of national energy goals requires cooperation and sometimes collaboration between public and private sectors, involving federal, state, and local governments; private industry; nonprofits; and, ultimately, the customer.

The goals should be to change the mix and texture of energy demand in order to establish the conditions for energy independence. This will involve research and development to design options, private industry competition within the framework of a clear national goal that takes precedence in corporate planning and marketing, and an informed market and customer base that acts deliberately to reduce individual and business energy costs and increase energy productivity. How does the federal program establish the conditions in which individual customers make decisions to achieve energy efficiency and independence and private enterprise responds with innovative energy systems? What kinds of federal activities—e.g., regulatory, program delivery, and economic/fiscal (price signals, interest decisions)—will act in concert to enable the country to preserve its private enterprise system but essentially change the demand and supply curves in energy?

Performance Budgeting and Energy Independence

Performance budgeting, the comparison of budgeted costs and program outcomes and benefits, is key to future program planning in the public sector. For instance, the evaluation of energy independence outcomes in the context of performance budgeting involves assessing the nationwide and global impacts of a program. The achievement of program objectives will require a *process* of establishing national goals, developing consistent federal agency roles and responses, setting up an outsourcing system, stewarding a competitive and innovative private enterprise response to the achievement of national goals, and a benchmarking and tracking system to ensure progress is being made.

The costs of investing in energy independence need to be looked at in the context of broad national policy objectives that go way beyond program management. Agencies must cooperate to perform such assessments. The new OFIM would be a "help desk" and technical support system to selected agencies involved in performance budgeting the energy independence process.

Coordinating Agency Program Setup

The way programs are set up determines their success or failure. Program integration management, or *integrated program and project management,* as it is sometimes called, involves selecting, coordinating, and synchronizing programs in an agency, and across associated agencies, so that all the key factors for success are optimized. Program managers can then see both the big picture and the details of program and project work, all at the same time. The program review process is set up so that management can make go or no-go decisions at each key transition point from one phase to another.

Integration involves analyzing project business value at the high level; mobilizing team performance and dynamics; monitoring projects to assure midstream adjustment and project recovery; resolving technical, resource, and interpersonal conflicts at every level; managing program interfaces and multitasking; identifying organizational constraints and exploiting them; keeping tabs on accountability; and reporting on ethical and waste problems.

In an integrated project organization, program and project managers do the following:

- See programs as capital investments made by the organization to achieve agency strategy.

- Analyze a portfolio of candidate projects using net present value techniques in combination with risk assessment and other tools.

- "Read" team dynamics as they relate to team leadership, motivation, effectiveness, decision-making process, and conflict resolution.

- Formulate and evaluate alternative completion plans to optimize program quality, cost, and schedule. Select and apply appropriate project planning and tracking tools, as well as project management leadership skills, to recover the program.

- Develop the appropriate program interface event management plan to integrate and manage the projects effectively, and resolve any complexities that result from multitasking within the networks or on the part of the project manager.

- Plan the development and implementation of suitable project management systems and methodology to support multiple projects. Identify best practices that lead to effective program management.

- Develop and manage an integrated program schedule. This includes the elements of the schedule, types of schedules, schedule development, and the processes of schedule management.

- Work with the structures of schedules based on the integrated master plan (IMP), integrated master schedule (IMS), supplier data integration, and the overall influences of the WBS.

- Establish requirements and a clear make/buy decision.

- Work with networking basics, primary elements of the network, what a critical path is, the critical path method, and program networks.

- Work with integrated product and process teams (IPTs), their outputs, and their relationships to each other, and recognize key product and process team tasks and their relationships in a program schedule. They can also analyze the performance aspects of an IPT.

- Work to transition from an integrated master plan to the integrated master schedule.

- Manage the integration of recurring production schedules with the non-recurring program-level schedules.

- Manage reporting of integrated schedule data to customers.

- Understand when and where to use earned value management (EVM), and how this system is used to benefit the project management effort.

- Design and manage an assignment matrix (AM).

- Serve as an effective cost account manager, and understand how cost accounting supports the cost account manager.

- Provide a working-level definition of a project/program baseline schedule to include the basic objectives of that schedule.

- Implement the planning and budgeting process within the resource loading activities of the schedule integration process.

- Produce a variance analysis report and the process of arriving at the conclusions cited in the report.

- Define and integrate risk management, and elaborate on the process of risk analysis and risk management.

- Provide the definition for low risk, medium risk, and high risk, while being able to integrate the likelihood of the occurrence of each with the consequences onto the tailored risk grid, and incorporate necessary mitigating elements.

Reviewing the current state of program management literature, it could be argued that the word "integration" is used in so many different contexts and applications that its usefulness is in question. Engineers integrate products and systems, managers integrate and coordinate their organizations, and planners integrate their plans. Yet despite its prolific use, the term still defines a critical aspect of successful organizations, projects, and teams working together toward program and project objectives. The integration of product and system means that the components come together to produce product performance and customer satisfaction. They come together because people make them come together; integration just doesn't happen—it must be proactively encouraged by the participants in the program management process.

Integration as a Leadership Function

OFIM would participate in the integration of programs across agencies. Organizations effectively integrate their program and new product development project work when their leaders encourage it. Programs and systems do not integrate unless key people at the working and project levels actually *think* integration. Thinking integration is a way of looking at your work as interdependent, as a part of the whole. Information is shared in an integrated organization simply because the key people know that shared purpose and shared information serves the customer better, faster, and cheaper.

Leaders prepare their organizations for integration by loosening bureaucratic barriers and encouraging cross-functional training and work settings. Information systems encourage integration. For instance, an electronic timesheet system is tied into networked Microsoft Project software so that project managers can see actual costs in real time. Leaders insist on these supporting systems because they know the value of information and sharing in building products that work.

Integration as a Wide-Ranging Quality and Process Improvement Standard

Integration is addressed in a wide variety of quality standards for corporate management and for program and project management, including the Project Management Institute PMBOK, the National Baldrige Quality Award, the PMI

OPM3 Maturity Model, and critical chain concepts. Along with increasing complexity in systems and projects, and the challenge of putting together the efforts of global outsourcing teams, the concept of integration becomes more and more important to achieve "cheaper, better, faster" project cycles.

For instance, the National Baldrige Quality Award criteria are used by many companies as benchmarks for best practices in integrating planning, operations, and project/process management. The Baldrige criteria address integration in terms of alignment and consistency of purpose and in measurement of outcomes. For instance, the 2005 criteria for health services organizations reads:

> This item examines your organization's selection, management, and use of data and information for performance measurement and analysis in support of organizational planning and performance improvement. This performance improvement includes efforts to improve health care results and outcomes (e.g., through the selection of statistically meaningful indicators, risk adjustment of data, and linking outcomes to processes and provider decisions). The item serves as a central collection and analysis point in an integrated performance measurement and management system that relies on clinical, financial, and nonfinancial data and information. The aim of measurement and analysis is to guide your organization's process management toward the achievement of key organizational performance results and strategic objectives. Alignment and integration are key concepts for successful implementation of your performance measurement system. They are viewed in terms of the extent and effectiveness of use to meet your performance assessment needs. Alignment and integration include how measures are aligned throughout your organization, how they are integrated to yield organization-wide data/information, and how performance measurement requirements are deployed by your senior leaders to track departmental, work group, and process-level performance on key measures targeted for organization-wide significance and/or improvement. (p. 40, 2005 Baldrige Award Health Criteria)

Translated to the program management environment, the Baldrige criteria stress the importance of selecting projects that implement business goals and plans, and making sure outcomes of multiprogram portfolios and business processes, such as project planning and control, are tied together through alignment with the business direction.

Tools in building an integrated federal program management system

Establishing cost reduction and performance as major program management objectives for all federal agencies.   The theme of program management systems is the trade-off between cost, time, and value. The focus of value is performance against program outcomes and benefits. The objective of the system is to optimize all three factors, since one can always be enhanced at the expense of the other two. The classic design is show in Fig. 8-1.

A program can be delivered early, but the process of crashing the program can increase the cost and create quality and value problems. On the other hand, individual project cost can be reduced, but this action may produce schedule

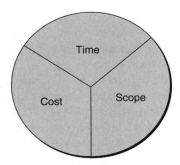

Figure 8-1   Program trade-offs.

and timeline problems, and again quality impacts. If value is to be increased beyond the scope of work and customer requirements, then the price will be paid in cost and time.

In building an agency system to support new program development project integration, there are 10 areas for process improvement. These are organization-wide or enterprise wide project management systems, program portfolio system development, integrated resource management systems, information technology, technical product development (including a stage-gateway review system), interface management system, project portfolio management, project monitoring and corrective action, change control, and program evaluation.

Federal government-wide program management system.   Here are the attributes of a fully mature federal program management system:

- *Integrated program management culture.* Assign Senior Executive Service career managers to program management positions and to top agency oversight positions in charge of long-term program benefits. The SES is the career cadre that can lead the transformation to better program management. Leaders develop their organizations to accomplish integration and coordination of agencies through an open exchange between the SES manager and interagency program management systems and communication. This system would involve the development of a federal culture of accountable program management, run by professionals. All training and development, and incentive systems, are built to encourage work that is accomplished through formal programs, projects, plans, and schedules that integrate cost, time, and quality.

- *Each agency has a program and project management council.* This council is made up of top management, functional departments and bureaus, and program management. It manages program reviews at the end of concept definition and full development, and makes the go or no-go decisions.

- *Generic program work breakdown structure.* The purpose of the generic WBS is to define a standard for how each kind of program is to be run. It integrates the work, which is project-coded to capture costs and task performance history. The generic WBS defines each task in a *data dictionary,* or task definition, that covers what the task expectation is and what its deliverable is.

- *Common scheduling system.* A common scheduling system across all agencies places all work in a program schedule software (e.g., Microsoft Project or Primavera, or equivalent) and assigns resource and estimate costs in order to control the work. Integration of all the work of the company is accomplished through scheduling, which is seen as a process of *committing resources to work.*

- *Program resource assignment.* Resources are assigned to programs and tasks so that the workforce is integrated into the work that is authorized and sponsored by the company. Projects are seen as investments in the agency's business plan. There is a major impetus in each agency to capture and control the work being performed in a resource assignment system.

- *Function, milestone, and task linkages and interdependency.* Multiple projects are consolidated into a program format, and tasks are linked to stress the interdependency of project work. No piece of work in the agency is left unconnected in order to ensure integration.

- *Program work authorization system.* Again, in order to ensure that any agency program work that goes on is authorized, all work is approved and directed by the program manager. The way work is approved is through the baseline schedule, which defines the authorized work.

- *Guidelines for program management plan.* The program management plan is defined in an agency program policy statement to guide the definition and control of the work. Therefore, the plan must include control points (e.g., stage-gateway reviews) that ensure that management authorizes movement from one phase or stage to another. Reporting and monitoring strategies, including the use of earned value to integrate cost, schedule, and quality performance, should be made explicit.

The plan should also address accountability, particularly in view of the recent legislative and regulatory requirements of the Sarbanes-Oxley Act. Compliance with internal control and accounting standards is no longer optional for project managers. In fact, the price of disconnected and inconsistently applied efforts throughout a project and its interfaces, and lack of financial tracking systems that provide for audits, could be business-wide. Compliance with Sarbanes-Oxley, therefore, is not a choice but a requirement, and the plan should state standards for estimating costs, tracking the costs and relating them to work performed, and the integrity of the closeout procedure and invoices to customers for work performed.

Program/portfolio planning and development system. An agency portfolio is its planned programs and investments that it will propose in the budget process.

- *Agency planning system and strategic objectives.* The integrated agency has a business and strategic planning process that produces a statement of strategic objectives that serves as a guide for all planning and budgeting. Such a system helps to shape the project portfolio and ensures that the company invests in projects that are integrated with the direction of the agency and its stakeholders.

- *Decision process.* Some kind of decision process supports integration because open decisions, if prolonged, can lead to waste and ineffective work. Decision trees are used to assess the commercial value of various decision paths involved in defining the task structure and sequence of approved projects.

- *Performance budgeting system.* Each agency must look at budget requests and outcomes to arrive at decisions that enhance cost effectiveness and achievement of program objectives and intended outcomes. A capital rationing system, or some way to allocate resources in line with the priority of relative strategic objectives, is part of agency integration. Once budgets are identified to carry out agency and business plans, projects are planned and prioritized in the portfolio system, and then costs are estimated. Finally, projects are fully funded according to their relative merit against business plans and available budget.

- *Risk management system.* Some kind of risk management planning system that identifies and assesses risks and generates risk contingency plans is necessary in an integrated program management system. The risk matrix is the format for developing risk information that is used in scheduling and controlling the work.

- *Program and program portfolio pipeline system.* A pipeline of approved projects is maintained so that as funds and resources become available, projects are quickly initiated. Project plans and schedules are produced for projects in the pipeline so that when authorized, they can proceed quickly.

Resource management system. Agency resources in an integrated program management must be managed, simply because there is value in targeting all resources and equipment on the right program and project work. A workforce resource pool information system can be established using Microsoft Project that records all assignments in order keep a running view of how people and equipment are being utilized.

A comprehensive system would include:

- *Workforce planning.* A program workforce planning system integrates the hiring and training of personnel with the needs of the program portfolio of projects. In others words, people, equipment, and systems are brought into the company to fill needs that are made explicit in the project resource allocation pool that reflects both current and planned work. Measures such as person-month needs by project are used to predict resource needs.

- *Staff planning.* A staffing system allocates staff to the priority project needs in order to fully integrate the core competence of the workforce with the priority needs of key programs. Staff are focused on assignments that are visible and reviewed regularly.

- Financial and accounting control is assured in a project management system that captures all project costs, both direct and indirect, and assures internal controls on project costs and equipment inventories.

  - Earned value: Reports on work progress and costs are used to calculate earned value so that the company knows how each project is doing in terms of schedule and cost.
  - Agency and industry standards: Appropriate industrial cost and work standards are used to control the estimated duration of scheduled tasks— e.g., using a trade association to schedule an industry-wide activity on which there are work and industrial standards.

Program information technology system.   A program information system that documents all program work in consolidated schedules and resource pools ensures that work is staffed, planned, and monitored in a uniform way. This allows comparison of project progress and supports decisions on where to focus resources.

The system would include:

- *Network system.* All program and project information (e.g., schedules, resource pools, project review and gate review data, configuration management documents) are kept on a company intranet to allow wide-ranging visibility.

- *Accessibility to key information.* Accessibility to information is controlled and focused on need-to-know criteria. However, customers are regularly informed on program and project progress through Microsoft Project Central web-based reporting systems that allow review of schedules without parent software.

- *Reliability planning.* Reliability planning targets products with failure mode effects assessments and functional hazard assessments, along with risk matrix documents, to consistently design and test reliability of product performance to customer requirements and specifications.

- *Workforce training.* Workforce training is designed to meet project needs as evidenced in work performance feedback reviews and lessons-learned exercises with project teams in close-out.

Program/service development process.   Integrated project management cannot be accomplished without integrated product development processes with strong stage-gate milestones.

A system would include:

- *Adopt Robert Cooper's stage-gateway process.* Program management is a process of managing time, cost, and quality, but the underlying strength of any project integration process is a strong, phased development process with clear controls on entry to the next stage. "Gate reviews" are documented, and generic work breakdown structures and data dictionaries are developed for all product and service development activities.

- *Technology support and testing.* Technical support that meets industry standards ensures that product integration and testing are verifiable. Designs are tested against specifications, specifications are tested against scope of work, and scope of work is traced to customer requirements and expectations.

Interface management.   Interfaces or linkages within the organization and outside would include:

- *Matrix organization.* Interfaces between functional agencies and departments (e.g., accounting, engineering, project management, and testing) are ensured through strong interface management. Separate departments and functions are brought together constantly through information and reporting systems and face-to-face review meetings at key program review points.

- *Program review meeting formats.* Review meetings are controlled by generic meeting agendas and data and information support from a professional project management office or staff. This way, review information is objective and consistent.

- *Procurement interface.* Because of the importance of contract and outsourced work, contractor personnel and processes are integrated with sponsor company personnel and processes. Common scheduling and reporting systems are designed.

- *Financial, accounting, and internal control interface.* New impetus for strong accounting and accountability reporting now requires that program and project managers capture costs and relate them to work performed and equipment purchased and in inventory. Ultimately, costs are compared to program outcomes in the performance budgeting process.

- *HR interface.* The interface with HR is important to integrate personnel and HR policies and procedures with project work and priorities. Performance reviews are left flexible, yet are important in assigning resources to future projects.

Portfolio management.   Portfolio management systems would include:

- *Top management visibility of programs and projects.* The whole set of projects in a multiple project system is managed consistently in an integrated project management system. All projects are monitored using common earned value and other measurement systems (e.g., balanced scorecard).

- *Uniform program management system.* A uniform approach to projects in the portfolio is assured through a professional project management staff and PMO support system.
- Pipeline management
  - Generation of projects: A systematic way of generating projects through brainstorming, budgeting processes, and business planning.
  - Evaluation of projects: A way of reviewing portfolio projects using net present value and cash flows, weighted scoring models to score projects against business objectives, and risk management.
  - Selection of projects: Projects are selected using a uniform set of measures.

Program monitoring and control system

- *Program management office (PMO).* Agencies create program management offices to support program managers. Monitoring is based on earned value reporting, and quality is ensured by a task planning system that relates percent complete to defined milestones in the baseline schedule.
- *Corrective action / risk management process.* Contingencies and corrective actions are based on remaining work and are forward-oriented. Contingencies are embedded in schedules to ensure that should risks occur, they have already been scheduled for and budgeted accordingly.
- *Escalation system for decisions.* Conflicts and differences within project processes are reviewed regularly by top management so that decisions are not delayed.

Change management system.   A change management system would include:

- *Change management system.* Change management system is important in new product development, since design changes are typical and frequent, and because each change in configuration or design must be documented.
- *Change order system.* All changes to a scope of work are submitted by project team members or the sponsor/customer to ensure that changes are reviewed and managed.
- *Change impact system.* Change impact statements are prepared for all substantial changes, with risk, schedule, cost, and quality impacts specified.

Program evaluation system.   A program evaluation system would include:

- *Document lessons learned.* Close-out includes a lessons-learned meeting and documentation of outcomes. The PMO is responsible for ensuring that lessons learned are integrated into future projects.
- *Financial auditing system.* A financial and program audit system is managed to ensure accountability and internal control of all assets.

Transparency

- Transparency requires a *serious* system to serve two interests: the interest of the general public for high-level information about a program, and the interests of agency program professionals and agency administrators, congressional oversight staff, and outside public organizations. A one-way White House website is not the answer to the transparency issue. Transparency requires interactivity and cross-communication at many different levels of a program, and the federal government has no such system in place.

Summary

This chapter addressed the root causes of failures in large, federal government programs, including lack of clear goals, lack of uniform processes, the lack of clear and measurable deliverables, and the lack of full funding. The chapter includes a proposed new Office of Innovative Federal Management (OIFM) to guide and coordinate federal program management activities in the agencies. The proposal also outlines a program management process to generate consistency and innovation in program design and implementation.

New Global Program Development

Introduction

The previous chapters addressed the global setting for program management, program portfolio and risk strategies, the global program manager, and the significance of partnerships. We found that the roles and functions of the program manager are different from those of the project manager. And we stressed the importance of portfolio planning and program and project selection because program managers want to run the *right projects right*. The previous chapter addressed a special program issue: how the federal government manages programs and what makes federal program management so challenging.

What Have We Learned about New Global Program Management?

Here are some conclusions from our analysis of program management:

1. Control is not always the goal in international programs.

   Program and project management are essentially Western concepts, grounded in our developed-country propensity to look at the world as rational and orderly. But perhaps global program managers are finding that this rational, orderly world does not always exist in other countries and cultures and that our Western, control-oriented approach to locking in on deliverables, scopes of work, tight work schedules, and controlling costs does not always fit.

   While program and project management tools are all focused on controlling time, cost, and quality, these aspects of a complex program are often determined by *softer* factors—e.g., client and stakeholder relationships and partnerships, the state of international relations, local politics and economics, and local trade dynamics. And global program teams in various countries such as China will likely be composed of members from several nations, with a Chinese national leading the effort. This will place a high premium on working in a multicultural team environment in which local nationals hold the leveraging positions.

Impact on program management. Global managers sometimes have to be comfortable with ambiguity and uncertainty in designing and managing programs in some parts of the world, including the Middle East and South America. This means that the traditional push toward controlling scopes of work, schedules, budgets, and quality standards often gives way to open-ended programs with no apparent end point.

2. A group of loosely related, short-term projects do not add up to a program in a global setting.

   Contrary to the conventional wisdom in the project management community, programs are not made up simply of multiple projects. Programs are not built and run bottom-up; they are top-down. In fact, programs are developed out of business plans and need broad support from administrative and support functions. Many projects do not make a program. A program is a complex of targeted resources.

   Impact on program management. There needs to be more emphasis on designing new programs around broad strategies—e.g., capturing a given new market for energy efficiency in developing countries—and planning for all the necessary support functions—e.g., logistics, IT, HR, and e-procurement.

3. Portfolio decisions to invest in new programs will have to revisit program benefits and revenue projections.

   Programs are built not only on income projections and rate of return (ROR) analysis, but also (and perhaps more importantly) on program benefits. This means that businesses will have to spend more time and resources on articulating and projecting program benefits over the longer term. For instance, a program aimed at developing and marketing a new wind source energy technology in a developing country will have to analyze the dollar value of future partnerships with local energy producers once the technology is adopted. For instance, Google's mission "to organize the world's information and make it universally accessible and useful" will have to be translated into long-term programs and benefits (e.g., market share and technological competence) for Google, as well as for the many users of Google-provided information.

   Impact on program management. Projections of benefits for global programs, particularly in developing countries, often involve speculating on customer behavior in unfamiliar societies, as well as deciding how broader social and economic benefits should be calculated.

4. Research suggests that trends in globalization are changing the dynamics of program management.

In 2009, the McKinsey Global Institute reported on a study of the underlying forces that shape the global business environment (Eric Beinhocker, Ian Davis, and Lenny Mendonca, in *Harvard Business Review*, July-August 2009). The authors of the study looked especially for discontinuities—e.g., changes in factors that will require a refocus on portfolio development. They found that the following forces could affect business and program success in a global setting.

a. **Resources feeling the strain:** With productivity down worldwide, basic resources, oil, water, and foods are in short supply.

Impact on program management. Projects oriented to resource development and efficiency (e.g., more energy-efficient buildings) represent good risks in program development. Schedule and cost becomes less important in light of the need for new program development on new energy systems and the building of local capacity to produce.

Programs that focus on new energy sources and sustainability are going to do better than programs that ignore those markets. New packaging and delivery markets will open up for programs equipped to increase the *greenness* of the program benefits to local countries and communities.

b. **Globalization under fire:** Movement of goods and services, and talent, across national boundaries is likely to slow as governments tighten controls. Finances for program development will be harder to arrange as countries delink from the global financial system in response to the worldwide recession in 2007 and beyond.

Impact on program management. Programs focused on achieving national energy self-sufficiency are likely to be popular. Labor for new programs will be more difficult to transfer into jobs worldwide, thus placing more emphasis on finding talent to support programs locally.

c. **Trust in business running out:** Trusting partnerships with other businesses and with customers will be more difficult in the future globally simply because the recent global recession altered the dynamics of trust in business.

Impact on program management. Program managers will find it harder to develop partnerships and consortiums with other businesses in the supply chain for program support. Stockholders will be intervening more in business decisions on program portfolios, asking questions about risks and returns.

d. **A bigger role for government:** The recession and government financial investment of public funds in private companies has changed the landscape for dealing with government in global programs. Now governments will play a larger role in program and project management decisions, and a premium will be placed on knowing the local public policies and regulations relating to targeted programs.

Impact on program management. Program managers are going to spend more time and resources on anticipating the actions and reactions of foreign governments, both national and local, to program development.

e. **Management as a science called into question:** The global recession of early 2007–2009 called into question the folly of relying on financial models that assumed economic rationality, linearity, equilibrium, and bell-curve distribution. Banks and other financial institutions approve large project-related investments on the basis of business planning models, including cost-benefit and net present value, but using these tools to select a program portfolio doesn't always work.

Impact on program management. Program development and portfolio selections based on rate of return, net present value, long-term revenue forecasts, and other financial models will be more vulnerable, given the failure of these models to predict major setbacks. This will place more importance in knowing local decision makers and developing trusting relationships based on achieving local recovery goals with new programs.

f. **Shifting consumption patterns:** Old assumptions about consumer spending and demand are changing as the world's productivity grows slower and slower. This means long-term business goals based on old rates of growth are no longer valid.

Impact on program management. Programs and project managers focused on consumer goods and demand in foreign nations will have to recalculate their longer-term income projections to match reality. This will place a premium on realizing small margins in program delivery and thus on cost controls.

g. **Asia rising:** The increasing importance of India and China and other Asian nations in the global economy suggests that they will be the focus for future program and project development.

Impact on program management. This means that global program and project decision makers will have to consider the impact of all their efforts on the Asian markets, both in terms of supply chains and customer base.

h. **Industries taking new shape:** New reliance on computer networking and mobile computer capacity is changing the structure of business models, with more attention being paid to enabling smaller companies to compete globally through the Internet.

Impact on program management. Programs will be increasingly supported by Internet tools and techniques and virtual teams, thus reducing the need for heavy logistics costs.

i. **Innovation marching on:** There will be more investment in research and development, despite the worldwide downturns, and more attention paid to program and project development in energy-related fields.

Impact on program management. This means program managers will focus more on rationalizing program and project portfolios and reevaluating foreign licensing agreements.

j. **Price stability in question:** Threats of deflation and unanticipated changes in supply pricing suggest that businesses will have to adjust their forecasts of margins and cost coverage.

Impact on program management. Program and project managers will rely more on their procurement and purchasing offices to find ways to take advantage of price decreases in key project supplies.

5. Global customers will increasingly take the shape of family businesses rather than traditional corporate structures, particularly in Asia, South America, and the Middle East, where nationalized businesses are actually family-owned and -operated.

This trend suggests that program and project managers in foreign nations will have to understand and prepare for a new structure for clients, stakeholders, and business customers: the family-owned business. This will mean more emphasis on trust and loyalty, and less on professional management techniques. It may also mean less emphasis on time and cost and more on quality.

Impact on program management. Program managers are going to have to deal with family members and bureaucrats who run state-owned companies in program design and delivery. There will a premium on contributing narrow functions and deliverables to a broad national purpose without interfering in family decisions and state approvals.

Along with this trend there will be more emphasis on entrepreneurial leaders and less on professional project managers, simply because foreign clients, particularly in developing countries in Asia will favor more aggressive risk-taking to maximize markets. This trend lessens the importance of traditional contracting, scheduling, costing, and quality controls, and envisions more risk-taking without project controls.

6. The world's governments are increasingly calling for transparency in program and project management.

This trend, started in the United States, provides an impetus to offset potentially bad or unethical business decisions with more public access to business information and data. But transparency can be a difficult goal, since different stakeholders will have different data and information needs. For instance, a partner contractor may need to see up-to-date schedule information and earned value, while the customer may want to see cost and schedule data targeted on remaining work. Meanwhile, a government oversight council may want to ensure adherence to safety regulations.

Impact on program management. Program managers are facing two opposing forces: one a move toward more family business clients, the other toward more transparency in program and project data and performance information. Managers will have to "read" their customers in order to arrive at a balance between techniques to inform stakeholders on program progress and costs, but at the same time making sure that proprietary information is not divulged to the wrong parties in a foreign country.

Generating Creative Program Concepts

So how does a business create new concepts for global programming? What makes Apple effective in marketing globally when Motorola fails in the same market with essentially the same products and services? The essential ingredient in successful new programs is not a disciplined portfolio process and analytic tools. Rather, a vibrant and energetic organization enables its creative people and its partners to think innovatively about what they are doing, know where the company is going and what it can do, and feel free to generate and integrate new ideas and new program concepts and processes.

In these companies, business associates are not risk-averse. They know *and feel* that their professional careers are enhanced by generating new program ideas in their domain, but they also know that their advancement is not solely determined by the success of a new project. This frees them from the fear of communicating bad news—if appropriate—or terminating a project because it turns out that it makes no sense in the market. The best ideas for new products come from both those who do the regular work of the company and from market research and new product teams.

This target organizational environment is a *culture of ideas,* the organizational *tone* that encourages free flow of feedback *into and out of the company* on current products, services, customers, and market operations from those closest to them. This does not happen naturally or easily because new ideas are not always welcome in companies working hard to develop and market current products and services, and because new ideas tend to challenge the current way of doing business—they change processes and strongly held paradigms.

New organizational structure for new programs

We live in a global economy in which the turnaround of structural and technological change is rapid and sometimes unpredictable. Business processes are changing, even as we speak, and traditional jobs and work settings are becoming virtual, e-driven, and mobile. This atmosphere of change creates the need to structure the organization so that people can create new products *before* they are anticipated by the customer, marketplace, or competition; and the company can design, develop, and produce new products better, faster, and cheaper and get them to market. Development life cycles have gone from years to months to days.

Traditional ways of organizing around product lines, markets, or geographic locations does not seem to encourage new thinking about products in a broader context. How does a company organize in order to avoid being captured by narrow views of the customer, determined by traditional organizational structure? The answer is that organizing around processes appears to be helping successful twenty-first century businesses to generate creative thinking because it is in process framework that companies can best *see* their customers in a system. Processes can describe various systems and interactions that lead to product success; thus, encouraging people to think in terms of process domain gives them a better chance of coming up with new processes and products to serve customers.

New products and outsourcing

Sometimes it pays for program managers to outsource creativity and front-end new projects and new product development to capture domains that the company cannot create or capture itself. That is, to the extent that outsourcing increases the probability of partners *seeing* new processes and products differently, out-sourcing increases the chances of successful new product introductions. This conclusion must be conditioned, however, on the presumption that the relationship with outsourced partners is more than simply a contract—that, indeed, the partner is a committed, long-term associate in a given market and, therefore, can be trusted to act in the interest of the partnership.

Restructuring for global program management

Many businesses are not organized to access and interpret global information or to work with international partners and experts in program design. Companies who can mobilize internationally without expensive foreign infrastructure will be at an advantage. This will require program managers to draw on the local talent and resources of partners and stakeholders who are located in the target nation, both to help design longer-term programs to meet local needs and to deliver and maintain the deliverables.

Examining competencies and outcomes

Program and project managers and their business associates will have to monitor the new dynamics of the global marketplace, both in choosing programs and in designing and delivering them. That process will require constant self-examination, focusing on in-house competencies and opportunities to outsource for more efficiency and effectiveness, and on opportunities to partner with foreign institutions of all kinds. For instance, a myriad of nongovernmental organizations (NGOs), roughly designed like U.S. nonprofits, figure importantly in working with program and project managers from foreign countries in multicultural, virtual program teams.

In addition, programs will have to be evaluated using conventional program evaluation techniques—e.g., compare actual program performance and deliverables against program benefit goals and objectives, and identify forces and process factors of success and failure.

The Broadband Case: An Application

Broadband communication systems provide sufficient capacity to carry multiple voice, data, or video channels simultaneously. Broadband enables, among other things, users to access streaming audio (Internet radio), streaming video (compressed images sent through the Internet), and streaming media (combines streaming video with sound). Broadband is a necessary element in any developing country's plans to equip its residents and businesses with global communication capacity.

Let's say that a communications business targets the African market for broadband communication in its strategic plan. In developing its implementation plans, the business would articulate and define its strategic objectives, then drill down to a single, globally targeted program, or possibly two or three.

Strategic global objectives

The three key strategic objectives related to global programming are:

1. To generate broadband options to selected African countries that have expressed an intent to provide such services to its population

2. To design and develop a low-cost broadband system and to find ways to provide broadband programs to rural areas at reduced costs in order to generate a market

3. To develop local competency in the targeted countries on broadband applications in online education, medical diagnosis and treatment, and e-commerce

Global program development

Let's say that the business identified its key program to be the development of broadband for online education in African countries. Other programs given high priority in the portfolio process include programs to deliver medical diagnosis and treatment, and to promote e-commerce.

The online program is funded and approved for implementation. This program would involve designing and installing broadband equipment and consulting services, along with software applications, to support online education in African countries. The business conducts market research in Africa and finds that several African nations have expressed interest in broadband and are providing subsidies for development programs. Surveys also discover that a variety of regulations in some African countries would affect the installation of broadband, including a censoring function on education content.

Global portfolio of projects

Next, the business would develop a portfolio of projects to support this program. Innovative ideas are solicited in the portfolio process, and several projects are generated and reviewed for inclusion in the approved program portfolio. These include the following projects:

- Project 1: Identify the potential population in targeted countries for online education, K–12, and higher education.

- Project 2: Design equipment and software to deliver broadband to 50 selected areas in five African countries.

- Project 3: Develop a curriculum for K–12 and higher education and integrate into software and course websites.

Global program manager

Next, a program manager is appointed from several candidates. The candidate selected has experience in delivering global telecommunication programs in South America. The business identifies this candidate as the best for the job, not only because of her experience, but also because she has personally undertaken a professional development program that includes learning the basic African languages and cultures, and she completed a "lessons learned" report on her telecommunications activity in South America.

Global program benefits

What are the benefits of this program, and how will these benefits be valued in the portfolio and program development process? The benefits of the program are:

1. Providing online education to low-income students who otherwise cannot attend school because of transportation, cost, and/or logistics reasons

2. Enabling local teachers and instructors to provide education services through online systems

3. Setting up a business model that the business can use for future broadband development for online education, providing the business with a test case, learning environment on what it takes to install such a system (e.g., cost, schedule, quality issues, regulatory issues, content censorship, etc.)

Calculating rate of return

Rate of return, or new present value of programs, involves projecting out income over a 5- to 10-year period, discounting for present value, and determining if the program should be undertaken. The business develops a listing of development costs, which comes to $30.5 million, including equipment, labor, contractor costs, and overhead. Income projections (net present value) from broadband user fees and leasing arrangements total $28 million over five years, resulting in a negative net present value of $1.5 million.

The decision, despite the negative rate of return in five years, is to proceed with the program, given its longer-term potential for creating a broadband market for the business and the broad benefits of the program for African country development. The business's top management determines that it is willing to invest in the long-term program in the hope that not only will the market develop, but also the company's own reputation for being willing to pursue a social and economic benefit in Africa along with the conventional profit motive.

Global partnering and strategic alliances

The business enters into a strategic alliance with the African Union Education Council, an association furthering African development funded by member countries. This alliance allows the business to access information and demographic data on various target areas for broadband development. In addition, a team of project staff is deployed to selected African countries to identify local talent for building the program teams.

Global transparency

Local governments and NGOs in targeted areas with a stake in broadband development are identified and structured into local broadband review boards. The business will use these boards to inform local leadership on its intentions and to get feedback on program and project progress.

Government relations

Each governmental jurisdiction in the selected markets will be briefed on the program. Top business associates are deployed to each jurisdiction to work out leasing and other arrangements, including necessary "payments" for local endorsement.

Regulatory issues

The business conducts a thorough search of extant regulations in broadband development in the targeted markets and develops a manual for program managers to deal with local requirements on access, safety, and installation and maintenance of equipment.

Culture and language issues

Special content and language experts are hired for each targeted area to translate negotiations and team collaboration and to anticipate particular cultural factors and forces that may shape the program.

Equipment production and delivery

Special attempts are made by the business to find local producers or suppliers to add to its supply chain for design and delivery of broadband equipment and systems. If local suppliers are not available, the business works with local governmental, NGO, and education providers to build a local industrial base for broadband management in the area. This base is intended to serve as the foundation for a local industry in the area that will take over the broadband activity as a licensee.

Education content censorship

A special issue arises in program development: the censorship of education content and curriculum. National and local governments in African countries express their intent to review and approve all online courses for consistency with local values and education programs. Recognizing that this requirement is part of doing business in Africa, the business works out a process of collaboration with local education institutions to use their courseware and content in designing its broadband online education program.

Summary

We can conclude that global program management is not very different from domestic program development in one sense; but in another sense, it is made more challenging because of a more complex societal, cultural, and political framework and added risks and uncertainties. Most of these issues can be anticipated and addressed if the business is committed to global work and is willing to do its homework. The key success factors in global program management may be the capacity to stay agile, to be comfortable with ambiguity, to be patient with outcomes and benefits, and to address the key role that trust and partnerships play in program success.

Program Risk Management Checklist

Integrated program risk management involves a whole set of activities that should be embedded into the program and project planning process. The following provides a risk checklist and associated exam and training questions raised by the checklist.

Action	What?	Why?	When?	Output?	Who?	Exam Question
			Business Culture			
Create risk management policy	Create business intent to manage risk	Confirm that it is important and back it up	Part of business plan; underlies project process	Policy statement on how the business will handle risk	Executive and program management level	What policies and procedures will encourage integrated project management? Write a policy statement that encourages integration across the organization.
Assess organization awareness	Find out how aware workforce is of risk and risk response impacts	Survey workforce	Every six months	Workforce awareness of risk management report	HR/project team	What indicators of awareness would you use to survey whether your organization is integrating its projects? What measures address integration?

(Continued)

ppendix A

Action	What?	Why?	When?	Output?	Who?	Exam Question
Business Culture						
Deliver training program	Design training around practice planning tools; use to introduce business risk	Workforce will implement if they understand tools	Every year with refresher	Certification	All PMs, project teams, and technical personnel	Identify the key training courses that you would provide to encourage integrated project management.
Reward effective risk management	Provide rewards for good risk management effort and effectiveness	Incentives motivate	During project	Compensation reward	Project managers and team members	What kinds of behaviors or actions that further integration would you reward? What rewards would you provide?
Business Strategy						
Risk component of business plan	Provide for a risk section in the business plan and communicate it	SWOT analysis: threats = risks; translate to product line risk exposure	Annual update of business strategic and business plan	Risk-based business plan; integrate with financial and profitability analysis	Executives and program managers	How would you include risk in your business plan; would it be a separate section, or integrated into the whole plan, or both? Why?
Strategic objectives	State objectives in terms of risk	Measurable strategy goals	Part of plan; communicate to workforce	Set of 10 long-term objectives	Executives and program managers	Write a strategic objective for an organization of your choice, incorporating risk.
Project Selection						
Do risk assessment of candidate projects	In developing business portfolio of projects, use risk as one criterion for project selection	Use PMBOK process; broad-brush risk assessment	Each time project portfolio pipeline is updated	Rank order projects using composite risk, alignment, cost, and revenue assessment	Program managers and functional managers	How would you evaluate candidate projects–in terms of whether they are integrated–in the project selection process?

Action	What?	Why?	When?	Output?	Who?	Exam Question
			Project Selection			
Weigh risk against revenues and alignment	Demonstrate that risk has been embedded in business and financial analysis	Trade off risk with opportunity for profitability and take advantage of business core competence	Each time pipeline is updated	Analysis, data, documentation	Project management office, project team, business planning staff	What measures would you use to evaluate projects on the basis of whether their planning included analysis of revenue risks? What about risks of misalignment, for example that the project is inconsistent with the organization's goals and strategies?
			Project Plan			
Requirements	State customer requirements in terms of customer risks	Risk that customer requirements do not reflect risk, or misunderstanding customer perspective and expectations on project risks	During initial concept phase; part of project plan	Requirements document stating customer requirements and risks	Project manager, functional manager, and customer	How would you assess whether a project integrated customer requirements into scope and project concept?
WBS	Include risk contingencies from risk matrix in WBS work activity	Because there is inherent risk in missing major parts of the deliverable in initial planning; WBS ensures coverage of major "chunks" of work	During development of the deliverable, the "work" should include initial contingencies identified in risk assessment	WBS in organization chart form and outlined in Microsoft Project	Project manager	How do you assess whether a project WBS has integrated all key work activities, for example that it is a truly integrated WBS?
Task list	Include risk tasks and contingencies in baseline schedule	Task list should include all anticipated contingency actions should risk events occur	After WBS is prepared, do task list and link; there is inherent risk that linkages will be too "hard;" allow for "soft" linkage	Task list in Microsoft Project GANTT chart or spreadsheet	Project management office and/ or project manager	How do you assure that a project task list is inclusive?

(Continued)

Action	What?	Why?	When?	Output?	Who?	Exam Question
			Project Plan			
Network Diagram	Show risk in network diagram with three scenarios: expected, pessimistic, and optimistic	Arrow diagram shows critical and noncritical paths; risks inherent in focusing on critical path when resource constraints in noncritical tasks may serve as bottleneck (theory of constraints)	During translation of WBS to GANTT chart, prepared to show dependencies and paths	Arrow diagram in Microsoft Project or other software	Project management office template or project manager	Assessment of risk options requires estimation of expected, pessimistic and optimistic alternatives, both in terms of weights and durations. How do you assure that these estimates truly reflect all the factors impacting on task duration and risk?
Calendar-based diagram	Relate network to time to begin to see schedule impacts and milestones	Histogram using arrows and calendar	During translation of WBS to GANTT chart	Graphics software or Microsoft Word document	Project manager	What is the value of a calendar based diagram in transitioning from a task list to a GANTT chart?
Risk-based schedule	Do risk-based schedule using Microsoft Project PERT analysis tool	Microsoft project GANTT chart showing calculated risk-based duration after weights and three scenarios are entered	During initial scheduling, then any time risk is identified and contingency prepared	Microsoft Project schedule file showing calculated durations for high-risk tasks	Project management office or project manager	How do you assure that changes in task durations from PERT analysis are integrated into the project team, for example, that project team members are made award of new duration estimates that affect them?

Action	What?	Why?	When?	Output?	Who?	Exam Question
		Risk Management Process (see PMBOK)				
Risk identification	Using input from business plan, identify and rank project tasks in terms of risk	Using data and information and past experience, rank summary tasks in WBS using risk matrix	During business planning, project and portfolio selection, and project WBS scheduling	Risk matrix	Project management office or project manager	How do you assure that the risk identification process is integrated with company business planning and SWOT (strengths, weaknesses, opportunities, and threats) analysis?
Risk assessment	Assess risks using risk matrix format	Complete risk definition, impact (schedule, cost, quality, business growth); make probability estimate (25%, 50%, 75% probability), severity on project outcome, and contingency	During business planning, project selection, and project planning and control	Risk matrix, updated monthly	Project manager	How do you assure that the risk assessment is integrated with the project WBS, for example that all tasks have been reviewed in the assessment?
Risk response	Prepare contingency actions and include in baseline schedule	Response is planning through definitive contingency plans and tasks, which are embedded in project schedule as regular tasks and triggered if risk event occurs	During project planning, responses and contingencies are designed to address specific risks and recorded; this is where the team anticipates what might happen to slow or delay the project, what can be done to prevent it or address it, and schedules contingency tasks into the project	Risk contingency actions	Project manager	How do you assure that the risk response plan is integrated with other project and business planning?

(Continued)

Action	What?	Why?	When?	Output?	Who?	Exam Question
			Risk Management Process (see PMBOK)			
Risk matrix	Prepare risk matrix as basis for scheduling	The risk matrix is the basic checklist item for risk throughout the process; it is the guide for action	During project planning, a basic risk matrix file is established and appears with all project planning and project review documents	Risk matrix following prescribed format	Project manager and team or task managers	How do you assure that the risk matrix data on each task are integrated across column headings for risk, impact, probability, severity, and contingency?
Decision tree	Do decision tree analysis to expected value of optional decisions	This is the way project managers anticipate decisions they will have to make based on risk and what alternative paths and expected values will follow each decision path	During project planning, decision tree analysis is applied to high-risk tasks	Decision tree diagram with expected values calculated	Project manager	How is the decision tree in assuring that all key decision risks are integrated with business risk and SWOT data?
			Integrate Risk into Project Manual			
Basic project manual	Ensure that risk is not treated separately, but seen as part of the way projects are planned and controlled	Because risk should not be treated separately from project management process; manual captures how risk is integrated into process	Business establishes a system of basic project manuals as part of "projectizing" the organization	Online and hardcopy manual, including basic project planning and risk management tools and templates	Project management office (PMO)	How do you assure that the manual calling for the integration of risk into project planning and operations is implemented?
Provide software tools	Train and provide software analysis tools in manual	Much of the risk analysis can be done through spreadsheets and decision tree analysis software; workforce needs to know how to use them	Business establishes a support system of risk management application software and trains appropriate staff	Software library	IT and project managers	How do you assure that training materials are integrated with other company software?

Action	What?	Why?	When?	Output?	Who?	Exam Question
		Product/Technical Process Development				
Define product/ technical development process in generic WBS	Ensure that business has defined the core, product development, and technical processes that it uses to produce products and services (e.g., engineering, construction, system development) standardizing where possible	Because project risk management cannot be successful unless both technical and product development/ testing requirements are conducted to control risk, and because management impacts (e.g., schedule and cost) are applied to the real industry processes that create customer value	Business establishes a generic WBS of technical processes; these are recorded and updated so that all project WBS and schedule information and risk data is taken from the generic model and tailored	WBS file	Functional managers	The generic WBS is intended to capture and integrate with all company business practices and processes, e.g. budgeting, accounting, performance evaluation, etc.; how do you assure that it does?
Project codes	Provide for coding actions in WBS so that costs can be captured	Because once you have identified all tasks and risk contingencies, you will want to capture costs against those codes to build a history of risk management and mitigation costs	When generic WBS is set up, codes are added at the appropriate level to capture costs	Coding system integrated with timesheets and accounting system	Accounting, project management	How do you assure that project codes assigned to various levels of the WBS are workable and integrated with normal personnel practices and timesheet systems?
		Identify Customer Risk Tolerance				
Assess Customer perspective on business and project risk and how much risk customer is willing to assume	Solicit customer input on customer risks and uncertainties	Because customer may have different and valuable insight on business and project risks that have been part of the customer expectations but not reflected in real planning	When requirements are being written	Customer risk analysis	Customer representative and functional and project managers, jointly	

(*Continued*)

Action	What?	Why?	When?	Output?	Who?	Exam Question
			Lessons Learned			
Risk audit	Do a project risk audit following selected projects to evaluate success in anticipating and managing risk	Because insights and documents that can lead to better risk management in the future will be lost unless a risk audit team builds a history of the project, how risk decisions were made, and how effective risk management was	At project close-out	Risk audit report to project manager	Project management office, audit staff, project and functional managers	How do you assure that risk audits are integrated with project reviews and lessons learned sessions?
Lessons-learned meeting and report	Prepare and communicate short report on what project team members and customers learned in the project that would reduce risks in a similar future project	Because the best lessons and insights are going to be lost unless someone facilitates a lessons-learned session and report	At close-out	Lessons-learned report, referencing systems, decisions, risk, outcomes, but no names	Project manager	How do you assure that lessons learned outcomes are integrated with future business and program planning?

Microsoft Applications for Program Management

Microsoft Project 2007 is not the only project management support package, but does offer useful tools to build and track complex, global programs of projects.

Project Consolidation to Create Programs

Since the difference between a program and a project is often simply in terms of complexity and scale, many projects can be consolidated into a program using the typical Microsoft GANTT chart. To do this, high-level milestones are chosen from project schedules and identified in a limited set of program milestones to be tracked at the program manager level in program reviews. See a Microsoft Project manual for details.

Project Server for Program Support

Microsoft Project Server provides tools for a program manager to use the Web to communicate, manage, and monitor programs globally. A Project Server integrates a web-based interface, called Project Web Access, with Microsoft Project Professional. A project administrator stewards the system as a database of project information uploaded by the program manager or project manager. Data security and access issues are handled by the administrator. Programs can be created using tools collaboratively with Windows SharePoint. Program managers can manage resource assignments by viewing and changing program or project enterprise resource and tracking the progress. Team members, wherever they are, can submit timesheets and report issues and risks in real time using Windows SharePoint. Program managers use Project Centers to access high-level project data and to communicate with team members.

A Word of Caution

Program managers are cautioned to careful when working with professional network specialists in developing a web-based, global program management system. The hardware and software configurations you select will directly affect the overall performance of Project Server in supporting global program management. The critical factors are to avoid underestimating the amount of traffic on the system globally and to anticipate any issues in reporting across nations and continents. Traffic can be estimated by anticipating all company executives, customers, team members, sponsors, stakeholders, outside regulators and/or government agencies, and other interested parties who might participate in reviewing and commenting on project information in the system, and then multiplying by two.

Global Project Management Strategies in a Program Management World

"Being busy does not always mean real work. The object of all work is production or accomplishment, and to either of these ends there must be forethought, system, planning, intelligence, and honest purpose, as well as perspiration. Seeming to do is not doing."
THOMAS EDISON

Project Management

The application of project management principles has dramatically increased in every industry. Some attribute a method of project management practice to ancient times when pyramids were constructed using large amounts of human resources and raw materials. There are many examples of the marvels of premodern project management, from coliseums to the Great Wall of China. Modern-day project management began to crystallize after World War II when people were looking for a systematic approach to carry out work effectively. There was a need to organize resources to obtain critical objectives within a specific timeframe. At this point, project management was mostly employed by engineers and those in the construction industry.

Over the years as the project management discipline solidified, you began to hear people from diverse industries and within different disciplines inside a company discuss projects. Today, from manufacturing to human resources to marketing, the project mindset is deeply imbedded in our corporate cultures. Project management is both an art and a science. The science is in the structure that is followed, but the art is in how you apply and manage it. Though many discuss projects, not all follow the project management disciplines laid out by the Project Management Institute. There is a link between project success and following the core strategies of the project management structure. Understanding the principles and being able to adapt those principles to best suit your corporate culture is one of the secrets to project success.

Thomas Edison accomplished many great inventions. These inventions were projects that themselves require initiating, planning, executing, controlling, and then closing phases of a project plan. Edison's hard work and dedication lead to 1,093 patents for different inventions. Many of these inventions you would recognize, like the light bulb, phonograph, and motion picture camera. These successfully completed projects have had a huge influence on our everyday lives.

In this appendix, we will explore the project management discipline, how it relates to the program management world, and how successfully completed projects can have a huge influence on the success of your business, professional, and personal lives.

Global Reference

The program world

As a project manager working within a program, you have the added responsibility to work within the parameters of the program. It is good practice to work closely with the program manager. In a global environment, you may be located in a different country, speak another language as your first language, and have different customs and beliefs than the program manager. However, it is important to essentially eliminate the "space" between the two of you. The elimination of space can also be thought of as a management continuum. You may never be able to physically remove the distance, or even the time barriers, that keep you from the program manager. You can lower these hurdles by being flexible, taking the time to fully understand what is required by you for the program, and communicating in a timely manner with the program manager.

It is very important that the program manager has a full understanding of your project objectives and status. It is vital to follow the guidelines that the program manager sets out for how they want to operate the program. It is critical that you, as the project manager, report to the program manager the status of your project in the form that is required by your program. If you work in a company that has a mature project management process, where you have histories from prior projects and programs, tools and techniques that are established, what your program's direction will be throughout your project will be clear. However, there are places that are not as mature, and the project manager must rely on the program manager for guidance on how they want to proceed. To understand better where your company fits in the project maturity model, you need to examine whether they leverage prior experiences, or if they share a common language and strategic vision. How does the company handle prioritizing work and resources? How does the company improve upon their processes? A maturity model is used to create an understanding of the company's strengths and weaknesses compared to other companies. Project management maturity is an organization's level of ability to accomplish project requirements with reliable methods and repeatable delivery of project management goals. To help determine the level of maturity your company supports, you can consider the following questions:

- Are the business processes within the organization repeatable?
- Is any project methodology followed?
- Is any project data captured?
- Are any software tools used?
- How are projects chosen and initiated?
- How are project roles determined?
- How is project information communicated?
- How are the executives in the organization kept informed on the projects?
- Are there certified project managers in the organization?

The more affirmative answers to these questions, the further down the project maturity path your company is located.

Explanation of Concept

Definition of a project

It is important to have a clear understanding of what constitutes a project as opposed to other forms of work. When you think of a manufacturing plant, you find that there is generally a conveyor belt, which advances down the line in a sequential manner that is strictly ordered and at each station adds or alters the product. When you reach the end of the line, you have a fully assembled product. You also see a similar process in food processing plants where ingredients are mixed together, formed, baked, and then packaged. The process of the assembly line doesn't end, but is continuous throughout the product life cycle. In other departments, such as accounting, you have the closing of the books every month. Normal operational business processes are not projects.

A project, on the other hand, is temporary in nature; it has a beginning, middle, and an end. The outcome of a project creates a unique product, service, or result. Where designing and creating a new computer is a project, manufacturing hundreds of them is not. Projects tend to have interrelated tasks, which generally have progressive elaboration in their nature. In progressive elaboration, you move through a project and each step builds upon the previous step in increments. The predecessor, or previous step, provides the ability for the next step or process to occur. If you consider building a house, you need to lay the foundation, and then frame the house, before installing the roof. Each step grows upon the one before it. It would be impossible to install the roof without the other necessary predecessor construction work being completed.

There are other attitudes to projects. Projects must have a customer, whether internal or external to the business. A project is formed to fulfill the customer's need. Since projects are temporary and you are creating something unique, there is a degree of uncertainty around them too. If you do this type of work all the time, it would not be considered a project, so there is risk in doing something that can be unfamiliar and new. Projects also use resources. Resources can be people with the required skills for the project work, money, workspace, technology, and raw materials, to name a few. Unfortunately, resources tend to be in limited supply and in high demand. As a project manager, you need to manage resources wisely.

It would be wonderful to say that all projects should have a well-defined objective. In a perfect world, projects are clear and understood by all parties. In the real world, when you start a project, you will spend a great deal of time defining and clarifying the project's scope and communicating it to the project's stakeholders. A project needs to have some definition when you start it, and to become further defined as it progresses.

PMI summarizes the definition of a project as:

A project is a temporary endeavor undertaken to create a unique product, service, or result.

This definition covers the major aspects of what a project is, but as you read, there are also finer details that need to be thought of when understanding the breadth of the meaning of projects. For the purpose of this book, we will use the following definition:

A project is an unprecedented effort that:

- *Occurs within a finite time period*
- *Possesses a unique set of interrelated tasks*
- *Uses resources effectively*
- *Contains a degree of uncertainty*
- *Accomplishes a specific customer-driven objective*

The purpose of a project

Projects are done for a specific purpose and become visible when a need is identified. The purpose of a project can materialize from a number of strategic reasons. There could be a market reason for developing a new product. There could be a competitor with a new feature or flavor that you have to match or exceed to maintain or enhance your market share. Or, through research, you discover an untapped market and design a product or service to meet the needs of that group.

There could be an organizational need to alter normal business processes and train staff on a new workflow to improve efficiency to increase revenues. One of your most important customers may request changes to the current product that you send them. These feature requests roll into your next product release cycle. Technical advances could lead to developing new products.

As Moore's Law states, computers double capacity, speed, and memory every 18 to 24 months. If Moore's Law is true, technical advances in computing alone change how we run our businesses and our projects every 18 to 24 months, and it is essential that businesses maintain this pace.

Finally, there are also reasons to create new projects based on legal or regulatory changes. It is important to understand the policies and procedures within the countries you are managing projects and programs. For example, lead paint was banned in the United States in the late 1970s, but is still in use today in other countries. What if one section for your project was manufactured in a plant outside the United States using lead paint and then the product was brought back to the United States and sold? Or you worked with a company that uses pesticides that are not approved in the country that the product will be sold in? There are a number of ethical and moral questions when it comes to managing a project. Ethics is discussed in greater detail in this appendix as it pertains to the role of a program manager.

Deeper Understanding—Strategy of Use

Project management process

There is a process to project management. The first step is to define the purpose of the project. It is just as important to define what will be omitted from the project as it is to detail what will be included. You can look at this process as defining a box, with your project being the box. It is important to state what is inside the box and what is outside of the box.

Project management is about managing expectations. A project charter is the first step in defining your project. The project charter should be a concise document, just a few pages in length, yet clearly define the agreed-upon work. A project charter is not your project plan. Further specifications about deliverables and work are in other documents.

Your project charter should be broad enough so that it does not need to be changed when the project changes. However, it needs to be specific enough to authorize the existence of a project and provide a project manager with the authority to apply organizational resources to the project activities.

The project charter format is as follows:

- Name of the project.

- Date and version of the document.

- Often, the project charter is revised a number of times until it is signed and authorized by all parties, so versioning is needed to make sure that everyone is singing from the same song book.

- Business case justifying the project and/or brief background of what led to the need of creating this project (no more than a paragraph or two).

- Project scope.

- The scope is a statement of the work that must be performed to successfully accomplish the project with the specified features and functions. This can include:
  - A couple of sentences describing the proposed end product
  - Identification of major components and phases
  - Specific customer requirements concerning workflow or functionality
  - Specific exclusions to the project
- *Project objectives.* The objectives need to be quantifiable so that they can be measured at the end of the project to determine the project's success.
  - These should be "SMART" goals.
    - *Specific.* The goal needs to be well defined and clear to anyone who has basic knowledge of the project.
    - *Measurable.* There is an old saying that says "what gets measured gets done." Setting a numerical target helps you measure your progress to make sure you are heading in the right direction.
      - Examples:
        The project will be completed no later than XX/XX/XXXX date.
        The project will not spend more than $100,000.
        The project will increase efficiency in the processing plant by 20 percent.
    - *Achievable.* The goal should be achievable, but that doesn't necessarily mean easy. Within reason, it should be possible to accomplish this goal.
    - *Realistic.* It is possible that you can complete this goal given your level of resources, constraints, assumptions, and risks.
    - *Time frame.* There is a reasonable amount of time to complete the goal.
  - Defining objectives.
    - Often, there is a large list of project objectives. This list can be further defined as primary and secondary objectives.
    - The primary objectives must be complete in order for the project to be considered a success.
    - Secondary objectives are additional benefits desired, but not critical, to the project. Secondary objectives are goals that would be beneficial and you would like to complete.
- Project stakeholders
  - Stakeholders are the people who are actively involved in the project or whose interests may be affected by the completion of the project.
  - All the critical stakeholders should be listed for the project and the role that they contribute to the project's success.

- Here is a sample list of project stakeholders:
  - Program manager
  - Project manager
  - *Activity manager(s).* Those team members who are responsible for managing a specific aspect or task in the project. These managers can come from the following areas: consultants, vendors, contractors, architects, and engineers.
  - *Project sponsor.* The person or group that provides resources for the project and exerts influence when the project manager is in need of help with organizational or other barriers.
  - *Project team members.* This is anyone who participates in the planning or execution of the project or is a source of information. Generally, there could be a number of people working on a project; however, this member list should focus on functional managers who direct others to fulfill the tasks.
  - *Subject matter experts (SMEs).* These are the people who will actively participate in the project and be at your status meetings.
  - *Customer/client.* Most will say that this refers to the people or organization for whom the project is being executed and who need to be satisfied with the outcome. In practice, I try to only have one name listed here. It is hard enough to make one person happy, let alone an entire department.
- Constraints, assumptions, and risks
  - This is a bulleted list of any known project constraints, assumptions, and risks. I generally list these items together because they tend to overlap within the three categories.
  - Constraints are limitations that reduce and/or restrict actions available to the project.
  - Assumptions are factors that are considered to be true, real, and certain. Concerns and risks are any issues or events that, if they occur, will affect the project. Part of the project plan is the risk plan, where you define how you will manage any risk that might happen in the project; in the charter, we just list them.
- Schedule expectations are listed, along with any critical dates and major milestones.
- Budget expectations are listed, along with possible limitation guidelines.
- *Approval signatures.* To make this an "official" document, it should be signed to confirm and acknowledge approval for the continuation of the project. The number of signatures should be minimal, but typically include project manager, project sponsor, and customer. Other names can be included. However, the larger the group, the more cumbersome and time-consuming this task becomes.

A copy of the charter and the other project plan documents should be stored in a shared server so that all project team members can access the information.

Work breakdown structure

Once the work is defined, it needs to be broken down into project deliverables called work packages. Here is where you need to be able to see the forest and the trees at the same time. The work packages are generally displayed graphically in an organizational chart format called a work breakdown structure (WBS). The WBS is the foundation of the project. The entire project plan and control is based on the WBS.

The WBS should be created with your project team. In the WBS, the top box is your project name. Your major deliverables are located in the first row. The entire scope of your project is listed in this first row. Below each major deliverable is a smaller segment of the work listed. You keep on breaking down project deliverables until they can't be broken down further. Basically, you break down the major initiative to smaller deliverables and finally end up with work packages. A rule of thumb is that the work packages are about 300 hours of work. However, depending on the size and scope of the project, the last row may consist of a different amount of work hours. If work is not listed in the WBS, it is not part of the project. The great thing about a WBS is that once you create one, you can use it as a template for future projects.

The WBS process triggers the team to think through all aspects of the project. The WBS holds your project's "big picture." A good project management practice is to bring out the WBS periodically, possibly once a month, for the core team to review on how you are progressing on major deliverables. It is easy to focus too closely on the finer details or the work that is happening today and lose track of the big picture. The WBS allows you to step back and holistically see where you are heading. As a project manager, you could print out the WBS on a plotter and bring it to your core team meeting, and then discuss each deliverable and cross out those that are complete and focus on larger issues with deliverables that are not meeting expectations.

Time management

From the WBS, you further define your project by creating a project schedule network diagram. The network diagram shows the relationships among the scheduled activities and how the tasks will flow from beginning to end. Here you take the major work packages and further break down the work effort into project tasks. This process will show project work chronologically. Each task is assigned an estimated duration to complete it. Also, the person who is responsible for completing the task should be assigned at this time. It is preferred that the person accomplishing the task help determine the estimated duration to complete it. Once the estimates are completed for each task and all the relationships among the tasks are set, the network diagram provides your best estimate of how long the project should take to complete.

Commonly, the network diagram information is entered into a project timeline software system to produce a bar (GANTT) chart of the tasks. After approval, this information creates the project's timeline baseline. Your baseline is a starting point to refer to during the project to determine if the project is on track. Any variances from the baseline need to be reviewed to determine the cause and to asses whether corrective action needs to be taken to return to the baseline or if the baseline needs adjusting. Using a baseline is one of the ways to monitor and control your project's progress.

Risk plan

A risk is an uncertain event that has a possibility of occurring and that will have an effect on the project. It is good practice to review your project deliverables and determine the risk for not achieving them or achieving them in a manner for which you had not planned. For each identified risk, the project team needs to determine how they plan on responding to it. Are you going to ignore, transfer, mitigate, or accept the risk? You will also have to factor your company's risk tolerance into the equation on managing risk. Are they risk-adverse, -neutral, or -seeking? Every project will have a different tolerance level, depending on the motives for creating the project. For more information on risk, please see Chap. 5, Global Program Risk Management.

Cost management

After tasks for the project are determined, you begin the process of creating cost estimates. A number of sources can go into building your cost plan:

- Project plan documents
  - WBS
  - Network diagram
  - Timeline
- Environmental factors
- Historical project information
- Organizational process
- Risk
- Resources
- You can use a number of estimating tools and techniques to help with cost estimating:
  - Analogous estimating, where you use actual costs from earlier comparable projects
  - Bottom-up estimating involves creating cost estimates for each work package or individual activity.

Once all the cost estimates are determined, you assemble your project budget.

Project team offsite

In practice, one way to accomplish building your initial project documents is to have a project meeting offsite. This meeting is held away from the normal office setting and blocks off half a day, a day, or a few days to accomplish this work. The amount of time required depends upon the complexity and estimated duration of the project.

The work that could be accomplished at the meeting is:

- Building the project charter
- Creating the WBS
- Developing your network diagram
- Constructing your initial timeline
- Outlining the project team structure, along with roles and responsibilities

Using facilitation processes, you work with your team through the deliverables that are decided upon. The offsite meeting should be managed like any other meeting. However, since this will be a bigger undertaking, there should be offsite preplanning meetings to determine the best way to create a smooth and productive process. The meeting agenda should be sent out prior to the event so that everyone knows what to expect. For your first offsite meeting, there might be a premeeting for the entire team to know what to expect at the offsite event. Make sure that all project stakeholders are invited to the meeting. If a key person cannot attend, someone should be sent in their place. At the offsite meeting, you will allow for more to be accomplished than producing your project documents. If the offsite meeting is done correctly, you will also:

- Speed up the process of getting the initial project documentation completed
- Set the tone for the project and the work will be accomplished
- Gain consensus on what is included and not included in the project
- Gain buy-in from the team who will lead the work to be accomplished
- Build team rapport
- Establish team roles and responsibilities

The project management process means planning the work and then working the plan.

Project life cycle

Projects are temporary; therefore, they have a beginning, middle, and an end. They have a limited life, and the cycle of the project's life usually goes through four phases:

- Define
- Select

- Execute
- Closeout

The project life cycle seems similar to the project process groups; however, the life cycle is a linear process. The cycle of a project varies from one project to another in duration, resources, and requirements. In the first phase, a project concept is identified. The concept is shaped and modeled into a defined idea. A need, problem, or opportunity is brought to light and recognized. The need requirement is clearly defined. The problem is then quantified, and a budget is determined. If a customer is looking for a solution to a need, problem, or opportunity, they may put together a request for proposal (RFP). In the RFP document, the customer is looking for vendors to reply on how they could best resolve the customer's concerns or provide the best solution to their new opportunity. The RFP comprehensively states what is required from the customer's point of view. The document also states the application process and the criteria for selection. The project concept is then reviewed by the project approval board. The project approval board can be a portfolio governance process, a program management process, or another form of business process. The project is reviewed to determine if it aligns to the company's strategy. If the concept passes, then the project is formally selected. A vendor may then develop a proposal for the customer. If the vendor is selected because their solution offers the greatest benefit at a reasonable price, a contract stating the working relationship and what is to be delivered is negotiated and signed.

Then the project work begins. Detailed project planning begins. Scheduling, monitoring, and control plans are created. Different resources are used. The focus is on the project objectives and achieving the results that are required by the customer.

After the project is fully executed, the project is evaluated to make sure that it met the customer's needs, and then it is terminated. Formal project closeout activities are performed. Resources are freed up, and feedback is requested of the customer.

Project process groups

The PMI defines five distinct process groups. These process groups define the activities that should be accomplished within that process group. The process groups are initiating, planning, executing, controlling, and closing. Often when process groups are described, they sound like unique, distinct, all-encompassing events and that once one is accomplished you move on to the next phase or process group. The process groups could be isolated events in the most simplest of projects. However, in most projects, the process groups overlap to some degree.

Initiating phase.   During the initiating phase of a project, a number of activities are accomplished. First, the project needs to be selected. This may be done by an external group prior to the project manager being selected or the project team being assembled.

Once a project is selected, it is important to look at former projects that are similar to the project you are working on. This historical data helps provide guidance and gives an insight to roadblocks that may occur. You want to learn from the lessons learned of previous projects. If you don't have historical information, you can interview experts in the company that have done similar projects.

After you gather this information, it is important to define the project objectives. What are we trying to achieve? What are we going to deliver? How long will it take us? How much will it cost? What are the roadblocks that we have to maneuver around? These constraints must be identified, defined, and planned for in a project. It is also important to define all the assumptions that you plan on being a factor in your project.

Other work that happens in this phase:

- Select the project.
- Gather background information.
- Verify the business need.
- Define project objectives.
- Identify constraints and assumptions.
- Define the responsibilities of the project manager.
- Determine high-level resource requirements.
- Define the project description.

Initiating is the first step in the project management process. It includes all the work necessary to create a project charter. Much of the work done in the initiating process is later refined in the planning process.

Planning phase.   Planning is important, and is often one area that is skipped over by anxious project teams ready to roll up their sleeves and get started on the "real work." There is a correlation between the amount of time spent in planning and the success of a project. Projects that spend little time in the planning phase generally take longer and require rework. This is due to the team not having a clear understanding of the scope of the project and what is required of them. The more experience a person has in doing projects, the more sophisticated the development of the project plan becomes. They really understand how critical it is to put effort into a clear and well-thought-out plan. You wouldn't build a house without a blueprint. Why would you start a project without a plan?

A number of documents are prepared during the planning phase of a project. The more complex a project is, the more details that should be included.

- The scope statement ensures that the project includes all the work and only the work required to complete the project successfully.
- The work breakdown structure includes all of the deliverables for the projects and groups them together in an organizational chart. Think of the deliverables

in the work breakdown structure as nouns. They are tangible deliverables. The GANTT chart should start with verbs, or action words, of what is going to get done.

- Team roles and responsibilities.
- *Time management plan.* Task estimates should not be longer than the amount of time that you can lose in a project. As a rule of thumb, no task should be longer than 10 days, but this is dependent on the length of the project. Each project manager should determine the maximum-allowed duration for a task's length. Tasks that take longer than the maximum should be broken down.
- Cost plan.
- Communication plan.
- Procurement plan.
- *Quality plan.* Quality must be planned in, not inspected in, and the impact of poor quality could be increased costs, low morale, lower customer satisfaction, and increased risk.
- Risk identification, quantification, qualification, and response planning.

Once all the planning work is completed, it can't be just put into a drawer and forgotten. It must be updated and revised as the project progresses. The project plan is used as a base for you to monitor and control the project. It also is a guideline to show how successful you have been in executing the project. At the end of the project, the plan is used in the closeout process to create lessons learned and histories, which can then be reviewed for future projects.

Execution.   Execution is where the rubber meets the road. All the planning and preparation goes into action. The team begins to execute on all the plans that were created. Here is where you:

- "Work the plan"
- Manage the project process
- Work on completing the work packages defined in the WBS
- Communicate project status
- Hold progress meetings
- Manage the team
- Manage project expectations
- Identify changes
- Assure quality
- Provide information and updates to the program manager

Controlling phase.   In the controlling phase, you are monitoring and controlling the status of your project. You use your project plan and check for variances.

To some degree, the controlling phase happens throughout all the phases of the project. Here your focus is on performance measuring and reporting. The project manager oversees the change control and quality process. Risks are monitored and risk triggers are identified that would cause the project to possibly have a response plan activated. All aspects of the project, including time, cost, and quality, are measured and analyzed to ensure the project progresses as planned. If there are deviations, corrective action is taken through the formal change control process.

Change requests are formal documents submitted by anyone on the project after the project has been approved. A change request can become a major issue if not formally documented. You will find that a decision could be made, thinking it is what is best for the project, then months later people will wonder why it was done. A change control board can review requests to determine if additional analysis is warranted. The change is then approved or rejected. The change control board may include the project manager, customer, experts, sponsor, and others.

Closing phase.   In the closing phase, all project activities are terminated and the product of the project is transferred to others. Project resources are released. Project procurement audits are conducted to make sure that the resources obtained were processed correctly. The results of the project are then verified to make sure that the team delivered all of the objectives for the project. Formal acceptance is given by the customer.

Often, there is a closeout meeting. In the meeting, project objectives are listed and the project team and stakeholders determine how closely the project met those objectives. Lessons learned are gathered at this closeout meeting too. The entire project plan is updated and stored in a central server system to be used as a history for future projects. The closing phase is critical to the long-term success of the business's project culture, and should never be skipped.

Throughout the phases

The project manager is leading the project effort throughout the phases of the project. They are solving problems and negotiating for resources. The project manager is communicating effectively with the project stakeholders and program manager. They hold productive and informative meetings. They also manage stakeholder expectations.

Project plan

Some people believe a GANNT chart is a project plan. While the GANNT chart is an important element in the project plan, it is not the only component. A project plan is a multipage document created by the project manager based on input from the team and stakeholders. The documentation includes many elements, from charts and tables to detailed descriptions and summaries. The project plan is the tool that helps the project manager effectively manage a project.

Though every project is different, a project plan needs to be designed and created to handle the nuances of the project and requirements of the stakeholders. Elements that could be present in a project plan include:

- Project charter
- Project management approach
- Scope statement
- WBS
- Responsibility and project team role chart/assignments
- Network diagram/major milestones
- Budget
- Schedule
- Resources
- Change control plan/system
- Performance measurement baselines
- Management plans (scope, schedule, cost, quality, staffing, communications, risk response, procurement)

The plan should be as complete as possible, although it may evolve over the course of the project. Maintaining the plan is an ongoing job of the project manager throughout the life of the project. The project manager must create a project plan that is believed in, approved, realistic, and formally documented. Many companies have standard procedures, forms, and guidelines for planning projects. These templates should be followed in planning the project. If not, the documents need to be created with input from the program manager and carried out by the project managers.

The plan needs to specify how you are going to manage every aspect of the project. This can come in different flavors, depending on the size of the project. The larger the project, the more formal you need to be with the plan, and every item needs to be documented and, in some cases, signed off on by your customers. The project manager must control the project according to the project plan. When working within a program, a project manager must:

- Follow the guidelines as set by the program manager on the level of detail and information that needs to be included in the Project Plan
- Monitor and control the project and roll up information to the program level
- Inform the program manager if any variances to the project occur
- Work with the program manager in risk response execution

How are projects selected?

Projects are selected based on many different methods. Basically, projects have to fit into the strategic plan of the organization. The selection of projects should be done through the organization's project portfolio process. The selection of the projects a company takes on is rarely straightforward and uncomplicated. There is often competing priorities and objectives, and decisions have to be made on the correct approach to best meet the strategic goals of the company. Some of the tools that are used to select projects include:

- A panel of people who analyze and review a new project idea
- Peer review
- Scoring models
- Economic models
- Benefit compared to cost

When working within a program, the project process may be different. A program of work may first be formalized. Then from the program, projects are created to best deliver on the objectives set forth in the program.

Knowledge and maturity

There is a link between the use of project management tools and processes and the success of a project. There are five levels in the project management maturity model. The first level has no formal processes or procedures. Projects are managed ad hoc. Often, projects are unplanned and informal, and those working on the project often improvise as they go along. Project outcomes at this stage are unpredictable. There could also be little organizational support for project processes, which are thought to be more effort than value. Often at this stage, projects never really end; they just die out, and deliverables are often not met.

As the level of maturity moves from level 1 to level 2, a basic grasp of project understanding forms. A project group might begin to define some processes that they will follow. These new processes are generally used for large projects.

Moving to level 3, the project management process is further defined and used on all large projects. There will be more extensive training at this stage. There is a consistent use of project tools and techniques for all projects.

As a company moves on to level 4, the process is defined even further and all projects, large and small, use the method. Project oversight is fully integrated with the business planning. At this point, senior management is involved in the process. The organization can effectively plan, manage, and control multiple projects.

At level 5, continuous process improvement is fully utilized. The organization becomes a project-centered organization. The business fully buys into the fact that the future growth of the company is dependent on the success of the projects it approves. The business learns from prior projects and improves processes. Along with the company focusing on a continuous improvement environment where they move up the project maturity ladder, the company needs to focus on what PMI defines as nine "knowledge areas" in project management.

The Nine Knowledge Areas

1	Project integration management	Project integration is the process that ensures project elements are coordinated effectively.
2	Project scope management	Project scope focuses on making sure that what is required for the project is the work that is completed and nothing more.
3	Project time management	Project time management is the work that ensures a project is completed on time.
4	Project cost management	Project cost management ensures that the project is completed within budget.
5	Project quality management	Project quality management ensures that the project will satisfy its purpose.
6	Project human resource management	Project human resource management focuses on how you manage and organize the project team.
7	Project communications management	Project communications management includes the proper creation, collection, distribution, and storage of information.
8	Project risk management	Project risk management identifies, quantifies, qualifies, controls, and responds to risk.
9	Project procurement management	Project procurement management is the process of acquiring supplies and services from outside the organization.

When you combine the nine knowledge areas with project maturity, you can gauge where you are in each of the knowledge areas. The following table further illustrates what typically occurs at each level of the maturity model using knowledge area examples. Each level builds upon the previous one. The tools listed are only examples of tools that an organization would use at that level of maturity. As you progress up the levels, the tools from previous levels are still completed but also include the next level's benefits. This is not a linear process. In some knowledge areas, you could be at a level 1, whereas at others you could be at a level 5.

	Level 1	Level 2	Level 3	Level 4	Level 5
Project management knowledge areas	No formal processes	Foundation	Formal methodology for all projects	Integrated process	Continuous improvement environment
Project integration management	No established processes	Project charter used for high-profile projects	Preliminary project scope statement and project management plan	Monitor and control project work and integrated change control	Close project; use information for future projects, building a history and feeding back into the organization's best practices

(Continued)

	Level 1	Level 2	Level 3	Level 4	Level 5
Project scope management	No established processes	Scope planning and definition and WBS to determine project objectives and business need high-profile projects		Scope verification and control	
Project time management	No established processes	Activity definition, sequencing, resource estimating, durations estimating, and schedule development Determine critical path		Schedule control	
Project cost management	No established processes	Cost estimating and budgeting		Cost control	
Project quality management	No established processes	Quality planning	Quality assurance	Quality control	
Project human resource management	No established processes	Human resource planning	Acquire project team and develop project team	Manage project team	
Project communication management	No established processes	Communication planning plan and communication requirements	Information distribution	Performance reporting and manage stakeholders	
Project risk management	No established processes	Risk management planning Risk identification Qualitative risk analysis Quantitative risk analysis Risk response planning		Risk monitor and control	
Project procurement management	No established processes	Plan purchases and acquisitions Plan contracting	Request seller responses Select sellers	Contract administration	Contract closeout

Maturity model benefits

The maturity model provides a number of benefits. First, it provides a roadmap for a continuous improvement model for the business. The model is a tool to help identify an organization's project-delivery strengths and weaknesses. It provides a way to set realistic goals for improvement. Here you can focus on improvements for selected projects, and as you learn from these efforts, you can expand to other projects. The model allows for a way to measure progress improvements to enhance capability.

When fully utilizing project management processes, you will find that your project management's effective use of time increases. Rather than focusing effort on putting out fires and reworking project deliverables, project managers can work on new initiatives. Project rework decreases because there is a clearer understanding of project requirements and a closer working relationship with customers.

Triple play of constraints

In baseball, a team accomplishes a triple play by making three outs in the same continuous play. Triple plays are infrequent because specific circumstances must be present to make the play possible. In project management, the triple constraints are circumstances that are always present in virtually any project, and a project manager needs to manage them continuously. To hit a home run with the quality of your project, you need to manage the constraints of scope, time, and budget. Time and budget is self-explanatory. Time is the length of time to go from the start of a project, through the project management processes, to project handover (Fig. C-1).

The budget is the planned financial cost of the project. A budget plan specifies how resources will be allocated or spent during a particular period. The plan shows the total amount of money allocated or needed for a specific purpose.

Project scope is the project's requirements. The requirements could be anything from workflow conditions, project specifications and design, to people and resources needed to complete the project.

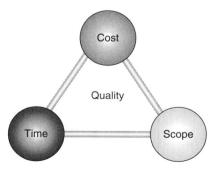

Figure C-1 Triple constraints of project management.

Scope, time, and budget make up the basic structure for your projects, and how these three "bases" are managed affects the quality of a project. These three foundational elements are so intertwined that a change to one will affect at least one of the other two. Constraints are major factors to consider when building your project plan. When you look at the project constraints, you have to ask yourself a number of questions, among them:

- What is included in the project, and what is not included in the project?
- Who do we have available to work on the project? What are their skill sets, and when will they be available?
- When does the project have to be completed by?
- What is the budget for the project? Is it realistic to meet the needs of the customers?

Along with these, numerous other questions need to be asked to better understand the project's constraints. Answering questions like the ones provided here help to build the foundation for your project. A project evolves from a blue-sky concept to something that is more down to earth and tangible.

Often, certain constraints hold true for any project that you work on in your company. These can be broad concerns that affect any manager, like the skill set of the staff, or limited resources. The project manager generally has to spend time with stakeholders defining the project to solidify the deliverables. In a project, it is critical to create a comprehensive roadmap of your starting location and your destination. This would become your scope framework. Then you have to estimate the time it will take to drive to your location and how much it will cost to get there. You also need to factor in a few roadblocks, road construction, and detours along the way. If you find that there are concerns with the time, cost, or scope, your roadmap needs to be reevaluated. The reevaluation processes is iterative throughout your project, and your customer must sign off on this roadmap.

Some constraints can be anticipated, and plans should be created to handle those situations. If you were building a house, an obvious constraint

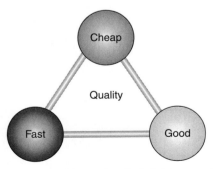

Figure C-2   The trade-off of triple constraints.

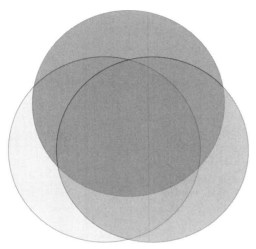

Figure C-3 Triple constraints blended together to create quality.

could be bad weather. Bad weather may not happen during the duration of your project, but it needs to be planned for, and time and cost need to be built into your project to handle such an occurrence. Contingency is factored into budgets and slack time into schedules to allow for unknown risks. Constraints are simply parameters to be managed by the project team. At the core or center of the triple constraint is quality. The balance among the three determines the level of quality the project will deliver. Quality becomes the residual effort of the balance of constraints (Fig. C-3).

Constraints are factors that limit the project team's options. A single project may have cost, time, human resources, and other constraints. For example, you might need a software developer with a specific skill set, but the best one in the company is already 100 percent allocated to another project. This would be a major constraint in completing your project successfully. Plans have to be made to prevent this from being an issue on your project.

Project quality comes from effectively balancing the constraints. There is a saying among project managers where they change the constraint names from time, cost, and scope to fast, cheap, and good, respectively (Fig. C-2). Then the project manager would be allowed to pick any two, but only two. The project can be fast and cheap, cheap and good, or good and fast. The result of these actions causes the following conditions: If it is fast and cheap, it will not be very good; if it is cheap and good, it won't be fast; and if it is good and fast, it won't be cheap. Projects are like walking a tightrope. You need to be in balance to make it across the rope to complete the project. Usually, you see the triple constraints in a form of a triangle. If you change one side of the triangle, it affects the other two sides. In reality, the constraints overlap a great deal. Time alone does not lead to a great project, nor does cost or scope. When the constraints are in harmony, the result is a project that meets the customer's objectives.

TABLE C-1   Project Team Roles

Project manager	Establishes project methodology
	Maintains customer relationships
	Oversees contract administration
	Manages project planning, control, reporting, and evaluation
	Helps with project marketing
	Advises on project engineering
	Advises on project manufacturing
	Manages purchasing and subcontracting
	Handles field operations
	Manages project finances
	Manages project closeout
	Reports to the program manager, program/project director, and/or PMO office
Sponsor	Is the company's executive responsible for the project
	Assists to get the project funded
	Facilitates project success
	Helps remove project barriers
	Supports the project manager
Customer	Source for project requirements
	Guides project manager on project risk concerns
	Receives the completed project
Project engineer	Owns the technical portion of any proposal
	Understands the customer's technical requirements and determines if company is technically capable
	Breaks down requirements to lowest level
	Ensures that the engineering schedules can be met
	Conducts design reviews
	Advises and supports project manager on technical issues
Functional manager	Owns the oversight of completing deliverables
	Assigns resources to projects
	Balances project workload with operational responsibilities
Contract administrator	Manages the legal aspects of the project proposal
	Leads all contract negotiations
	Explains contract language
	Submits contract deliverables
	Identifies scope changes and customer delays
	Initiates request for final payment
	Retains all relevant records
Stakeholder	Provides the project team with project requirements
	Participates in the project
Program manager	Supports the project manager
	Guides project manager on the deliverables for the project
	Ensures project alignment to the organisations objectives

Project manager

The project manager is the person who has project oversight and is the primary person accountable for the success of the project. Depending on how a company is structured—whether it is a functional, matrix, or project-based organization—a project manager will have varying degrees of authority to match the accountability. If authority and accountability do not match, the project manager needs to work with the sponsor of the project to resolve the issue.

The project manager's main responsibilities are to ensure that:

- The customer is satisfied

- The work is completed inside the parameters set for the scope, time, and budget

- The quality required by the customer is met

The project manager provides leadership in planning, organizing, and controlling the work effort. This person will also coordinate the activities of the project team by delegating the task work, and does not try to take the entire project work upon their own shoulders. The project manager works to gain commitment from the project team and buy-in from the stakeholders.

If the project manager's organization has an established project management office (PMO) with the structure that has a formal project methodology, it is up to the project manager to follow the prearranged structure. However, if a structure does not exist, it is up to the project manager to define how the project is going to be run, with sign-off from the critical stakeholders and program manager. Table C-1 provides an example of a project team with the roles and responsibilities outlined. If a project manager hasn't run a project before, it would be good to follow the policies and procedures according to PMI. Every project and organization is unique, and it is often necessary for the project manager to tailor the project structure to fit the needs of the project, team, and organization.

The project manager needs to build relationships with customers and stakeholders. They oversee the project planning, control, reporting, and evaluation. The project manager is a jack of all trades; however, they do not have to be the master of any of them. They need to help market the project, work with engineering to develop requirements, and work with manufacturing to build a process to develop the project.

- Establish project methodology

- Customer relationships

- Contract administration

- Project planning, control, reporting, evaluation, and direction

- Marketing

- Engineering

- Manufacturing

- Purchasing and subcontracting
- Field operations
- Financial
- Manage project closeout

Role of the project manager.   A project manager is responsible for coordinating a project and making sure that the desired outcome is reached on time and within budget. Being a project manager is a large responsibility. Picture a juggler. First, they may start out with a couple of balls in one hand, and then move to many balls using both hands. The juggler's eyes are always on the different balls. Project managers usually do not have the luxury of building up the number of balls that they are juggling in a project. Often, they have to manage all nine balls right away. Why nine balls? The nine balls represent the nine areas of expertise that a project manager must focus on when managing a project. The nine areas were discussed earlier and are called knowledge areas by PMI. These nine areas are project integration, scope, time, cost, quality, human resources, communication, risk, and procurement. A project manager must keep their eyes on the nine balls, making sure that none of them fall.

The project management discipline evolved out of the need to coordinate resources to achieve predictable results. It is up to the project manager to coordinate resources effectively to deliver the objectives of the project.

A project manager must possess "hard" and "soft" skills. The hard skills lie in understanding the project management principles and being able to execute the discipline in a manner appropriate for the organization.

The soft skills are often overlooked when discussing project management. These include leadership, interpersonal skills, communication skills, conflict management, and delegation.

A project manager provides leadership to a project. They are the ones that lead the effort to plan appropriately, organize effectively, and monitor and control judiciously. They are responsible for coordinating the activities of the team and managing the delegation of tasks and resolving conflicts. Project managers gain commitments from their stakeholders and manage their expectations throughout the project. They realize that they can't manage a project just in project meetings, but rather roll up their sleeves and communicate with stakeholders at various times and build relationships with them. Project managers, by their very nature, need to be inquisitive and ask questions and not wait until information comes to them. Project managers are proactive.

Project managers must also have the ability to accept uncertainty and manage change effectively. Usually when you start a project, you will not receive comprehensive project information. Detective work is required to uncover project deliverables, customer needs, and stakeholders that need to be involved. Most projects start with a level of ambiguity, and while working on a project, the haziness lessens. However, unplanned events arise. A project manager must remain flexible and agile to handle the ebbs and flows of the different currents that may wash upon a project's shore.

Only in the smallest of projects is the project manager working alone. On most projects, a project manager is managing a team and needs to work to build that team and keep them focused on the project's goals and deliverables.

Projects are done to satisfy a customer's need; therefore, a project manager needs to be customer-focused. The project manager also must keep the customer informed. The project manager must first understand their customer and the customer's needs for the project. Overall, projects are undertaken due to a business need. It is critical that the project manager adheres to the priorities of the business. In a program, the project manager must work closely with the program manager to make sure the deliverables are meeting the program objectives.

Project manager best practices:

- Hold regular project status meetings with appropriate team members.

- Meet with stakeholders in groups and individually. There could be times that some issues need to be discussed privately.

- Provide team with consistent feedback on project's status.

- Ensure that you communicate in the mode that is best for the receiver and the message.

  - Messages can be:

    - Formal or informal

    - Delivered verbally or in writing

	Formal	Informal

Verbal	Presentation
	Project status/
	  review meetings
		Hallway
		  conversations
		Stop by the office
		  or work area
		Phone call

Written	Project plan document Status report	E-mail Instant/text message

- Be available, friendly, and willing to discuss any issues that may be presented to you.

- *Documentation.* Document, document, document! You need to document the project plan, decisions made, and any changes. The documentation must be clear enough so that when you go back to review it many months later, it is clear and understandable to all of the stakeholders.

- Documented information should be distributed in a timely manner to the appropriate parties for review. Documents should be stored in a shared storage area so that the entire project team can access them.

- Project managers can't manage from their desks; they need to be around the work areas and talking with stakeholders. You can find out a lot about the status of the project in informal settings.

Level of project management effort. During the various stages of a project, different levels of effort are required from a project manager. In the initiating phase of a project, a project manager is often not part of this process until near the end if it. Usually, the initiating phase is handled by the PMO or portfolio management process to determine whether this work effort should become a project for the organization given the limited resources that it has. A project manager is often not assigned until the end of initiating phase or the beginning of the planning process.

The planning process is where the majority of the project manager's work should focus, as depicted in Fig. C-4a. However, some project managers and teams are so excited to get to work on a project, they often skip this process or spend little time in planning, as depicted in Fig. C-4b.

In a poorly run project, the execution phase of the project never seems to end (Fig. C-4b), unlike in a well-run project (Fig. C-4a). Throughout the project, the project manager needs to maintain and update the project plan.

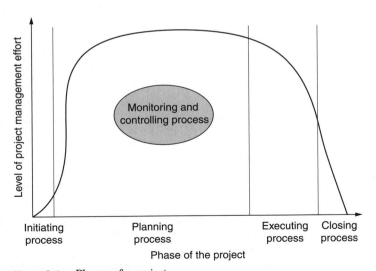

Figure C-4a   Phases of a project.

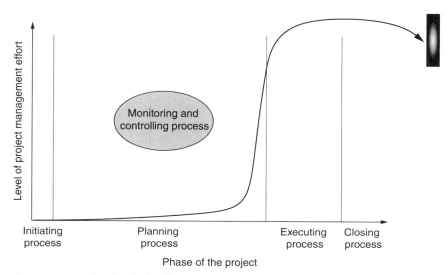

Figure C-4b   Typical level of project management effort in a poorly run project.

In a well-run project, you will see that the monitoring and controlling process is at the center of the diagram and should be at the center of all your projects. Monitoring and controlling a project starts as soon as you begin a project and doesn't end until you close a project. In a poorly run project, the monitoring and control circle is off to the side. Some project teams do not focus on monitoring and controlling at all, or only as an afterthought. Often it is too late to focus on monitoring and controlling when you get to project execution. The project often spins out of control, and a project manager spends a great deal of time "putting out fires."

At the execution phase of a project, the project manager directs the project team on work packages and lets them get on with their work. At times in this phase the project manager will have a slightly smaller workload since the heavy lifting of the planning phase is done.

Unfortunately, in poorly run projects, project managers will spend a great deal of time working and putting out fires. In a well-run project, there will be a formal closeout and celebration for the completion of a successful project. In a poorly run project, the project hits a road block or another project comes along to take the resources, leaving the project floundering and never delivering upon the deliverables it set out to accomplish.

In different phases of a project, the project manager will have a slightly different focus. As discussed earlier, a project goes through five process groups, and the project manager has different responsibilities in each of these groups. In the planning phase of a project, the project manager should spend the majority of their effort in this process group. In this process group, the project manager will lead the effort to create the project scope, WBS, schedule, cost communication, and risk plans. They will work to determine the quality standards for the project. With all this information, they will then build a budget.

They will also create a formal change control process. The change control process can be in the form of a document that a person fills out, requesting a change. Then the change control committee reviews the changes to make a determination as to whether the change should take effect. The change needs to be weighed against other predetermined project work to determine if there are any conflicts with what was planned in the project.

In the execution process, the project manager will manage the change control process, direct the technical team to execute the work as defined in the project plan, manage the project interfaces to execute the work as defined in the project plan, acquire and manage project resources, execute the project plan, make sure work packages are completed as planned, and distribute information to make sure the project is at the quality required by the customer. They will also hold project meetings effectively. Prior to meetings, agendas will be produced; following the meeting, minutes will be distributed. They will also actively identify variances to the project plan. A manager cannot wait for concerns to come to him or her—they must be proactive and seek out issues.

The monitoring and controlling process is usually depicted in most books as the fourth process, which may be confusing. The monitoring and controlling process happens throughout the entire project, although to a lesser degree at the beginning and to a greater degree at the end. During the monitoring and controlling process, the project manager monitors risk triggers and monitors project activities and makes sure that they are following the plan. They also keep the stakeholders informed of the project status; produce dashboard project status reports on schedule, cost, scope, resources, quality, and risk; control cost and changes to the project; manage the project team's performance and provide feedback and help to resolve issues to enhance performance; and manage stakeholder expectations and resolve issues.

The closing process is one process group that often is overlooked. Closing is a critical process and can be a valuable tool to improve future project success. In the closing process, the project manager formally terminates all project activities. There is a formal handoff of the project results to the customer. Resources are released, including equipment, facilities, money, and people. A lessons-learned document is created and shared with other project managers. The lessons-learned document is reviewed in the initiation stage of future projects to help judge the resources and processes needed for it to be successful. This document is then archived with the other closeout documents.

Overall, there are a number of essential items that the project manager should keep in mind. In the real estate business, there are three things to focus on: "location, location, location." In project management, it is critical to focus on "communication, communication, communication." It is critical to understand the customer's concerns and speak their language. You always want to give your customer what they want—no more, no less. Always provide an accurate picture of the status of your project. You do not want to be the boy who cried wolf; however, you do not want to under-represent critical issues either.

You want to use the tools and techniques set out by your program office. Deviating from them will cause confusion and provide issues to the office, who often use the information to roll it into other documentation for upper management. Monitor your project closely, and actively seek out concerns. Identify risks and discuss them with your team. Create a daily log of your activities. Small decisions that are made on one day will be forgotten a few weeks or months later. It is important to keep a record of all your activities, like a captain's log. Mentor your team. In the end, the project's success or failure rests on your shoulders. I once had a manager advise me to treat a project and its resources as if they were my own. In a sense, the resources do become your own, and it becomes your responsibility to deliver what you promised. Your reputation as a project manager is built upon previous successes.

Sponsor

A sponsor is a person or group of people in the performing organization that helps fund the project. The sponsor is the company's executive ultimately responsible for the project. The sponsor also helps remove project barriers and supports the project manager throughout the project.

Customer

The customer is the person or group receiving the results of the project, and is the source for project requirements. The project manager works with the customer to make sure expectations are met according to the project plan. Thus, it is important for the project manager to manage customer expectations.

When project concerns arise, the project manager reviews options with the customer to determine the best project outcome.

Project engineer

Highly technical projects will have a project engineer or software engineer consulting on the technical aspects. The project engineer is responsible for the technical feasibility and success of the project. This person must understand the customer's technical requirements. Also, they must understand the technical aspects where the project is going to reside. The project engineer conducts design reviews, and is a member of the core project team.

Functional manager

The functional manager is generally the manager of the group of people that will be conducting the work for the project. The functional manager owns the oversight of completing projects and assigns resources within their own department. They also balance the project workload with operational responsibilities. Functional managers are generally on the core project team.

Contract administrator

Projects with a lot of legal ramifications should have a contract administrator who, in most cases, is a lawyer. The contract administrator would be responsible for all of the legal aspects of a project proposal. This person would also lead contract negotiations and explain contract language to the core project team. The contract administrator would submit all of the project contract deliverables. They would also work with the customer if there are changes to the scope or other delays and explain how that affects agreements. In addition, the contract administrator would handle the request for final payments. The contract administrator would not necessarily be on the core team, but would be the financial advisor and would be called upon whenever there was a legal matter to manage.

Stakeholders

Stakeholders have a stake, or interest, in the project, and their interests may be positively or negatively affected by it. However, not all stakeholders play an active role throughout the entire project, or even complete tasks on the project. Stakeholders in projects include the program manager, project manager, sponsor, customer, and project team. Basically, anyone who participates or could be affected by the project is a stakeholder. Stakeholders have influence on a project. Different projects will have a different mix of stakeholders, depending on the needs of the project.

It is critical that a project manager identify anyone who is necessary for the project's success. It is not always easy to identify a stakeholder. The project manager needs to ask who will make a contribution and who will be affected by the project. The sponsor is the person or persons who provide the financial resources for the project and removes high-level obstacles and corporate issues, and helps with resource allocations. The project team is the group of people who complete tasks on the project. Often, the project team is broken into a core team, who are the leads on key areas in your project, then there is a larger team that accomplishes the tasks of the project.

It is critical for your project's success to accurately identify all of the stakeholders. Nothing is worse than coming close to the end of execution and finding that you did not have a necessary functional project requirement because you had not met with and gathered information from a key yet unknown stakeholder. Stakeholders are not static. They evolve and change throughout a project, especially in larger projects. People leave the company or industry, or are promoted to different areas of the business. Also, the needs of the business or market can change. Thus, it is important to periodically reassess your stakeholder list. When you are working with a stakeholder, you need to make sure you provide them with:

- Information about the project status and make sure their needs are still being met

- Involvement in the project through assigning them work, using them as experts, reporting to them, and involving them in changes and creation of lessons-learned documentation

- Project management information so that they know what to expect from the project management process

- Formal acceptance documents for them to sign-off on

Stakeholder management is proactive work. The processes discussed here help with managing stakeholder expectations.

The practice of project management has evolved into a discipline with standardized processes and procedures. As organizations adapt the project management methodology, they become more successful in delivering project objectives. This appendix illustrates some of the common understandings of the project process. As projects are aligned under a program, an added dimension of effort is supplemented to the process. A successful project or program manager should have a strong understanding of the project management standards to constructively achieve the organization's strategic program outcomes.

Index

Note: A page number followed by a *t* or *f* indicates that the entry is included in a table or figure.